Lecture Notes in Intelligent Transportation and Infrastructure

The series “Lecture Notes in Intelligent Transportation and Infrastructure” (LNITI) publishes new developments and advances in the various areas of intelligent transportation and infrastructure. The intent is to cover the theory, applications, and perspectives on the state-of-the-art and future developments relevant to topics such as intelligent transportation systems, smart mobility, urban logistics, smart grids, critical infrastructure, smart architecture, smart citizens, intelligent governance, smart architecture and construction design, as well as green and sustainable urban structures. The series contains monographs, conference proceedings, edited volumes, lecture notes and textbooks. Of particular value to both the contributors and the readership are the short publication timeframe and the world-wide distribution, which enable wide and rapid dissemination of high-quality research output.

More information about this series at http://www.springer.com/series/15991

Mohammad Ayoub Khan · Fahad Algarni ·
Mohammad Tabrez Quasim
Editors

Smart Cities: A Data Analytics Perspective

Springer

Editors
Mohammad Ayoub Khan
College of Computing and Information Technology
University of Bisha
Bisha, Saudi Arabia

Fahad Algarni
College of Computing and Information Technology
University of Bisha
Bisha, Saudi Arabia

Mohammad Tabrez Quasim
College of Computing and Information Technology
University of Bisha
Bisha, Saudi Arabia

ISSN 2523-3440 ISSN 2523-3459 (electronic)
Lecture Notes in Intelligent Transportation and Infrastructure
ISBN 978-3-030-60921-4 ISBN 978-3-030-60922-1 (eBook)
https://doi.org/10.1007/978-3-030-60922-1

This Springer imprint is published by the registered company Springer Nature Switzerland AG
The registered company address is: Gewerbestrasse 11, 6330 Cham, Switzerland

This work is dedicated to my loving parents who let me do whatsoever I liked because they always have trust in me. I owe everything whatever I have today. This work is also dedicated to my loving wife, son and little angle Amira Fatimah who have been a great source of inspiration to lead farther in the life.

Mohammad Ayoub Khan

Preface

The role of the Internet and wireless sensors network has changed greatly as a result of development in different communication technologies. Today, the Internet is connected to billions of people and sensors. The smart city has recently emerged with underlying infrastructure that is connected to Internet of things (IoT). The smart city has a focus on data analytics facilitated through the IoT platforms. The IoT architecture and technology need to be modified to address the needs of different fields of application in smart cities like telecommunications, transport, health, and the environment. Smart cities will provide next-generation urban development solutions that include environmental sustainability, personalisation, accessibility, efficient use of resources, improved governance, and better quality of life. The Smart City: A Data Analytics Perspective provides research insights on recent tools and methodologies of smart cities.

This book is distinct from other books with detailed case study and data of best practices. The book can be used as a technical guide for a research in smart cities at the postgraduate level. The book focuses on the interplay of data analytics; design and infrastructure between communities of information; and IoT technologies that combine ideas, methodologies, techniques, and tools in practical terms.

Organization of the Book

The book is organized into three parts and altogether 15 chapters. Part I, titled "Fundamentals of Smart Cities and Data Analytics" and contains chapters "Fundamentals of Smart Cities"–"Governance in Smart City: An Approach Based on Social Network". Part II, named "Algorithms in Smart Cities" and contains chapters "An Overview of the Machine Learning Applied in Smart Cities"–"A Multi-level Ontology to Manage Service Level Agreements in Smart Cities". Part III, titled "Smart Transportation and Mobility" and contains chapters "Smart Transportation: A Reference Architecture for Big Data Analytics"–"Validation Frameworks for Self-Driving Vehicles: A Survey". Part IV, titled "Case Studies in Smart Cities" and contains Chapters "Big Data for Smart Cities: A Case Study of NEOM City, Saudi Arabia"–"An Assessment of Smart Urban Furniture Design: Istanbul Yildiz Technical University Bus Stop Case Study". A brief description of each of the chapters follows:

Chapter "Fundamentals of Smart Cities" presents fundamental concepts of smart cities. This chapter presents importance of underlying technology infrastructure of smart cities such as wireless sensor network (WSN), Internet of thing (IoT), and data analytics. Authors mentioned that big data analytics has an important role to play in smart cities as large amount of data is continuously received from many sensors, autonomous machines, or intelligent IoT devices.

Chapter "A Context Aware Big Data Analytics Service-Centric Process Modeling Approach" presents discussion on the dynamic process modeling approach based on context-related knowledge of the data analytics services. Authors have proposed a novel framework to design the context-aware process model. This chapter also presents some examples for determining how the framework works in a real-life scenario.

Chapter "Smart Energy: A Collaborative Demand Response Solution for Smart Neighborhood" presents a detailed discussion on smart energy. This chapter is based on collaborative demand response toward a smart neighborhood. The presented solution enables households to collaborate achieving an overall welfare such as load flattening to reduce the cost. The analysis shows that the proposed solution can obtain a significant trade-off between utilities' and households' welfare.

Chapter "Governance in Smart City: An Approach Based on Social Network" presents discussion on citizen-centric and government policies which is based on the demand of the citizen. This chapter presents the importance of social network to elevate the governance process to next level. Authors emphasized that social media analysis can provide knowledge for good governance.

Chapter "An Overview of the Machine Learning Applied in Smart Cities" presents a detailed discussion on the role of machine learning in smart cities. The chapter has focus on the fundamental, taxonomy, and algorithms addressing the key points with a precise bibliographic background.

Chapter "Clustering Techniques for Smart Cities: An Artificial Intelligence Perspective" presents a detailed clustering technique for smart cities. Clustering is a leading method used for solving scalability problems and providing a robust operational network in highly dynamic environments such as IoT-based smart cities. In recent years, the techniques of clustering have been used to revolutionize artificial intelligence and machine learning.

Chapter "A Multi-level Ontology to Manage Service Level Agreements in Smart Cities" presents an ontology for managing service level agreement in smart cities. The service agreement has been discussed at data consumers' level. The presented ontology responds to the challenge of managing customer information and providing a service autonomously in response. The chapter also presents an application of the ontology which has been illustrated using the smart city waste management.

Chapter "Smart Transportation: A Reference Architecture for Big Data Analytics" presents smart transportation systems with perspective of big data analytics. The chapter focuses on the underlying infrastructure for public transport, mobility, and logistics. Also, chapter discusses about the dataset, high volume, speed, and heterogeneity characteristics of the data in smart cities. This chapter has proposed reference architecture to address the challenges of big data analysis in smart transportation. The validating of the model has been shown using case studies.

Chapter "Designing for Urban Mobility: The Role of Digital Media Applications in Increasing Efficiency of Intelligent Transportation Management System" presents design for urban mobility to improve traffic efficiency. This chapter addresses issues such as transportation services, traffic congestion, and road navigation and also has use cases such as DiDi and Amap.

Chapter "Validation Frameworks for Self-Driving Vehicles: A Survey" presents a detailed survey on self-driving vehicles. This chapter focuses on the safety and reliability concern in self-driving vehicles. To test all the possible scenario, a validation framework for testing self-driving vehicles in simulation has been proposed.

Chapter "Big Data for Smart Cities: A Case Study of NEOM City, Saudi Arabia" presents role of data analytics in smart cities. This chapter presents a comprehensive discussion on the fundamental of data analytics, challenges, and a case study of NEOM smart city in Saudi Arabia.

Chapter "Smart Cities Pilot Projects: An IoT Perspective" presents discussion on many pilot projects around the world. This chapter focuses on key IoT technologies that cover various constituting domains of smart city, viz., smart health, smart transport, smart building, and smart industry. The chapter summarizes the outcomes of these pilot projects and also recognizes the major technologies, tools, challenges, and future opportunities in these domains.

Chapter "Performance of Smart Cities Concerning the Use of Internet of Things: A Case Study of Four Indian Himalayan Cities" presents discussion on the rapid urbanization of Indian cities. Authors have listed the problem that needs special attention to solve it. This chapter also presents the performance investigation of Indian cities and the analysis shows how the important things have been ignored by the policy-makers based on different evaluation criteria.

Chapter "Sustainability and Smart Cities: A Case Study of Internet Radio", "Sustainability and Smart cities: A case study of Internet Radio" presents an interesting paradigm of entertainment and sustainability. This chapter presents how the hybrid Internet radio is necessary for the sustainability of smart cities.

Chapter "An Assessment of Smart Urban Furniture Design: Istanbul Yildiz Technical University Bus Stop Case Study" has an interesting focus on the ergonomics. This chapter evaluates smart urban furniture design standards for the smart bus stop. The focus of the design standard is based on safety, form and character, function, value, maintenance, sustainability, and technology. The use of Istanbul has been detained along with the testing observation performed at Yıldız Technical University.

Who and How to Read This Book?

The potential audience of this book is three groups of people: (i) undergraduate students and postgraduate students conducting research in the areas of smart cities; (ii) researchers at universities and other institutions working in these fields; and (iii) practitioners in the R&D departments of urbanization, smart cities, Internet of things, and many more. This book differs from other books with detailed case studies and best practices details. The book can be used as an advanced reference for a course taught at the postgraduate level in urban planning, smart cities, data analytics, and Internet of things.

Bisha, Saudi Arabia

Editors
Mohammad Ayoub Khan
Fahad Algarni
Mohammad Tabrez Quasim

Contents

Fundamental of Smart Cities and Data Analytics

Fundamentals of Smart Cities 3
Mohammad Tabrez Quasim, Mohammad Ayoub Khan, Fahad Algarni, and Mohammed Mujib Alshahrani

A Context Aware Big Data Analytics Service-Centric Process Modeling Approach 17
Anindita Sarkar Mondal and Samiran Chattopadhyay

Smart Energy: A Collaborative Demand Response Solution for Smart Neighborhood 43
Sawsan Al Zahr, Elias A. Doumith, and Philippe Forestier

Governance in Smart City: An Approach Based on Social Network 63
Paramita Dey and Sarbani Roy

Algorithms in Smart Cities

An Overview of the Machine Learning Applied in Smart Cities 91
Reinaldo Padilha França, Ana Carolina Borges Monteiro, Rangel Arthur, and Yuzo Iano

Clustering Techniques for Smart Cities: An Artificial Intelligence Perspective 113
Mohamed Tharwat and Ahmed Khattab

A Multi-level Ontology to Manage Service Level Agreements in Smart Cities 135
Cathryn Peoples, Mamun Abu-Tair, Bin Wang, Kashif Rabbani, Adrian Moore, Philip Morrow, Joseph Rafferty, Sally McClean, M. Zoualfaghari, Z. Cui, and P. Kulkarni

Smart Transportation and Mobility

Smart Transportation: A Reference Architecture for Big Data Analytics 161
Camilo Castellanos, Boris Perez, and Dario Correal

Designing for Urban Mobility: The Role of Digital Media Applications in Increasing Efficiency of Intelligent Transportation Management System 181
Stephen T. F. Poon

Validation Frameworks for Self-Driving Vehicles: A Survey 197
Francesco Concas, Jukka K. Nurminen, Tommi Mikkonen, and Sasu Tarkoma

Case Studies in Smart Cities

Big Data for Smart Cities: A Case Study of NEOM City, Saudi Arabia 215
Tanweer Alam, Mohammad Ayoub Khan, Natheer K. Gharaibeh, and Malik Khlaif Gharaibeh

Smart Cities Pilot Projects: An IoT Perspective 231
Syed Rameem Zahra and Mohammad Ahsan Chishti

Performance of Smart Cities Concerning the Use of Internet of Things: A Case Study of Four Indian Himalayan Cities 257
Mahfuzuar Rahman Barbhuiya, Navneet Munoth, and Ranjeet Singh Rajput

Sustainability and Smart Cities: A Case Study of Internet Radio 281
Haifaa Omar Elayyan

An Assessment of Smart Urban Furniture Design: Istanbul Yildiz Technical University Bus Stop Case Study 297
Gökçen Firdevs Yücel Caymaz and Kürşat Kemal Kul

About the Editors

Dr. Mohammad Ayoub khan is working as Associate Professor at University of Bisha, Saudi Arabia with interests in Internet of Things, Blockchain, RFID, Wireless Sensors Networks, Ad Hoc Network, Smart Cities, Industrial IoT, Signal Processing, NFC, Routing in Network-on-Chip, and Real-Time and Embedded Systems. He has more than 14 years of experience in his research area. He has published more than 70 research papers and books in the reputed journals and international IEEE conferences. He is contributing to the research community by various volunteer activities. He has served as conference chair in various reputed IEEE/Springer international conferences. He is a senior member of professional bodies of IEEE, ACM, ISTE, and EURASIP society. He may be reached at ayoub.khan@ieee.org.

Dr. Fahad Algarni is working as Assistant Professor at the College of Computing and Information Technology at the University of Bisha, Saudi Arabia. He received his Ph.D. from Clayton School of Information Technology, Monash University, Melbourne, Australia. He achieved his Bachelor's Degree from the Department of Computer Science at King Abdul-Aziz University with First Class Honors. He also graduated with Master's Degree in computer networks from La Trobe University, Melbourne, Australia. He has published over 15 journal articles, book chapters, and conference papers in various internationally recognized academic databases. His research interests include Wireless Sensor Networks (WSN), Cloud Computing, Systems' Reliability, IoT, and IT Security. He may be reached at fahad.a.algarni@gmail.com, fahad.alqarni@ub.edu.sa.

Dr. Mohammad Tabrez Quasim is working as Assistant Professor at University of Bisha, Saudi Arabia. He received his Ph.D. (Computer Science) from Tilkamanjhi Bhagalpur University, India and Master of Computer Application from Punjab Technical University, India. His research interests include but are not limited to: IoT, Big Data, Cloud Computing, Wireless Sensors Networks, and Blockchain. He has more than 10 years of experience in his research area. He has published more than 30 journal articles, book chapters, and conference papers in various internationally recognized academic databases. He is contributing to the research community by various volunteer activities like Associate Editor, Editorial Member, Reviewer of many Scopus/ISI journals, and Program Committee Member of IEEE/Springer international conferences. He may be reached at tabrezquasim@gmail.com, mtabrez@ub.edu.sa.

Fundamental of Smart Cities and Data Analytics

Fundamentals of Smart Cities

Mohammad Tabrez Quasim, Mohammad Ayoub Khan, Fahad Algarni, and Mohammed Mujib Alshahrani

Abstract The concept of smart cities is getting popularity day by day. Many countries have started adopting the idea of smart city to improve the quality of life by achieving the recommended level of sustainable development. The smart city is a technologically advanced area which could understand the world by analyzing the data in order to improve living conditions. The underlying technology infrastructure of smart cities are wireless sensor network (WSN), Internet of thing (IoT), RFID, and 6G are among others. Along with the technology, the role of machine learning and data analytics can't be ignored. The smart cities generate massive amount of data from the monitoring equipments and sensors. Big data analytics is one of the important technologies which is capable of improving intelligent urban facilities. In smart cities large amount of data is continuously received from many sensors, autonomous machines or intelligent IoT devices. The accurate prediction depends on the approaches of data analytics and machine learning techniques. This chapter presents fundamental of smart cities, vertical in smart cities and data analytics approaches.

1 Introduction

The smart cities have been most addressed issues since last few years, due to the urbanization of the world's population. During the 1950s, only 30% of the world's population lived in cities; but gradually till the end of 2014, the urbanization level had reached 54%, and the United Nations predicts that by 2050 this figure reach to 66% [1]. However, the speed of the urbanization in the developing countries like Asia and Africa is more rapidly growing than other regions of the world. Since the cities are not only centers of human activities, rather also the center of economics, environmental and societal demand are magnified. The process of urbanization may

M. T. Quasim (✉) · M. A. Khan · F. Algarni · M. M. Alshahrani
College of Computing and Information Technology, University of Bisha, Bisha, Saudi Arabia
e-mail: tabrezquasim@gmail.com

M. Khan et al. (eds.), *Smart Cities: A Data Analytics Perspective*,
Lecture Notes in Intelligent Transportation and Infrastructure,
https://doi.org/10.1007/978-3-030-60922-1_1

also cause of different important and significant economic, social and demographic transformation [2].

There has been a great revolution in the information and communication technology (ICT) due to advancement of hardware and software design in last few years.

This revolution in ICT provide a great opportunity to solve different issues related to urbanization. The use of ICT in cities to enhance the different city activities is generally term as "cyber Ville", "digital city", "electronic city", "flexi city", "information city", "telicity", "wired city", and "smart city" [3].

However, the smart city is a concept and still there is not a clear and consistent definition of the concept among academia and practitioners. In a simplistic definition "A smart city is a sustainable, innovative city that uses information and communication technologies and other means to improve quality of life". The efficiency of urban operations and services, and competitiveness in smart cities ensures that it meets the needs of present and future generations with respect to economic, social and environmental aspects. A broad overview of various components needed in a smart city is shown in Fig. 1.

The rapidly increased world population is a great challenge for the current world. The population has increased significantly in the last decades and so has the expectation of living standards. It is predicted that around 70% of the world population will live in urban areas by the year 2050 [3]. At present cities consume 75% of the world's resources and energy which leads to the generation of 80% of greenhouse gases [3]. Thus, in the next few decades there can be severe negative impact on the environment. This makes the concept of smart cities a necessity. The creation of smart cities is a natural strategy to mitigate the problems emerging by rapid urbanization and urban population growth. Smart cities, in spite of the design costs associated,

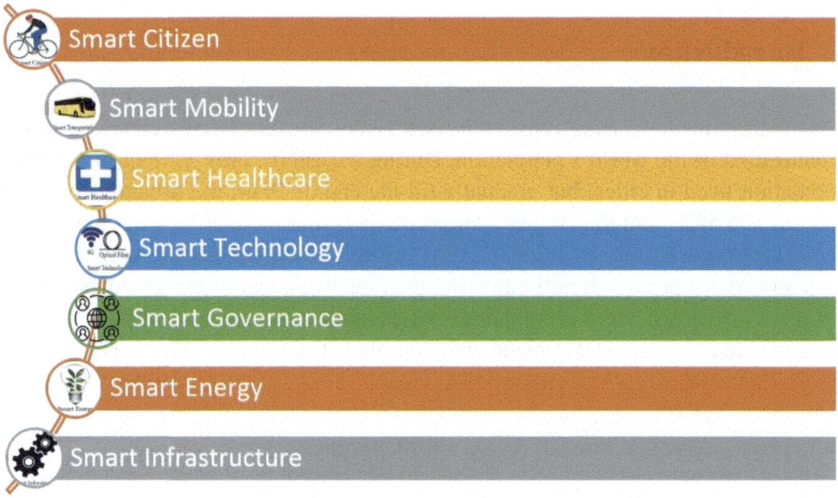

Fig. 1 A broad overview of smart cities components

once implemented then it can reduce operational cost by reducing energy consumption, water consumption, carbon emissions, transportation requirements, and waste. However, the concept of smart cities around the world is slightly different in term of their characteristics, requirements, and components. Some international organization such as the international organization for standardization (ISO), make a standard for globally understood specification to ensure quality, efficiency and safety. These standards are very important in monitoring and controlling the different technical and functional aspects of the smart cities. The ISO 37120 is one of the smart city standard example developed by IEEE which define 100 city performance indicators which include 46 core and 54 supporting indicators [3]. Some selected indicators are the following: economy, education, energy, and environment, which can be used by city civic bodies to benchmark their service performance, learn best practices from other cities as well as compare their city against other cities. The standards like industry 4.0/5.0 and industrial Internet of things are also supporting infrastructure for smart factories [4, 5]. Here, it is also important to mention that the research interest in smart cities among researchers are also growing very fast. We have extracted some data of last five years from the web of science (WoS) using keyword "smart cities". The result has returned 15,660 published papers from different discipline as shown in Fig. 2. From this figure, we can observe that most of the papers are from computer science disciplines as compared to others. The computer science and engineering research has great role to play in smart cities.

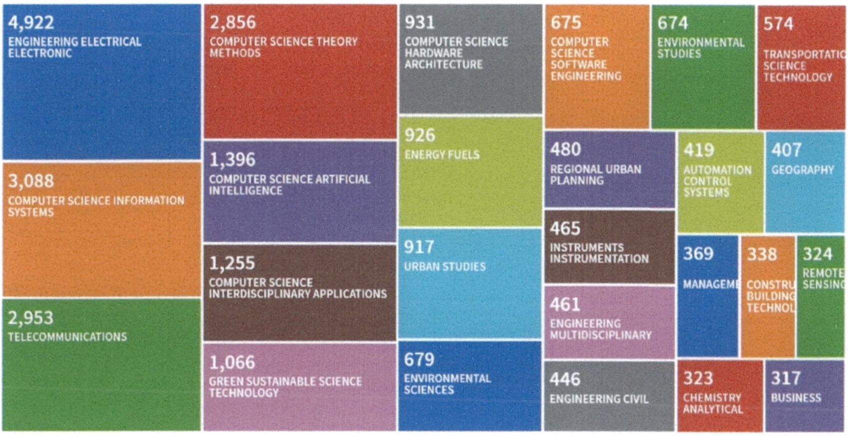

Fig. 2 Number of papers published in last five years indexed in WoS

2 Smart Cities: Components, Characteristics and Role of Big Data Applications

The different components and characteristics of the smart city are summarized in Fig. 3. The components of a smart city include the following: smart infrastructure, smart buildings, smart transportation, smart energy, smart healthcare, smart technology, smart governance, smart education, and smart citizens. Among others the smart healthcare is a demanding requirement for any country [6, 7]. Different smart cities have different levels of components, depending on their focus [8].

Economy, environment, governance and society are the four key themes for smart city. Here smart city society theme means that the city is for its citizens. The smart city economy theme implies that the city is capable to boom with continuous job growth as well as economic growth, whereas the environment theme indicates that the city must be able to sustain its different functions for its current and future generations. The smart city infrastructure includes information, physical and communication technology and different services. The physical infrastructure of the smart city is indeed physical in nature it includes railway track, buildings, roads, power supply line, water supply system, etc. The ICT infrastructure is actually the most essential component of the smart city which controls different operations and components. The smart city service infrastructure is actually based on physical infrastructure and

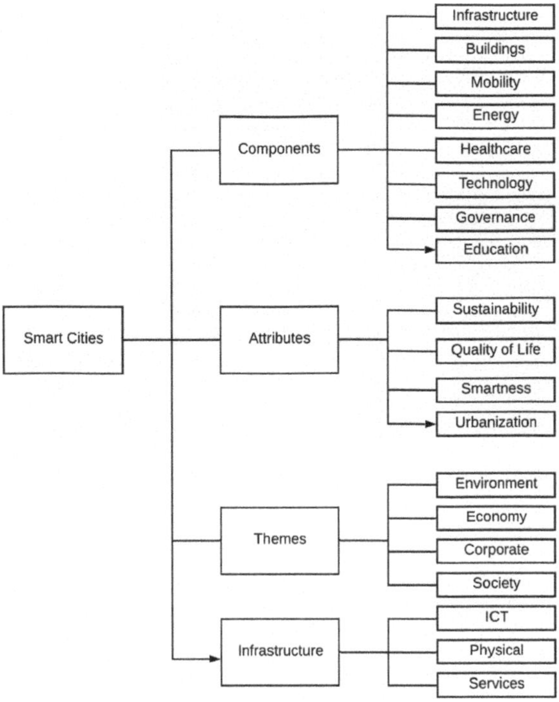

Fig. 3 Components and characteristics of smart cities

have some ICT components. Some examples of service infrastructure are smart grids and mass rapid transit system.

2.1 *Applications of Big Data in Smart Cities*

Big data Analytics can collect, analyze as well as mine data about smart cities projects in such an intelligent way to generate information to improve various smart city facilities. To turn, big data would enable decision-makers to prepare for any growth of smart city infrastructure, facilities, and places. Smart cities and big data applications could be categorized into two groups: disconnected big data apps and connected big data apps. Connected big data apps are distinctive as they depend on immediate feedback and quick evaluation to make a decision or intervention within a brief as well as the very precise timeframe. At certain instances, when a recommendation could not be reached in the timeframe, it will become meaningless. This is critical that all information required for this assessment is made available promptly and also that the study is carried out in a comprehensive and timely manner. In a consequence, big data apps in real-time typically want additional specifications. After designing broad data-based smart city apps, it is important to explain a range of criteria arising from the uniqueness of smart city demands and big data features. The criteria are defined along with the basis of the nature of big data apps and the complexities of developing such apps for the smart city. Many of these conditions are technical, whereas others are linked to the resident's knowledge and the function of authorities [9].

The smart city provides an opportunity for connecting people and places that use emerging technologies that help to improve urban planning and design [10]. Smart cities incorporate technological technologies to enhance public infrastructure and people's living experience. Municipal governments utilize IoT cameras, infrastructure, and software to collect relevant information, like congestion, energy consumption, and air pollution. This information could then be used through technical solutions to enhance community infrastructure, covering services, transportation, and public security. A broader picture of flow of information and data analytics in smart cities is shown Fig. 4.

Emerging communication technologies play a key role in smart cities by providing information gathered across internet technologies. The IoT functions through interacting among smart devices when sending and receiving information that needs internet, mobile networks as well as another way to communicate. Essentially, smart cities are using IoT gadgets to collect and store information effectively for implementation in a specific location. The smart city sensors and smart devices obtain information from various smart city access points configured in a region and then examine it for the great decision-making process. Making sure residents protection has become a primary concern for every community so it is critical to secure citizens in every kind of situation. In order to prevent future issues inside the area, statistical modeling may help to analyze historic and geographic information to determine

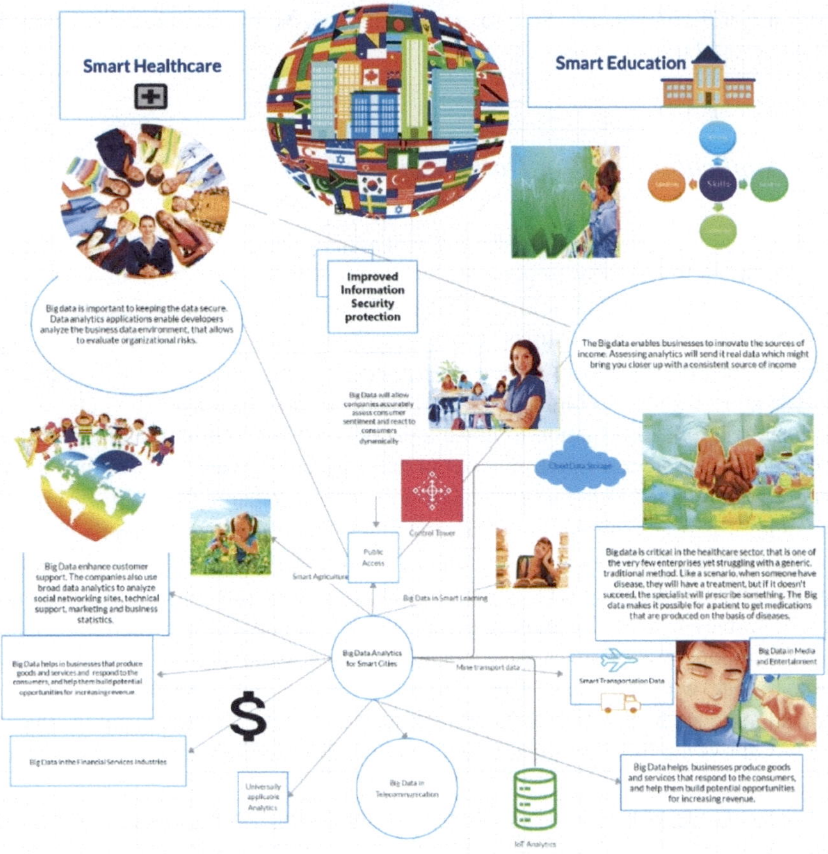

Fig. 4 Big data in smart cities

whenever and wherever incidents are likely to take place. The substantial amount of success could be seen as the required information transform the city into something much better location.

2.1.1 Big Data in Smart Learning

Big data in the field of education provides excellent possibilities for institutions such as mLearning, curriculum development, enhancing the evaluation process, offering career advice to learners and introducing new educational strategies [11–13]. Because of the importance of big data in education, organizations may monitor academic achievement along with several disciplines for the students and the global standard, and create specific strategies to help learners improve. Data analysis of the scores of learners in numerous topics could assist to direct learners easier and the learning schedules could be applied appropriately. Along with the basis of the success in the corresponding courses and the field of study, educators would be able to advise learners about what direction in life they must pursue. Learners would also be capable of recognizing determinants of the success of students as well as coming up with successful approaches by big data analysis. Many learners can only learn through listening, and many learners could learn through understanding. However, some students may learn through watching the videos and others may use some other techniques. However, some learners are forced to contribute to a given model that may be opposed to their kind of learning techniques, and that is something that discourages their development. It can also have a dramatic effect on the student's educational success. The big data analytics provide various educational proposals and would then allow learners the right to select their desired thinking strategy that is why there are big data used. Using the means of big data analytics, organizations would be able to construct various learning strategies that will potentially allow them to make it much more enjoyable.

2.1.2 Big Data in Smart Health

The big data in health services is an excellent fit. This enhances the health sector more than before. A volume of data that the health industry has to concern with would be overwhelming. The days have gone when medical providers have been unable to take advantage of such information. Including seeking a cure from diseases, the big data has brought everything under its umbrella, and investigators have seen too many lives-saving results. Connected wearable technology has recently continued to gain prominence and are the current trends across individuals of all age categories. The smart healthcare creates huge quantities of real-time data in the context of warnings that help to save people's lives [14, 15].

2.1.3 Big Data in Smart Tourism

The big data could help humans in the tourism industries in a variety of many respects, helping everyone to reach better proof-driven choices. This allows predicting potential activity fairly efficiently, to refine business models, to sell information better specifically and to enhance customer satisfaction. Smart cities have had an enormous wealth of experience which can change the future of tourists. But, merely getting knowledge does not have to be prepared to do so efficiently. Smart cities and travel participants would need access to services that will allow them to derive implementable inferences. Several of these findings would allow them to support not only tourists, as well as locals [16].

2.1.4 Big Data in the Smart Retail Economy

Modern technologies have also been a mechanism for socially and economically transformations. Renewable technologies, innovations, and communication channels have brought great changes to the community, to the financial system, and citizens during the recent economic booms. That is not different today. The current economic model of the digital economy is now the basis for commercial activity. People begin experiencing the emergence of a modern technology era. The latest smart ecosystem has an unparalleled effect, which is both overwhelming and satisfying. Retail sector is facing the most extreme rivalry among everyone. Dealers are continually looking for forms that offer businesses a strategic advantage over anyone else. Consumers have always been, in fact, the true winner of sensations for the retail sector. In order for distributors to succeed throughout this dynamic environment, they will have to understand better potential consumers. When they understand the expectations of their clients as well as how to better satisfy their expectations, they understand best [17].

2.1.5 Big Data in Media and Entertainment

The big data in the media and entertainment sector also often allows companies to achieve confidential information through consumer conduct, as well as helps produce customized advertising. The Big data thus held the key to making entertainment and media businesses more competitive. Media and entertainment organizations would use big data to determine which user information, goods, and apps they need. Due to the study of consumer information, software changes became more economically efficient and timely efficient. They would never realize that functionality the customers need or want to deliver when they don't look into the details, it might offer them a significant advantage, increase market share and customer satisfaction [18].

2.1.6 Big Data in Telecommunication

Telecommunications companies are gathering vast quantities of data from phone history reports, cellphone uses, internet infrastructure, database records, bills, and social networking sites, offering a lot of details regarding their consumers and the infrastructure, and how telecommunications companies could use this information to enhance their company. Industries with big data could monitor areas with the minimum and maximum data traffic rates but also do what is needed to ensure trouble-free network access. The big data, as well as other sectors, have allowed the telecommunications industry to recognize its consumers relatively effectively. Telecommunications companies are now supplying consumers with the most personalized services available. The big data would be behind the digital explosion that we are now witnessing. Global telecommunications industries are the essence of the industrial revolution taking place over time. To the ever-increasing popularity of smartphones, the telecommunications company has been overwhelmed with huge amounts of storage. Using methods of big data analytics, firms are willing to have consumers with seamless access, thereby removing all the infrastructure obstacles that consumers are faced with [19].

2.1.7 Big Data in Automobile Industries

The idea of smart devices is not only restricted to wearable devices but has been at the core of automotive manufacturing. Besides the IT sector, the automobile industry has benefited most from the big data. This is revolutionized the way all pass about. There is an opportunity for autonomous vehicles to make our roads safer, as 90% of the number of deaths on the highways is due to the mistakes of humans. Those automobiles require information to be genuine. Vehicles become completely fitted with sensors that monitor everything including location, distance, movement, and steering; traffic lights, traffic danger as well as emergencies. Through making use of certain information, the automobile is in a position to infer and to carry out possible responses without human mistakes [20].

2.1.8 Big Data in the Financial Services Industries

The big data is a common new slogan in the field of computing and analytical approaches for gathering and analyzing vast quantities of data. Since before big data came to prominence, the financial sector had already dominated the technological sector. The financial institutions were among the first to embrace big data. Advancements in computer resources, together with dropping costs, are making massive information initiatives progressively technologically and technically feasible. To general, the emergence of cloud services puts the expense of large amounts of data processing beyond the scope of several smaller companies that do not need to make major reinvestment in their technological assets. Its functionality of every financial institution depends greatly on its information and maintaining the information is one of the most challenging tasks facing any financial company. Information was the second biggest resource after money for everyone. Internet banking and transactions are one of the trendiest phrases and big data seems to be at the center of this one. The big data is responsible for important areas of financial companies including fraud prevention, risk management, statistical arbitrage, and service and support [21].

3 Technology in Smart Cities and Challenges

The underlying infrastructure of smart cities are connectivity, mobility, sensors, big data analytics and machine learning techniques as shown in Fig. 5.

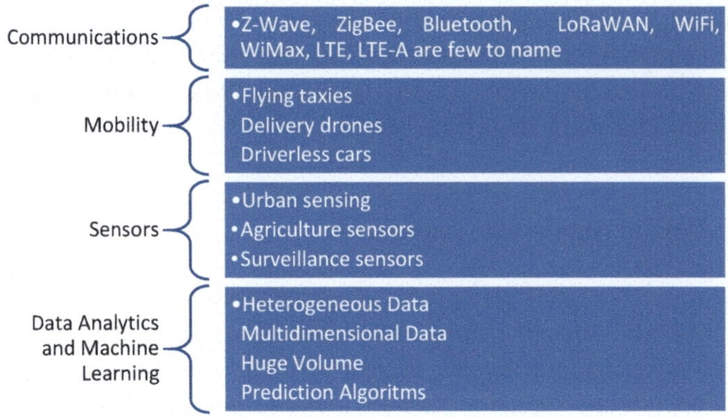

Fig. 5 Technology in smart cities

3.1 Communication Technologies

The connectivity is a key technology to smart cities. Local governments are turning concrete cities into artificial intelligence and sensors converging physical and virtual worlds through a large network of connected applications and devices via an Internet of things enabled infrastructure. In order to solve urban problems, cities now may take advantage of real-time city functions, including traffic control, public transport, surveillance cameras, sensor systems, air quality monitoring, etc. The smart city projects tackle key areas such as urban ecosystems, smart energy and resource efficiency, with intelligent data analysis and predictive diagnosis systems. The smart city planners have initiated several sustainability solutions to develop non-greenfield communities as part of their smart national initiatives. The enabling technologies includes Bluetooth, Z-Wave, ZigBee, LoRaWAN, WiFi, WiMax, LTE, LTE-A are few to name [22].

3.2 Mobility

The strategy of the smart cities project is to build a connected infrastructure that can generate data to manage traffic flows with heterogenous sensor-enabled physical devices. In addition, recent trends in the intelligent mobility revolution have shown an ever-increased interest in autonomous technology with the greatest investment opportunities as below:

- Electric and renewable energy-based mobility
- Driverless cars
- Artificial intelligence, big data and cloud-based automobile [23]
- Flying taxis

- Drones based delivery of supply chain
- Shared mobility
- Mobility as a service (MasS)

Nonetheless, the promise of smart mobility is an audacious dream that involves both regulatory and technological collaborations between government and business. This is also important to mention that people should be educated enough to adapt to the new concept of efficient mobility [8, 22].

3.3 Sensors

The sensors are the building block to capture the data in smart cities. The sensors collect the information regarding mobility, energy consumption, air pollution, cultural information, tourism information and many more. The computing and sensing infrastructure is integrated into smart cities for control and decision-making [8, 24]. The key infrastructure of smart cities is a network of sensors, cameras, wireless devices, and data centers, allowing municipal authorities to provide essential services more quickly and efficiently. Sensors that will communicate continuously with the people to manage public services in real-time to improve the quality of life through traffic management, waste collection, disposal, irrigation systems, parking, alerting the municipal authorities for better governance. Smart cities also use sustainable building materials and reduce energy consumption, which is much easier for the environment. It is important to mention that the cost of the sensors, energy consumption, resilience, adaptability, self-learning capability, deployment, user experience, cognition, and maintainability plays crucial role for the success of the smart city.

3.4 Data Analytics and Machine Learning

A greater awareness of the environment, CO2 emissions and a variety of other factors allows policymakers realize the importance of people-centered cities. The new design requires huge analytics on the gathered data. The variety of data such as geospatial data, crime records, weather data, mobility data, traffic and any more need to be sensed and analyzed. Technologies such as AI and machine learning allows analytics to analyze data using complex algorithms to link to different sources, to reveal useful information to urban planners. The smart city planning does not only concern the present, it also concerns the future. The urban city planning improves living standards for the people but an influx of people can put pressure on resources and disrupt the quality of life. While making policies and programs for smart city, the planners need to take account of the future.

In the predictive analysis algorithms, what data analytics can do is forecast the future. The urban planners and experts can get a precise number of people coming into the city and plan accordingly using analytical algorithms for the services needed.

The rising number of IoT devices used in all cities has the potential to change our perception of how cities work. The continuous research and development in machine learning, analytics, visualization provides effective tool for city planners to help make sense of these data flows. In turn, cities are developing new data integration and sharing platforms which break down traditional transactions and provide access to all stakeholders and potential providers of solutions.

The advantages of big data are therefore an important part of many smart city approaches as follows.

(1) Predictive transport analysis could reduce congestion and enhance performance of public services.
(2) City resources can be utilized more efficiently by up-to-date analysis for public security, social care and other important services.
(3) The most vulnerable households and appropriate buildings for retrofitting can be aimed at energy efficiency programs.
(4) Open data platforms can enhance citizen participation, innovation, and creativity among developers and other service providers.

3.5 Smart City Design: Challenges and Opportunities

There are many complex and diverse challenges for building smart cities such as cost, efficiency, sustainability, communication, safety, and security are few to name. Different factors, including natural environment, government policy, social cultures and economy, control these design challenges. The important factor to design any smart city is the cost involved that includes design and operational cost. The cost of design for smart cities is a one-time cost, however, operation cost is recurring. To make a smart urban realization possible, design costs must be small. The small cost will also facilitate the sustainability of cities, with minimal burden on the city. The optimization of design and operational cost is one of the major challenges in smart cities. The operational cost and sustainability can be reduced by increasing operational efficiency. Reducing pollution and urban waste is necessary to improve sustainability and efficiency and reduce operating costs.

The smart cities also have to confront with population growth to ensure long -term sustainability and reduced operating costs. The increasing population may decrease the efficiency of the smart cities. In addition to efficiency, the cities should be resilient to failures and natural disaster. The smart cities should be able to recover quickly from the disaster and failures.

Lastly, the security of the data and underling infrastructure is among major challenges. Public safety is, above all, an important design challenge for smart cities as the health of the public is crucial which may raise budgets for design and operation.

4 Conclusion

In recent years, smart cities became an important area of research. In this chapter we have presented the components of a smart city which include the smart infrastructure, smart buildings, smart transportation, smart energy, smart healthcare, smart technology, smart governance, smart education, and smart citizens. Different smart cities have different levels of smart components, depending on their vision. However, in every smart city there are many complex and diverse challenges. The design costs must be small to make smart urban development feasible. The low cost will also help to make cities more sustainable, with minimal burden on the nation. The role of machine learning and data analytics can't be ignored in conjunction with other technologies for building a smart city.

References

1. United Nations, Department of Economic and Social Affairs. World urbanization prospects: the 2014 revision. New York (2014). https://esa.un.org/unpd/wup/Highlights/WUP2014-Highlights.pdf
2. Abu-Lughod J, Hay RJ (2013) Third world urbanization. Routledge Kegan & Paul, Abingdon
3. Mohanty SP, Choppali U, Kougianos E (July 2016)Everything you wanted to know about smart cities: The Internet of things is the backbone. IEEE Consumer Electron Magaz 5(3):60–70. https://doi.org/10.1109/MCE.2016.2556879
4. Abuhasel KA, Khan MA (2020) A secure industrial Internet of Things (IIoT) framework for resource management in smart manufacturing. IEEE Access 8:117354–117364. https://doi.org/10.1109/ACCESS.2020.3004711
5. Khan MA, Abuhasel KA (2020) Advanced metameric dimension framework for heterogeneous industrial Internet of things. Comput Intell 1–21. https://doi.org/10.1111/coin.12378
6. Khan MA, Quasim MT, Alghamdi NS, Khan MY (2020) A secure framework for authentication and encryption using improved ECC for IoT-based medical sensor data. IEEE Access 8:52018–52027. https://doi.org/10.1109/ACCESS.2020.2980739
7. Khan MA (2020) An IoT framework for heart disease prediction based on MDCNN classifier. IEEE Access 8:34717–34727. https://doi.org/10.1109/ACCESS.2020.2974687
8. Mardacany E (2014)Smart cities characteristics: importance of buit environments components. In: IET conference on future intelligent cities, London, pp 1–6. https://doi.org/10.1049/ic.2014.0045
9. Al Nuaimi E, Al Neyadi H, Mohamed N, Al-Jaroodi J (2015) Applications of big data to smart cities. J Internet Serv Appl 6(1):25
10. Khan Z, Anjum A, Soomro K, Atif Tahir M (2015) Towards cloud based big data analytics for smart future cities. J Cloud Comput 4(1):1-11
11. Hammad R, Ludlow D (2016) Towards a smart learning environment for smart city governance. In Proceedings of the 9th international conference on utility and cloud computing, pp 185–190
12. Tabrez Quasim M et al (2019) 5V's of big data via cloud computing: uses and importance. Sci Int(Lahore) 31(3):367–371
13. Tabrez Quasim M, Meraj M (2017) Big data security and privacy: a short review. Int J Mech Eng Technol 8(4):408–412. https://www.iaeme.com/IJMET/issues.asp?JType=IJMET&VType=8&IType=4
14. Pramanik MI, Lau RYK, Demirkan H, Abul Kalam Azad M (2017) Smart health: big data enabled health paradigm within smart cities. Expert Syst Appl 87:370–383

15. Khan MA, Algarni F (2020) A healthcare monitoring system for the diagnosis of heart disease in the IoMT cloud environment using MSSO-ANFIS. IEEE Access 8:122259–122269. https://doi.org/10.1109/ACCESS.2020.3006424
16. Xiang Z, Fesenmaier DR (2017) Big data analytics, tourism design and smart tourism. In: Analytics in smart tourism design, pp 299–307. Springer, Cham
17. Bradlow ET, Gangwar M, Kopalle P, Voleti S (2017) The role of big data and predictive analytics in retailing. J Retail 93(1):79–95
18. Schlieski T, David Johnson B (2012) Entertainment in the age of big data. Proceedings of the IEEE 100, no. Special Centennial Issue (2012): 1404–1408
19. Wang Z, Wei G, Zhan Y, Sun Y (2017) Big data in telecommunication operators: data, platform and practices. J Commun Inf Netw 2(3):78–91
20. Lv S (2019) Design of the automobile marketing system based on the big data. In: International conference on big data analytics for cyber-physical-systems, pp 1713–1719. Springer, Singapore
21. Fang B, Zhang P (2016) Big data in finance. In: Big data concepts, theories, and applications, pp 391-412. Springer, Cham
22. Yaqoob I, Hashem IAT, Mehmood Y, Gani A, Mokhtar S, Guizani S (2017) Enabling communication technologies for smart cities. IEEE Commun Magaz 55(1): 112–120, January. https://doi.org/10.1109/MCOM.2017.1600232CM
23. Quasim MT (2015) Artificial intelligence as a business forecasting and error handling tool. COMPUSOFT Int J Adv Comput Technol 4(2), February-2015 (Volume-IV, Issue-II)
24. Quasim MT, Khan MA, Abdullah M, Meraj M, Singh SP, Johri P (2019) Internet of things for smart healthcare: a hardware perspective. In: 2019 first international conference of intelligent computing and engineering (ICOICE), Hadhramout, Yemen, 2019, pp 1–5. https://doi.org/10.1109/ICOICE48418.2019.9035175

A Context Aware Big Data Analytics Service-Centric Process Modeling Approach

Anindita Sarkar Mondal and Samiran Chattopadhyay

Abstract In the dynamic process based big data analytics environment, user's service-related demand, services provided by service unit, and big data volume and velocity are associated with data processing tasks. For proper assignment of appropriate process based services, modeling approaches takes an important role. The process model needs to be context-aware for handling the dynamic behavior of these components to support the versatile dependency on it. The existing process modeling approaches are not able to support these circumstances. This creates a gap between the design time and run time behavior of the process model. To solve this issue, here we propose a modeling approach where it considers context-related knowledge of the data analytics services and a state transition diagram to control the activity flow at run time by determining the workflow pattern and assign the corresponding services. The process model will be much flexible that can handle the dynamic behavior of a process-based data analytic environment by controlling the execution of processes. Here, we also provide a framework to support our technique to design the context-aware process model and provide some examples for determining how it will work in a real-life scenario.

Keywords Big data analytics · Context-aware · Process model · Service unit

1 Introduction

In the competitive business world, asset maintenance is the key factor to increase the manufacturing product. This challenge makes the necessity to analyze the asset of health-related big data for making the decision about the health of an asset. Through

A. Sarkar Mondal (✉) · S. Chattopadhyay
Department of Information Technology, Jadavpur University, Kolkata, India
e-mail: sarkar.anindita5@gmail.com

S. Chattopadhyay
e-mail: samirancju@gmail.com

M. Khan et al. (eds.), *Smart Cities: A Data Analytics Perspective*,
Lecture Notes in Intelligent Transportation and Infrastructure,
https://doi.org/10.1007/978-3-030-60922-1_2

the process modeling approach, the working procedure of the involved processes is monitored, controlled, and simulated in a structured way to achieve the business goal [7, 19]. From this perspective, it needs to consider the knowledge of the processes, it's provided services and dependent components. In recent years, it is more important to support the executable process-based data analytic environment and user satisfaction rather than informing of what process doing what task [22].

In the process based data analytic environment, there is the involvement of multiple types of analytic process engine instances[1] to reach a particular data analytic goal for supporting the process-based services. Each analytic process engine instance holds a specific data analytic goal. The goal of this chapter is to configure a new process modeling approach by considering the usability of the data analytic process engine instance. Through this process modeling approach, we consider when and at what respect, which process engine instance will invoke to reach the data analytic goal.

Therefore, here we consider the context-related knowledge of the dynamic data analytic environment to configure the process model. The context related knowledge [25] mainly describes the collection of information of the data analytic components (e.g.., user's choice, available data analytic process engine instances, involved processes and their sequence to reach a data analytic goal, incoming data volume from the data sources) involve in this context. It makes beneficial to the process modeler of the data analytics system to take the decision which process engine instance will be executed and the execution pattern of this instance.

In process model design, it needs to consider the relationship between data analytic components with the user's choice. This relationship is represented via context-related knowledge. Different tools (e.g.., jbpm [2], stata [4], pegasis [3], taverna [5], WSBPEL[2]) are used to configure activity flow of data analytic operation. IBM WebSphere [1] is used to configure the artifact-based process model and context4BPEL [32] is used to control the activity flow on the base of the context event. To handle the big data, Oozie[3] is used to provide the apache Hadoop jobs. But these modeling approaches are not flexible enough to handle all the components (i.e., big data principle, user's demand, and service unit) of a dynamic process-based data analytic environment.

1.1 Contribution

The contribution of this chapter is presented below.

1. A modeling approach for the context-aware process model. Through this model process engine instance is selected and arranged on the basis of present data analytics context.

---

[1]It is a kind of business process engine, running under the data processing purpose.

[2]https://www.oasis-open.org/committees/tc_home.php?wg_abbrev=wsbpel.

[3]http://oozi.e.apache.org/.

2. A framework to describe the working procedure of this modeling approach. Here, it shows how this modeling approach is able to minimize the gap between design time and run time behavior of data analytic processes.

1.2 Organizations

In the next section, we have presented our motivation and scenario of the proposed work. Section 3 describes the related work of the presented topics. Detail terminology has been presented in Sect. 4. Representation of context knowledge in the form of process service decision tree which is one of the contributions is presented in Sect. 5. Another major contribution, state transition diagram of control units has been presented in Sect. 6. In Sect. 7, some use case scenarios have been given to explain the proposed approach. At last, a small conclusion and future work are given in Sect. 8.

2 Scenario and Motivation

Here we consider the task of asset maintenance via process-based data analytic service [23] under the scenario of smart city domain [16] and in the business world. For instance, consider the involvement of two type stakeholders as shown in Fig. 1: (A) Asset Manager,[4] name of such agency or company, responsible for maintenance of assets like chiller and heater. For maintaining assets, he needs to manipulate the monitored data of chiller or heater [21] in the process based data analytic system. (B) Service Provider, name of the company, or organization who is responsible to provide the information of services along with the service units. This process-based data analytic system is mainly concerned about the process engine instance to perform the data analytics job.

Therefore, the service provider is responsible to provide the process engine instances as service units. To perform this data analytic job, the different user has different business and economic demands of process engine instances which are controlled through Quality of Result model (QoR model) and Ascertain of the Process model (AoP model). QoR model shows service-related demand like execution time, execution cost, processed data size, and output data quality, etc. [31]. AoP model shows to reach the data analytic objectives which processes should have to execute and in which order. On the other side, the service provider mentioned the information of the data analytic process engine instance through the data process engine information model.

Here we give a simple example to describe our motivating scenario. We assume that the asset manager needs two processes one is a sampling (P1) and another is

[4]Here asset manager acts as a user of the data analytic system.

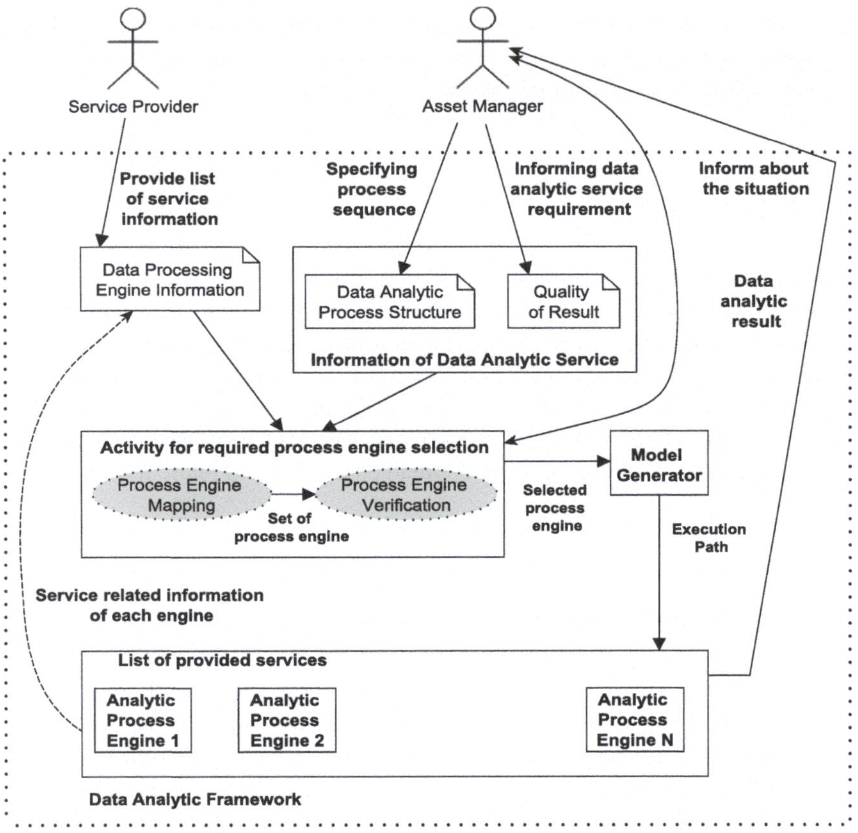

Fig. 1 View of the process management system to configure the context aware big data analytics service-centric process model

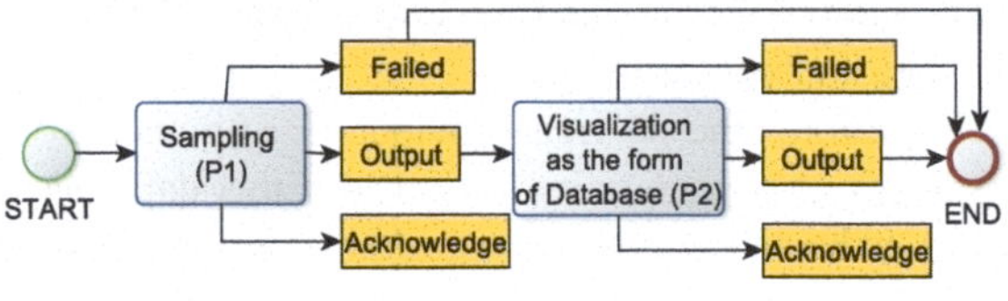

Fig. 2 Example of ascertain of processes

visualization as the form of database (P2) is shown in Fig. 2. At the same time, the asset manager also mentioned the QoR model corresponding to each process mentioned in the ascertain of a process model. An example of such QoR model of Fig. 2 is given in Table 1. The table shows the QoR value to process 0 TB to 2 TB text data. Also, the service provider provides the process engine information metrics to inform about the service-related information of each process engine instance to the asset manager. An example of such process engine information metrics to process per TB text data is given in Table 2.

Table 1 Example of quality of result metrics

Process name	Execution time (min)	Execution cost (euro)	Data accuracy (%)
P1	<10	<81	>85
P2	<11	<45	>85

Table 2 Example of process engine information metrics

Process engine name	Execution time (min)	Execution cost (euro)	Data accuracy (%)	Analytic objective
PE1	8	50	90	Sampling
PE2	9	40	86	Sampling
PE3	2	82	99	Sampling
PE4	11	22	87	Visualization as the form of database
PE5	10	22	85	Visualization as the form of database

Here we mentioned certain situations to explain the challenging factors, faced by the process model designer, in the above scenario:

Situation 1: User-defined business and economic demands are going to make the restrictions for designing the executable process model. For example, by considering the service requirement (i.e. execution cost, execution time, data accuracy) of P1, it is going to match with the analytic process engine instance PE1 and PE2 for 1 TB data volume. Also, there have another processing engine instance PE3 that is able to fulfill all the service demand except data accuracy. Therefore the process engine instance will be selected between PE1 and PE2.

Challenge: Here we observe that there have a number of process engine instances under the same data analytic objective but all are not able to perform this task. Therefore the challenge is, how the system automatically decides the proper service unit for a particular process to satisfy the user's expectation.

Situation 2: One of the main big data challenges is changing big data volume over time. This is another factor for designing the process model of the data analytic framework. Considering the situation 1, PE1 and PE2 are selected for process P1 at 1 TB input data volume. But when the input data volume is 2 TB, only PE2 will be able to fulfill all the demands of users.

Challenge: Here we observe that the process engine instance is changed to support the big data volume. At the run time, how the system defines the proper process engine instance and also decides the workflow pattern for a particular process to support the change of data volume with time.

Situation 3: The system should be flexible enough to choose a process engine instance before stopping the execution even after the lack of proper process engine instance. For example, for the process P2, both the process engine instance PE4 and PE5 are not applicable to fulfill all the user's demands. Therefore, if the execution time is more important than data accuracy to the users, PE5 will be selected else PE4 will be selected at the place of P2. **Challenge**: How the system will be intelligent enough to provide this flexibility to the users.

These above challenging factors expressed that there needs a proper process modeling approach that is able to make the relationship between user and application to handle these situations. Here we consider certain principles which are going to solve the above situations by incorporating them in the process model:

Context-aware: In [15], the author has defined the context as the execution environment which is changed with the user, user's social status, and resources. He also mentioned that context is any information which helps to make the interaction between human and application by characterizing the situation of entities (e.g.., person, place, object). In the above situations, we noticed that the context depends upon the user's demand, data volume, and available resources. Therefore, to show the holistic view of these contexts, a new modeling approach is provided.

User's Demand: For the data analytic service organizations, the main role is played by the user in the service collaboration as shown in the above situations. In this case, users can come from a technical or non-technical background. His main issue is designed services are able to fulfill his business and economic demands. Through the quality of the result model [31], the author mentioned how users can control the data analytic job. The modeling approach should be able to describe that the user's demand is associated with service selection.

Service-centric: Technologies are increasing day by day. Here, the main problem is how to use these provided technologies to build a better service from a business and economic perspective. So, the user can involve only the needed services that he wants to include instead of considering the complex technical details of the processes. For the service collaboration, it is needed to interact between the service provider and the user to confirm the available services and required services. This modeling approach consults the selection and arrangement of data analytic process engine instances to build up the process model.

Big Data Volume and Velocity: Nowadays, the main challenge is to handle the big data. To provide better asset management service, the asset should be monitored in real-time. In this way, it will be possible to take the pre-action regarding the asset. This process modeling approach considers this problem and provides a possible solution at run time.

In this chapter, our main aim is to propose a modeling approach for process-based big data analytics framework which is context-aware, service-centric, able to fulfill the user's demand and big data volume and velocity to handle the unpredictable situation.[5] In such a way, it will be able to provide better asset management services via monitoring the assets in real-time.

3 Related Work

In this section, we point out certain process modeling approaches that are helpful to build up this novel process modeling approach. We arrange them according to their working procedure. Here we also point out certain paradigms of process modeling which help to model this new process modeling approach.

3.1 Activity Oriented Approach

This type of modeling approach considers data and control flow of the associated processes as a main building block of the process model design. Through WS-BPEL,[6] invoke the business processes via the web services. In the article [32], the author extends WS-BPEL to the context-aware workflow management system (Context4BPEL) for controlling the activity flow in the new process-oriented domain. Here context event is used to start any technical processes. Other related tools, like BPMN [2], follow the procedural process modeling structure, and YAWL [6] follows the declarative process modeling structure. In these tools, tasks are communicated via the exchange of messages or events and also tasks are not dependent upon the data principles. Through this process modeling approach, our most important aim is to make our users as much satisfied as possible. Therefore, for this reason, we involved the users only those situations which are needed to provide flexible services without pushing them to provide unnecessary information like how a process engine will work. But the working procedure of these type of tools does not support this condition.

3.2 Artifact Based Approach

Artifact based approach is an integral part of service-centric process modeling. In [24], the author mentioned such an artifact-based approach that focuses on the pre and postcondition of business services without considering what types of goods are going to produce. For supporting this type of job, there have some information entities

[5]We said them unpredictable situation because the system is not able to predict the timing of the occurrence of a particular event (e.g.., changing of data volume).

[6]http://docs.oasis-open.org/wsbpel/2.0/OS/wsbpel-v2.0-OS.html.

to hold information about business process goals, known as Business Artifacts [12]. These artifacts are used to design the business process model to determine when and which services will be executed. Through this modeling approach, different stakeholders can communicate with each other regarding the services [13]. IBM WebSphere [1] follows the techniques of service-oriented architecture to design the services by considering the artifact-based process modeling approach. It supports the interconnection between different processes and monitoring of the tasks.

Artifact based modeling approach considers the information entity (presented by business artifacts) which is used to hold the information about the services, how and what way they are going to reach the business goal. In such a way, if another person comes and takes charge of these services then they will be able to know the whole working procedure by using the business artifacts [13]. Therefore the business artifacts represent the pre and post conditions of the processes. In our proposed process modeling approach, the context related knowledge is used to represent the information of what services are able to handle the data volume by fulfilling the user's demands. This type of information model does not support by the structure of business artifacts (i.e., class and attribute related structure). In this proposed process modeling approach, we apply the concept of an artifact to represent the information of each process.

3.3 *Workflow Pattern and Case Handling*

Workflow pattern [8] is used to describe the business requirements in the workflow specification. The process modeling approach, based on workflow pattern focuses on the control flow of the activity. Now a days, there have different number of tools like [2–6], which follow the concept of workflow pattern based modeling to design the process model. This modeling approach considers the context to determine the structure of the process model in certain situations like merge, split, choice, sequence, and cycle which are defined at design time. At the run time, there have the possibility to do not occur in this type of situation. According to our proposed approach, we consider the run time to decide the process model structure depending upon the data analytics environment.

Case handling [9] shows a new paradigm of the business process modeling. It firstly represents the necessity of human support in each service of the process model. For the business process modeling, it shows that the process model should be more focused on "what can be done rather then what should be done". Adaptive case management [30] follows the concept of case handling to perform the process modeling job in an unpredictable situation. This job is done by knowledge workers where they consider business rules, business events, and users (e.g.., regular customer, V.V.I.P. customer) driven pattern to solve the problem. Oracle BPM [7] tool supports the adaptive case management to provide the customer with expected services. According

[7]http://www.oracle.com/us/technologies/bpm/overview/index.html.

to this modeling approach, knowledge workers find out a service unit (e.g.., people, activity) which is able to solve the occurring case at run time. But it is not enough to handle the incoming case, it also has to support the user's expectation (e.g.., time, quality, cost) regarding service unit selection. In such a way, it makes better satisfactory results to the users. Our process modeling approach considers this concept to build up the process model and also considers the users' demand for service unit selection.

3.4 Context-Aware Approaches

In the definition of context, Dey [15] mentioned that "the context is any information which is used to characterize the interaction between the human, application and the surrounding environment". The author of the article [14] represents how a context-based system works in the service-based world. It represents a strong relationship between the environment and adaptation which are a key factor of the context-based system. In [27], the author proposes the concept of context-aware process design to support the need for flexibility in business processes. In the article [20], the author mentioned that at the service collaboration time, context awareness makes an increase in adaptation of services.

The context related knowledge ([25, 28]) is used to design the role driven business process model. It is designed on the basis of four categories: time, location, organization, and resource to manage the context-related information in a graph-based structure [28]. Here the author describes how context-related knowledge is used to select the entities of the business process model. The context model for BPM (CM4BPM) is used to write the various context-related rules and it's a related query on context using context-related knowledge [25]. Here the context is limited in the relationship between actor and role. Liptchinsky [17, 18] applies context-dependency rules in the social network for process collaboration purposes. In the executable data analytic environment, the working procedure of the processes depends upon the big data principle, involved processes (e.g.., data analytic process engine), and the choice of involved actors (e.g.., user). For this reason, the actor and role relationship are not able to solve this type of problem. Our proposed process modeling approach considers the relationship of all the components of an environment and takes the final decision for modeling the processes.

The context-awareness for service adaptation is already applied in different domains such as e-health, location finding, advertisement services [10, 11]. Here, the service is mainly web service which is going to collaborate with devices on the basis of device identity (e.g.., mobile, personal computer) and user identity to design the user-friendly interface. Omni [10] presents a context-aware location-based service-oriented architecture that maps with an advertisement, routing services, finding products, and management of daily tasks to provide a smart lifestyle of citizens. Context-Aware Workflow execution Framework (CAWE) [11] are used to control the user's interface via the context-aware web service adaptation. In this type of context-aware process modeling approach, the context is focused upon the user's requirement. To provide a satisfactory result, it needs to consider the system's state with user demand. For example, in a data analytic environment, context should be aware of the system's state like incoming data volume, available service unit to process the incoming data amount. Our process modeling approach considers all the components associated with the data analytic environment (e.g.., the system's state, user's requirement) to produce a satisfactory service.

4 Context Related Knowledge Elicitation

To configure the proposed process modeling approach, we consider context-related knowledge and state transition diagram as the main key concept. Here context-related knowledge is represented as the form of a decision tree. The state transition diagram is used to decide the possible action at execution time on the basis of context-related knowledge and occurring event. In this section, we describe the components of context-related knowledge as the context of our process management system. The state transition diagram has been discussed in the subsequent section.

In this section, we will first describe the information of process engine instances to present how it is related to the data analytic framework. Also, the structure of process engine instances is represented to understand the provided services by an instance. This will help to understand how the service provider and asset manager can present the service-related information of a process engine instance to distinguish between different engine instances. After this, we present how the asset manager represents their requirements related to the selection of process engine instances to determine the elasticity of the needed services. At last, we mention how the service provider presents the service-related information of the provided process engine instances and requires details.

4.1 View of Multi-level Data Analytic Framework

According to the hierarchical view of the data analytic framework presented in Fig. 3, the main component of this framework is the data processing engine. The example of the tools is used to configure the data process engines are Hadoop, jbpm, esper. To distinguishing the process engine instances are made by their providing service quality under the same data set. The same processing engine instance, for example, Hadoop engines are different by their running workflow model for a particular data analytic objective. This workflow model is executed by performing a certain number of tasks as a process service unit sequentially. For performing the task of the workflow model, it takes the help of different tools that run under different virtual machines.

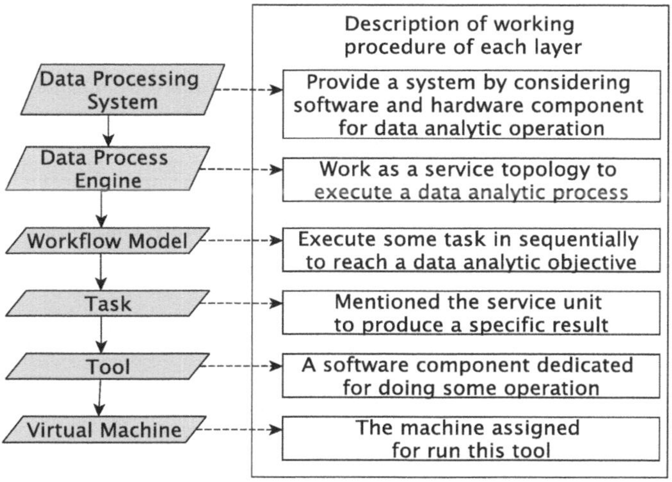

Fig. 3 Hierarchical view of the components of Data Analytic Framework

Under a specific data analytic objective, the working procedure of the data process engine is varied with the data type, data amount, execution time, execution cost, and output data quality.

Here we consider the most three properties which are followed by a process modeling approach to support the data analytics services under this process based data analytic framework.

1. Data Dependency: determines to reach a data analytic objective, what types of data are needed as input, and what type of data will be produced after performing the data analytic operation.
2. Control Flows: It represents the sequence of processes, executed one after another to reach a data analytic goal.
3. Business and Economic Demands: For data manipulation purposes asset manager has a different requirement regarding execution time, execution cost, and output data quality.

4.2 Capturing Elastic Specifications

In the data analytics environment, the asset manager act as a user. The user has some business and economic demands regarding the contexts of data analytic processes present in the form of a service model. It concentrates on the utilization of the service of a particular process regarding execution cost, execution time, and output data quality. Asset Manager provided a service model has two parts:

Quality Of Result Model

Through the quality of the resulting model, a user mentioned the data analytic service-related demand. There have three dimensions of the quality of the resulting model:

1. Execution Time: determines the user's expected time limit to complete the job.
2. Execution Cost: the cost limit user's want to expend for this data analytic job.
3. Data Quality: determines the expected output data quality. Under the data quality, there have different sub-dimensions like data accuracy, data completeness, data consistency, data reliability.

Definition 1 Quality of Result model is one type of ontology-based service model by which the asset manager describes the data analytic service elasticity to get his own expected data analytic result.

This model holds a specific unique id to identify the quality of the resulting model of each process of a particular user. It considers the elasticity dimension for the quality of the resulting model is arranged as the classes. There have four types of data property to represent the content value of these classes:

hasPriority: The filtering operation of the process engines is done on the basis of the priority value of the classes. For a class has priority 1 means first filtering will be done on the basis of this class.
hasDataCondition: User's provided data value of each class has some condition like greater than, less than, equal, greater, and equal, less and equal.
hasDataValue: It represents the elastic data value of this class.
hasDataUnit: It represents the union of the data value like a minute, euro.

There have also another class *ProcessedDataSize* which is used to represent up to what limit of data size this elastic information will be valid. Data properties of this class are, hasDataUnit, hasLowerProcessedDataSize, hasUpperProcessedDataSize. The ontology-based diagram of this model is presented in Fig. 4. The usability of this model for the elasticity specification in data analytic services is presented in listing 1.

Listing 1 Example of the content of QoR model for data analytic process

```
ExecutionCost less than 50 euro, hasPriority=1
Execution Time less than 20 %, hasPriority=2
DataAccuracy greater than 10 %, hasPriority=3
ProcessedDataSize in between 0 TB to 2 TB
```

Ascertain of Process Model

The quality of the result model describes the business or economic demands of the asset manager but it properly does not describe the information of the data analytic objective. Through this model, the user is going to represent the type of processes

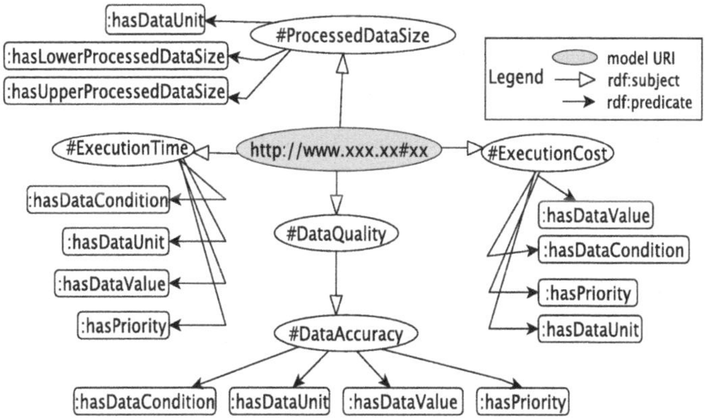

Fig. 4 Ontology based diagram of Quality of Result model

Fig. 5 Ontology based diagram of ascertain of process model

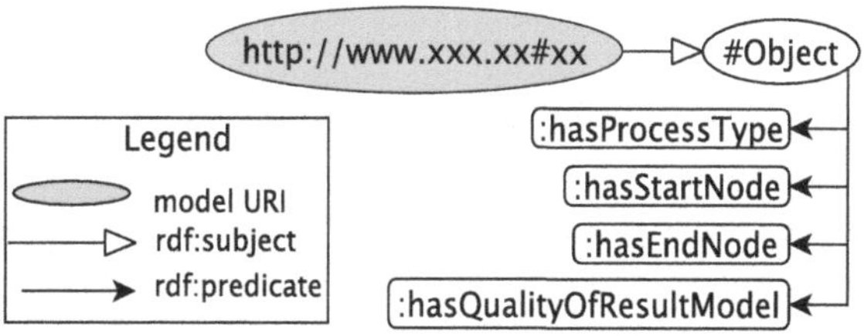

with an execution order for performing the data analytic job. Through this model asset manager mentions that which quality of result model will be valid with which process type. In Fig. 5, we present the ontology-based diagram of Ascertain Of Process Model. In this chapter, we do not consider the data variety (e.g.., text, image, audio, video) which is involved in data analytic operation. This approach part is focused on the process type (e.g.., Querying, sampling, clustering) and the control flow of processes under a different context. Therefore, here we consider process type as a key factor for process engine instance selection.

Definition 2 This type of ontology-based service model is provided by asset manager to describe the sequence of processes with the specification of the quality of result model applicable with which process. In this ontology model process, the type is placed as classes.

To represent the context of these classes we consider four property: hasProcessTypeName, hasQualityOfResultModel, hasEndNode, hasStartNode. In listing 2, we present the example of the data property value of the ontology-based ascertain of the process model of the process sequence presented in Fig. 2.

Listing 2 Example of the content of Ascertain Of Process model

```
@ Sampling Process (URI= "http://processsequence.com#Sampling")
(Data Property= hasProcessTypeName: Sampling
hasQualityOfResultModel:
http://DataAnalyticServiceDemand.com#ResultQualityOfSampling
hasStartNode: START hasEndNode:
http://processsequence.com#Visualization)

@ Visualization as the form of DataBase Process (URI="http://processsequence.com
    #Sampling") (Data Property=
hasProcessTypeName: Visualization as the form of DataBase
hasQualityOfResultModel: http://DataAnalyticServiceDemand.com#
    ResultQualityOfVisualization
hasStartNode: http://processsequence.com#Sampling
hasEndNode: END)
```

4.3 Capturing Service Specifications

Service Registry acts as a repository which mainly stores the personal contact details of the service provider and technical details of the provided services [29]. Under the technical details of the provided services, it mainly discusses the quality of service parameters of the services like execution cost, execution time, generated data quality, process type, processed data type. In our chapter, we represent the quality of service parameter of the process engine instance as the form of the Data Process Engine Information model.

Data Process Engine Information Model

In order to design the process model on the basis of available processing engine instance, we need to collect the technical information of services which is provided by the service provider. This service-related technical information should be presented in a proper structure way. Therefore, it can help to configure the dynamic process model which will satisfy both required services and available services.

Definition 3 The service model provided by a service provider known as the data process engine information model, is used to describe the generated data quality and service quality of each data analytic process engine instance in process-based data analytic framework. This is also an ontology-based RDF model. Using this model, each and every process engine instance should have a unique id to make the separation of technical details from other engine instance.

The process engine information model describes the process type (e.g.., Querying, Clustering, Sampling) of all the provided process engine presented in Fig. 6. It points the technical details of each data processing engine instance as the form of data

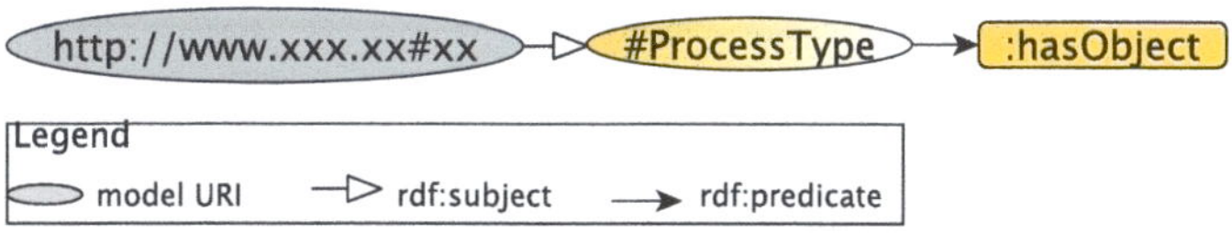

Fig. 6 Ontology based diagram of data process engine information model

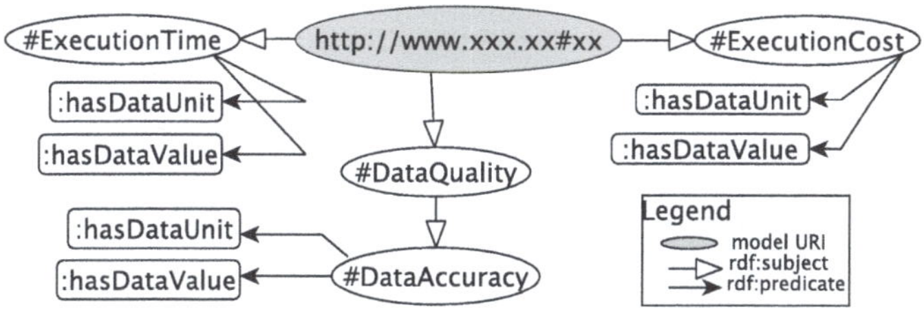

Fig. 7 Ontology based diagram of data process engine service model

process engine service model, presented in Fig. 7. The example of the process engine service model of process engine PE1 under process type Sampling is presented in listing 3.

Listing 3 Example of the value of Data Processing Engine Service Model of PE1

```
Execution Cost 6 minute/TB Execution Time 60 euro/TB
Data Accuracy  60 %/TB
```

5 Context Related Knowledge Representation

For the data analytic operation, the behavior of the process model depends upon the context of the data processing system. In the previous section, we have mentioned certain components that are used to represent the context of executable data analytic environment. In Fig. 8, we have shown the relationship between a different component of the context. This semantic relationship is helped to build up the context-related knowledge.

The context related knowledge should support the selection of service units in the place of a particular process mentioned in the user's defined ascertain of the process model. To support this job, it should be structured in such a way that it is able to show the knowledge which is collected by analyzing the information of this context. Here we follow the decision tree technique to provide this type of structure. According to this technique, the context related knowledge is represented as the form of process service decision tree.

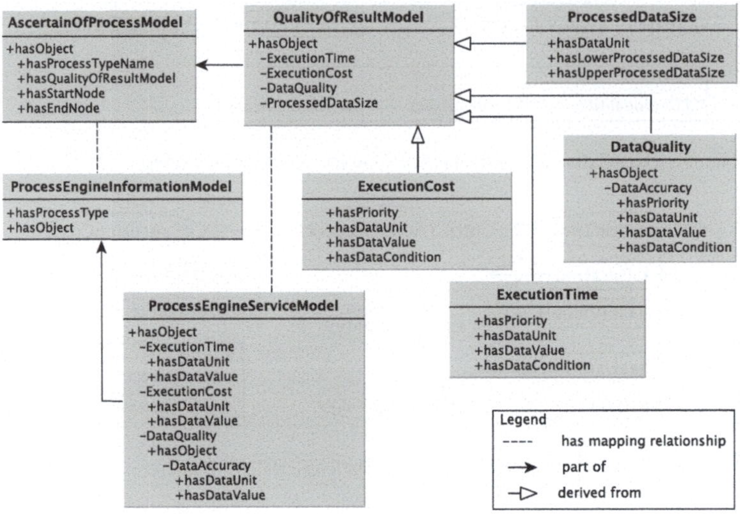

Fig. 8 UML diagram to represent the relationship between different components of the context of data analytic environment

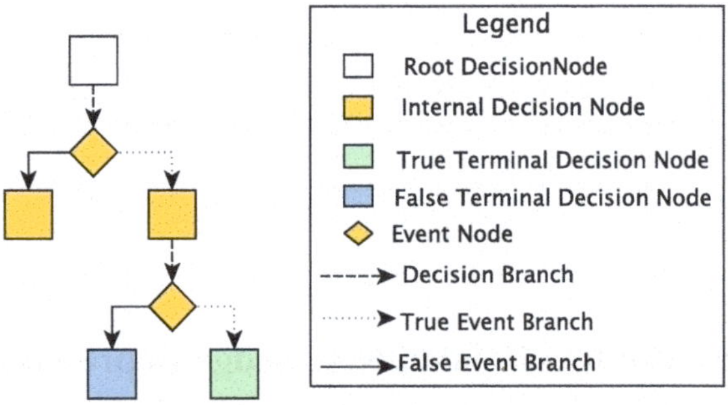

Fig. 9 Structure of process service decision tree

To construct the process service decision tree, we follow the principles of this proposed context-aware process model where user's demand is associated with service-centric context-dependent data analytics job. Therefore, the process service decision tree is designed based on the user's demand.

Under the decision tree, we consider the user's demand as the event node, and the process engine instance is represented via the decision node as shown in Fig. 9 and explained below. According to the usability, we divide the decision node in four-way:

1. Root Decision Node: Represents the set of process engine instances which able to reach the same data analytic objective as mentioned in the required process.
2. Internal Decision Node: This node holds those specific process engine instances which going through the filtering of the user's provided service-related demand. Here we do not consider the set of process engine instances able to fulfill all the demands or not.

3. True Terminal Decision node: It holds the set of process engine instances that are able to fulfill all the user's demands.
4. False Terminal Decision Node: It holds the set of process engine instances which able to fulfill all the user's service-related demands except the last one.

In the case of a decision node, the branches extended from this node and ended to the event node are named as decision branches. For the event node, it has two branches:

- True Event Branch: the decision node under this branch represents the set of process engine instances that are able to fulfill the service demand of this event node.
- False Event Branch: This branch denotes those decision nodes which hold the set of process engine instances that are not able to fulfill the service-related demand mentioned in this event node.

The height of this process service decision tree is decided by the number of service-related demands presented by the user in the quality of the resulting model. Through the QoR model, the user presents the importance of each service demand in hasPriority data property. Here low priority value (e.g.., 1) represents the most important service demand than the service demand presented in higher priority value (e.g.., 2). In this decision tree, the filtering operation is done by the service demands holds from lower priority value to higher priority value. Therefore, the tree level of a particular event node is decided by the priority of the service demand presented in that event node. The process service decision tree generation algorithm is presented in Algorithm 1.

6 Modeling Framework

Through this modeling framework, we are going to demonstrate how the context-aware process model is going to link with changing the environment and its corresponding action. In this framework, we represent the environment as a context instance and the possible action is represented as a state. This context instance represents the knowledge extraction from the decision tree. Here, firstly we describe the graphical representation of the proposed context-aware process model in Fig. 10. According to this model, we can observe that there has a control unit before any required process.

This control unit is going to decide the pattern of the process engine instance on the place of this process and also decides the executing process model pattern in the executable platform. The control unit works as the form of the state machine. It follows the concept of state transition [26] to move from the current state to the next state after satisfying the corresponding context. Here for the transition purpose, the binary value is used to determine the occurred event to satisfy that context (i.e., represented by 1) or not (i.e., represented by 0). The state transition diagram of the control unit is presented in Fig. 11.

Algorithm 1 Algorithm to generate the Process Service Decision Tree

```
Input: QualityOfResult Model (QoR), processTypeName, ProcessEngineInformation Model
  (PEIM)
Output: Decision Tree (DT)
begin
count ← 1
DT.Level ← count
for each PEIM.hasProcessType do
  if processTypeName == PEIM.hasProcessType.Name then
    DT.RootDecisionNode ← PEIM.hasProcessType.hasObject
  end if
end for
DT.DecisionNode ← DT.RootDecisionNode
for each DT.DecisionNode.ProcessEngine (PE) do
  for each QoR.hasObject.hasPriority do
    if (QoR.hasObject.hasPriority==DT.level) then
      DT.EventNode ← QoR.hasObject.[hasCondition + hasDataValue +
      hasDataUnit]
    end if
    for each PE.hasObject do
      if (PE.hasObject.[hasDataValue + hasDataUnit] == DT.EventNode) then
        DT.DecisionNode ← PE
        DT.Level ← (count + 1)
      end if
    end for
  end for
end for
end
```

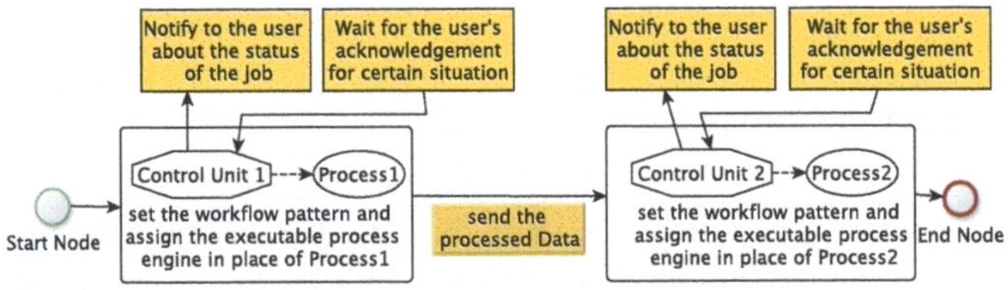

Fig. 10 Graphical representation of context aware service-centric process model

Here we consider certain states which are occurred in the process model at design time are as follows.

State A: Represents the generated process service decision tree.
State B: Assign the corresponding process engine instance name.
State C: Notify to the users with a notification message.
State D: Failure with a message where it has occurred.
State E: Do nothing. If the event is not in any context then the state machine switches to State E.

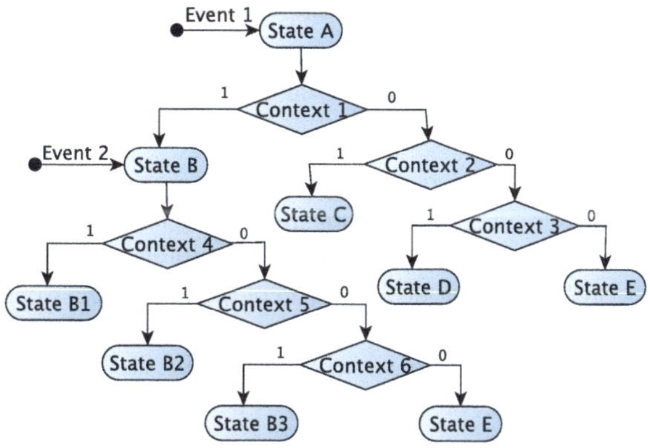

Fig. 11 State transition diagram of control unit

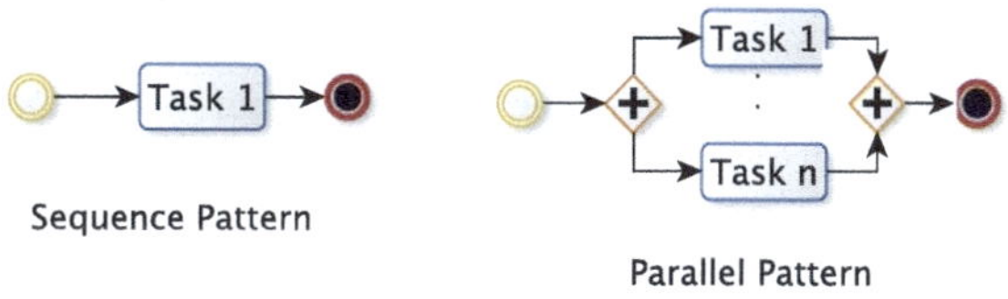

Fig. 12 The graphical representation of the two Workflow pattern: Sequence Pattern and Parallel Pattern

In this chapter, the state is considered to represent the possible action taken by the control unit. The control unit takes the help of users to design the process model. When the state machine switches from one state to another state based on context then the present state determines which action needs to perform at that time. The working procedure of each state is presented in the Algorithm 2.

Here we consider the process fragmentation on the basis of selected analytic process engine instance. To get the ultimate context-dependent process model structure, we apply the concept of workflow pattern [8]. Here we use two types of workflow pattern as shown in Fig. 12:

1. Sequence Pattern: Applying this pattern, incoming data is processed by a single process engine instance.
2. Parallel Pattern: The incoming data is processed by two or more than two process engine instances and they are parallelly connected.

Under state B what will be the pattern of the process model, it is defined by certain states. These states occurred in run time.

State B1: It represents the process structure mentioned in the sequence pattern.
State B2: Supports the parallel pattern as the process structure.
State B3: Failure by the message due to the increase of data volume.

In this framework, we consider certain contexts that are used to represent the behavior of the situations in the data analytics environment as given below. These situations represent the knowledge that is collected from the process service decision tree.

Context 1: If there have any process engine instances able to fulfill all the demands of users then the state machine moves from State A to State B.

Context 2: By considering the situation where the service provider has no such process engine instance except the one process engine instance able to fulfill all the demands except the last demand and the difference between the expected value with required value is within 0.0 to 0.1. At this condition, the state machine moves from State A to State C.

Context 3: At the design time, we consider the situation where the matching operation is performed in between 1 to (n–1) and for the nth demand, the mismatched value is more than 0.1. In this case, the state machine moves from State A to State D. Depending upon the incoming data volume, here we also consider certain contexts:

Context A: If only one specific process engine instance is able to fulfill all the user's demands to process this amount of data than the state machine moves from state B to State B1.

Context B: If more than one process engine instances are needed to process a certain amount of data under the demanded services then the state machine moves from State B to State B2.

Context C: If there have no such process engine instance able to process this amount of data with providing all the user's demanded services then the state machine moves from state B to State B3.

On the description of event type here we consider two types of event:

1. Event 1: Here we consider the matching operation which is performed at design time to find out the corresponding process engine instance of a particular process.
2. Event 2: This event occurs in run time where the incoming data volume is varied from time to time. But here we consider the incoming data volume is in between the user's mentioned data volume.

7 Illustration Examples and Architecture

To demonstrate our process modeling approach, here we represent some real-world use cases that are involved in an asset management job. These use cases represent the usability of our proposed process modeling approach. It also mentioned the complex relationship between the context (i.e., the components associated with the executable data analytics environment) and data analytics processes. It shows how the process engine instance collaboration depends upon the context.

Algorithm 2 Algorithm to invoke the state on the basis of decision tree manipulate result

```
treeManipulation(DecisionTree (DT))
begin
if (DT.TrueTerminalDecisionNode == Null) then
  if (DT.FalseTerminalDecisionNode == Null) then
    invoke state D
  end if
  DT.Level ← DT.FalseTerminalDecisionNode
  tempDataVal ← DT.Level.EventNode.dataVal
  tempDataVal ← (temp + 0.1)
  tempEventComponent ← DT.Level.EventNode.object
  for each DT.FalseTerminalDecisionNode do
    if (DT.FalseTerminalDecisionNode.tempEventComponent.
    dataVal == tempDataVal) then
      invoke state C
    end if
  end for
end if
invoke state B
end
```

7.1 Service Centric Without Considering the Technical Details

Goal: For some confidential asset monitoring like data asset monitoring. The user does not want to take the help of another person but also he is from the non-IT background. In that situation, he can design the asset monitoring job by applying this technique. If he chooses some service value that is not supported by any service unit (e.g.., analytic process engine) then the service provider will inform the users where the mismatch has occurred.

Activity: The involved process is dedicated to providing the way, how a non-IT background user can able to design the required process model for getting the required services. The working procedure of this process is presented as the flowchart diagram in Fig. 13.

7.2 User's Service-Related Demand

Goal: Here we consider the scenario of the chemical factory where the chemical chamber needs to monitor on the basis of Users (e,g., chemist, technician) demand, like the time duration to receive the monitoring result. The collaborative process engine instances should be able to follow the instructions of the user.

Constraints: Here all the sensors are working all the time. Therefore the generated data volume is fixed with time.

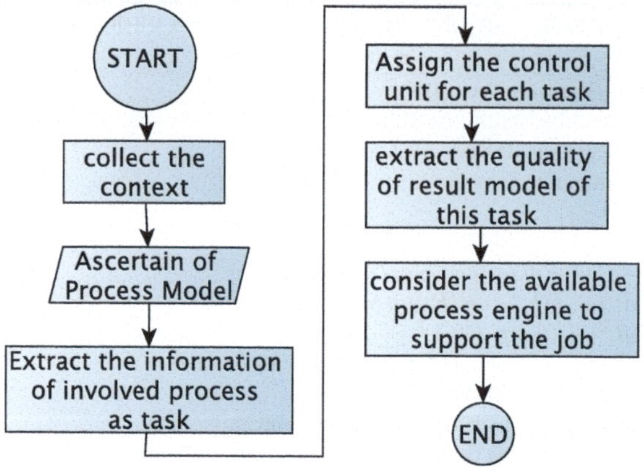

Fig. 13 The flow chart diagram of usecase "Service centric without considering the technical details"

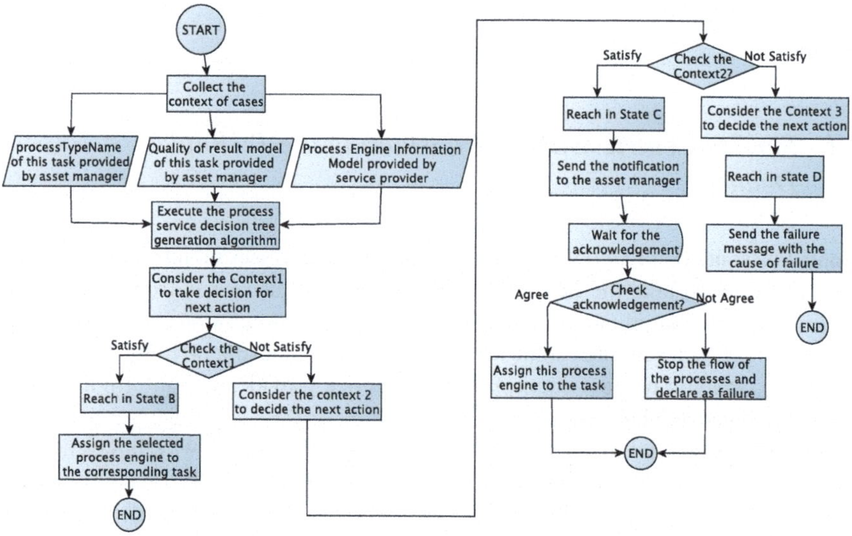

Fig. 14 The flow chart diagram of usecase "User's service-related demand"

Activity: To handle this type of case, the control unit executes a specific process for providing the corresponding service. This specific process is dedicated to assigning the proper process engine instance to this assigned task. Here the context is a user's provided quality of result model for this task and service provider provided process engine information model and process engine service model. The flowchart diagram of this process is presented in Fig. 14.

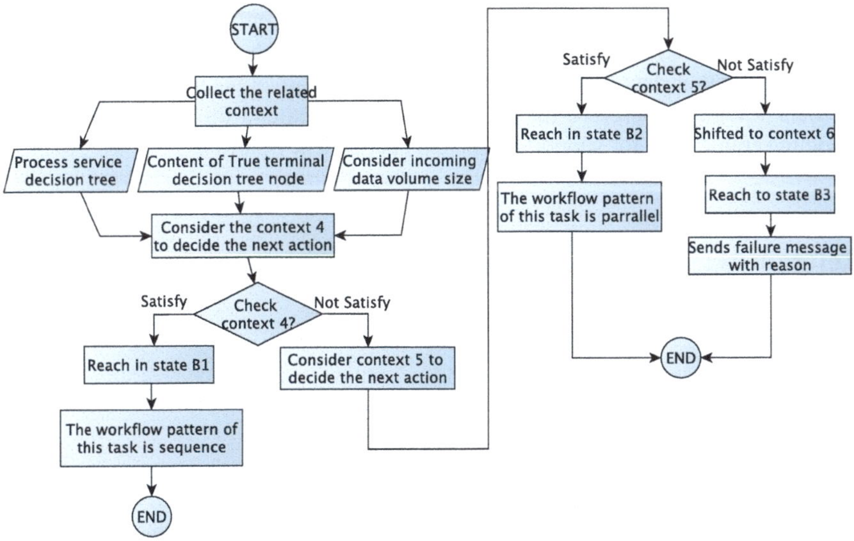

Fig. 15 The flow chart diagram of usecase "Big Data Volume and velocity"

7.3 Big Data Volume and Velocity

Goal: In a building there has more than one number of chillers. But all the chillers are not working at the same time. Therefore the generated data volume is changed with time. The chillers need to monitor in real-time due to the avoiding of short circuits.

Constraints: Here we consider the user's demanded services are matched with the services provided by available process engine instances.

Activity: The dedicated control unit for this task executes a specific process to provide the required services. The aim of this process is to assign the workflow pattern of the monitored task on the basis of the service model provided by assigned process engine instances. In such a way, the task can be able to handle the big data velocity (changing data volume with time). The flow chart diagram of this process is presented in Fig. 15.

8 Conclusion and Future Work

In this chapter, we propose a context-aware service-centric process modeling approach. Under this approach, we are going to consider the user's business and economic demands in business process modeling. This process model is build up by considering the context-related knowledge. Therefore it is able to adapt with the changing executable environment of process-based data analytic services.

In this proposed process modeling approach, here we only consider the big data volume and velocity and the whole experiment is done on the text-based data. We do not consider other data dependencies of the process model means big data variety, big data veracity. Therefore for the future, we will extend this proposed process modeling framework for supporting the data dependency part.

References

1. Ibm websphere. http://www.ibm.com/software/websphere
2. jbpm. http://www.jbpm.org/
3. pegasus. http://pegasus.isi.edu/
4. stata. http://www.stata.com/
5. Taverna. http://www.taverna.org.uk/
6. yawl. http://www.yawlfoundation.org/
7. van der Aalst WMP (2013) Business process management: A comprehensive survey. ISRN Software Engineering 2013:1–37
8. Aalst, W.M.P.V.D., Hofstede, A.H.M.T., Kiepuszewski, B., Barros, A.P.: Workflow patterns. Distribted Parallel Databases **14**(1), 5–51 (2003)
9. van der Aalsta, W.M., Weskeb, M., Grünbauer, D.: Case handling: A new paradigm for business process support. Data Knowledge Engineering **53**(2), 129–162 (2005)
10. Almeida, D., Baptista, C., Silva, E.D., Campelo, C., Nunes, C., Costa, B.D., Andrade, W., Cabral, J.: Using service-oriented architecture in context-aware applications. In: Proceedings of the VII Brazilian Symposium on Geoinformatics (GeoInfo 2005), pp. 15–29. Campos do Jordao, Brazil (2005)
11. Ardissono L, Furnari R, Goy A, Petrone G, Segnan M (2012) The context aware workflow execution framework. IJAACS 5(1):58–76
12. Bhattacharya K, Gerede C, Hull R, Liu R, Su J (2007) Towards formal analysis of artifact-centric business process models. In: Alonso G, Dadam P, Rosemann M (eds) International Conference on Business Process Management (BPM 2007), vol 4714. Lecture Notes in Computer Science. Springer-Verlag, Berlin, pp 288–304
13. Cohn D, Hull R (2009) Business artifacts: A data-centric approach to modeling business operations and processes. Data Engineering Bulletin 32(3):3–9
14. Coutaz J, Crowley JL, Dobson S, Garlan D (2005) Context is key. ACM. Communications 48(3):49–53
15. Dey AK, Abowd G, Salber D (2001) A conceptual framework and a toolkit for supporting the rapid prototyping of context-aware applications. Human Computer Interaction 16(2):97–166
16. Falconer G, Mitchell S (2012) Smart city framework: A systematic process for enabling smart+ connected communities. Cisco report, Cisco Internet Business Solutions Group
17. Liptchinsky V, Khazankin R, Schulte S, Satzger B, Truong HL, Dustdar S (2014) On modeling context-aware social collaboration processes. Information Systems 43:66–82
18. Liptchinsky, V., Khazankin, R., Truong, H.L., Dustdar, S.: A novel approach to modeling context-aware and social collaboration processes. In: Advanced Information Systems Engineering, *Lecture Notes in Computer Science*, vol. 7328, pp. 565–580. Springer (2012)
19. Lu R, Sadiq S (2007) A survey of comparative business process modeling approaches. In: Abramowicz W (ed) Business Information Systems, vol 4439. Lecture Notes in Computer Science. Springer, Berlin Heidelberg, pp 82–94
20. Matthias B, Schahram D, Florian R (2007) A survey on context aware systems. Int. J. Ad Hoc Ubiquitous Comput. 2(4):263–277
21. Miller, C., Nagy, Z., Schlueter, A.: A seed dataset for a public, temporal data repository for energy informatics research on commercial building performance. In: 3rd Conference on Future Energy Business & Energy Informatics, pp. 1–2. Springer, Rotterdam, Netherlands (2014)

22. Moore, C., Mines, C., Clair, C.L., Miers, D., Hamerman, P.D., Hagen, P., Musto, S.: The process-driven business of 2020. Forrester research report, Future Strategies Inc. (2012)
23. Murguzur, A., Schleicher, J.M., Truong, H.L., Trujillo, S., Dustdar, S.: Drain: An engine for quality-of-result driven process-based data analytics. In: Business Process Management, *Lecture Notes in Computer Science*, vol. 8659, pp. 349–356. Springer Publishing (2014)
24. Nigam A, Caswell NS (2003) Business artifacts: An approach to operational specification. IBM Syst. J. 42(3):428–445
25. Oumaima, S., Selmin, N.: Context-awareness for adequate business process modelling. In: RCIS, pp. 177–186. IEEE (2009)
26. Parnas, D.L.: On the use of transition diagrams in the design of a user interface for an interactive computer system. In: 24th ACM National Conference, pp. 379–385. ACM (1969)
27. Rosemann, M., Recker, J.: Context-aware process design: Exploring the extrinsic drivers for process flexibility. In: M. Petit, T. Latour (eds.) Proceedings of the Workshops and Doctoral Consortium, pp. 149–158. Namur University Press (2006)
28. Saidani, O., Nurcan, S.: Towards context aware business process modelling. Tech report, Université Paris 1 (2007)
29. Schahram D, Martin T (2005) A view based analysis on web service registries. Distributed and Parallel Databases 18(2):147–171
30. Swenson KD (2010) Mastering the unpredictable: how adaptive case management will revolutionize the way that knowledge workers get things done. Meghan-Kiffer Press, Tampa, Fla
31. Truong, H.L., Dustdar, S.: Principles of software-defined elastic systems for big data analytics. In: 2014 IEEE International Conference on Cloud Engineering, pp. 562–567. IEEE Computer Society (2014)
32. Wieland, M., Kopp, O., Nicklas, D., Leymann, F.: Towards context-aware workflows. In: CAiSE 2007: Workshops and Doctoral Consortium, pp. 1–15. Stuttgart, Germany (2007)

Smart Energy: A Collaborative Demand Response Solution for Smart Neighborhood

Sawsan Al Zahr, Elias A. Doumith, and Philippe Forestier

Abstract Demand response aims to improve the reliability of the power system by mitigating the impact of the growing penetration of renewable energy sources. It aims to influence, through time-variable pricing or incentives, the ability and willingness of consumers to modify their consumption in line with utilities' needs. In a residential context for instance, under time-variable pricing, an advanced home energy management system can take advantage of demand flexibility to optimize the cost of energy consumption while meeting household comfort constraints. However, uncoordinated demand response programs may lead to undesirable outcomes such as rebound peaks at the neighborhood level. In this chapter, we propose a novel collaborative demand response towards a smart neighborhood. Our solution enables households to collaborate achieving an overall welfare, e.g. load flattening, while still meeting their own optimization objective, e.g. cost reduction. For this purpose, we formulate and solve the problem of robust collaborative neighborhood, where robustness measures are applied to mitigate the collaboration uncertainty, i.e. whether households collaborate to achieve overall welfare or focus on their optimization objective. Numerical analysis highlights that our solution can achieve a significant tradeoff between utilities' and households' welfare. Compared to non-collaborative scenarios, collaboration will reduce the peak power by about 42% to 46% while increasing the consumption cost by about 1.06% to 2.75% per household.

S. Al Zahr (✉)
Télécom Paris - LTCI, Institut Polytechnique de Paris, 19 Place de Marguerite Perey, 91120 Palaiseau, France
e-mail: sawsan.alzahr@telecom-paris.fr

E. A. Doumith
TICKET Lab, Antonine University, B.P. 40016 Hadat-Baabda, Lebanon
e-mail: elias.doumith@ua.edu.lb

P. Forestier
EDF R&D - ICAME Dpt., 7 Boulevard Gaspard Monge, 91120 Palaiseau, France
e-mail: philippe.forestier@edf.fr

M. Khan et al. (eds.), *Smart Cities: A Data Analytics Perspective*,
Lecture Notes in Intelligent Transportation and Infrastructure,
https://doi.org/10.1007/978-3-030-60922-1_3

1 Introduction

Urban areas account for 10% of the Earth's surface, a proportion that keeps growing. These areas consume about 65% of the available primary energy and account for about 70% of the greenhouse gas emissions, mainly due to the provision of energy for lighting, heating, cooling, and transport. Since 2007, more than 50% of the world's population has been living in urban areas; this rate is expected to reach 58% and 70% by 2025 and 2050,[1] respectively. With this trend, urbanization will face various challenges, such as overpopulation, climate change, environment quality, access to energy, etc. Faced with these challenges, cities must rethink how to provide population with basic services in a sustainable fashion.

The sustainable development program adopted by UN member states in 2015 defines 17 goals to be achieved by 2030.[2] One of these goals is to develop sustainable cities and communities. Beyond the obvious issues related to energy and environment, a sustainable city is also concerned with its social, economic, and cultural organization. At the same time, the development of information and communication technologies (ICT) has fostered the concept of smart city wherein the strategic use of digital technologies and data analysis allows the optimization of urban planning and management, with the aim of increasing the quality of urban services and/or reducing their costs, and ultimately promoting the development of sustainable metropolises. The smart city has been conceptualized by Giffinger et al. [1] according to six features, namely smart economy, smart governance, smart mobility, smart environment, smart living and smart people.

In this context, we are interested in the connection between smart city as a background and sustainability as a goal. More specifically, we are interested in the opportunity smart city bring to achieve one major goal of sustainable development, namely affordable and clean energy involving a substantially increasing share of renewable energy sources in the global energy mix.

What are the challenges arising from such a change in the power system? What are the solutions to face these challenges? How could the concept of smart city and digital technologies such as the Internet of Things (IoT) and data analysis enable the development of such solutions? What impacts could these solutions have on households and their environment? These are the questions we attempt to address in this chapter through a demand response solution in the context of a residential neighborhood.

The remainder of this chapter is organized as follows. In Sect. 2, we describe some of the challenges arising from substantial penetration of renewable energy sources. Moreover, we draw an overview of demand side management as one promising

[1] The world population is expected to reach 9.8 billion people by 2050 according to the UN report "The World Population Prospects: The 2017 Revision", published by the UN Department of Economic and Social Affairs. It provides a comprehensive review of global demographic trends and prospects for the future.

[2] See the Sustainable Development Goals on the link below:
https://www.un.org/sustainabledevelopment/sustainable-development-goals/.

solution to facilitate the deployment of renewable energy sources. In Sects. 3 and 4, we present our demand response solution. Indeed, we present the model and formulation of the robust collaborative smart neighborhood problem. Numerical results are presented and discussed in Sect. 5. Finally, we draw our conclusions in Sect. 6.

2 Demand Side Management

Power generation has long been derived primarily from nuclear and fossil fuel sources. As these sources are weather-independent, they provide a highly controllable form of power production. Along with the increasing penetration of renewable energy sources (RES), a growing attention is being paid to the capacity of power grids to handle the demand-supply equilibrium. Indeed, because of the intermittent nature of RES, the ongoing transformation in the power system has a significant impact on the power grid's performance in terms of reliability, quality of supply, voltage and frequency fluctuations, etc. Moreover, the average power consumption and the peak power[3] have steadily increased in recent years. This is mainly due to economic growth and prosperity, accompanied by the need for further comfort. For example, the increasing adoption of electric vehicles, which require high charging power, has significantly modified the pattern of household power consumption. Indeed, the electric vehicle load can be an order of magnitude higher than the average household's load.

Investing in more power generation capacity to overcome the above challenges would be a costly solution, both for utilities and consumers, and would most likely be used for only a few hours a year. Demand side management (DSM) will help coping with RES variability while avoiding or postponing heavy expenditures. DSM covers various approaches ranging from energy response (e.g. use of energy efficient appliances/construction materials, etc.) to sophisticated demand response (e.g. interruptible/curtailable rates, critical peak pricing, etc.).

2.1 Impact of Demand Response Programs

In contrast with energy response, demand response (DR) refers to short-term changes in the power consumption targeting to match the consumers' demand with the power production. Changes consist in decreasing/increasing the power consumption over specific time periods with respect to the power grid's needs.

DR has been recognized as an important approach to reduce demand side uncertainties and can therefore significantly improve the reliability of the power system

[3]Peak power, also referred to as peak demand, critical demand or maximum demand, is generally caused by power usage spikes in households, industries, and commercial facilities.

[2]. Another positive effect of DR is the ability to compensate for supply shortages through load reductions in the event of generation interruptions or transmission line failures. This reduces the need for investment in compensatory controllable energy sources, typically fossil fuel generation. Avoided investments in additional power plants not only save money, but also result in avoided CO_2 emissions. Therefore, DR programs can bring significant benefits to energy systems and society as a whole. However, they will also entail significant costs associated with their implementation, in particular if advanced measurement, communication and remote control infrastructures are developed to facilitate automatic demand response. For this reason, assessing the outcomes of a DR program is a key step in determining its value from the perspective of both regulators and market players.

To date, there has been little consistency in quantifying the benefits of DR programs in residential areas. Indeed, existing analyses focus primarily on the method of quantification, assumptions about consumer participation and responsiveness, and market characteristics. However, a proper quantitative assessment of the benefits of DR programs requires first an estimate of the changes incurred by demand patterns, and then a thorough analysis of the effects of these changes on intrinsically complex power systems. This assessment should be carried out over a sufficiently long period of time to capture the full range of market circumstances and evaluate the long-term impacts. It is worth noticing the difficulties in predicting changes in demand patterns. First, customers' decisions to enroll in DR programs and to respond during critical emergencies and critical events can be quite complex. On the one hand, residential consumers do not have a formalized notion of their usage needs and budget. On the other hand, they are not used to making electricity decisions on a daily or hourly basis. As a result, the response of consumers is uncertain and can be influenced by multiple factors, such as the consumer's lifestyle, education, and awareness. Second, adopting smart charging strategies that shift the time of day when electric vehicles draw power from the grid, can improve the overall benefits of DR programs. Not to mention that electric vehicles can be considered as a valuable source of load flexibility through vehicle-to-grid (V2G) technology. Finally, while the impact of DR programs in peak demand reduction is well known, the overall effects on energy consumption are less explicit. Actually, no studies have investigated the increase in energy consumption after a DR event. For instance, under residential air conditioner direct load control program, when the air conditioner is switched back on, it may consume more energy to bring the house back to the target temperature in a short period of time.

Although there is no appropriate quantitative assessment of DR benefits, it is commonly agreed that DR programs yield benefits for both consumers and utilities. On the one hand, they help consumers reducing their energy consumption cost. On the other hand, they help utilities reducing their capital investments and costs for operation and maintenance. Moreover, DR programs can improve the safety and reliability of the power system by reducing the likelihood and consequences of forced outages that result in financial costs to utilities and inconvenience to customers. They also reduce the need to invest in new power plants to meet peak demand that occurs for only a few hours a year. In addition to these benefits, it is also recognized that

DR programs provide utilities with a major opportunity to mitigate some of the relative unpredictability of RES, thereby facilitating their integration in the energy mix. Finally, DR programs bring other benefits which are not easily quantifiable. For example, power generation now accounts for a significant share of CO_2 emissions. By facilitating the integration of renewable energy sources, DR programs play an important role in achieving clean energy and climate goals.

We conclude that the DR triggers many positive developments. These include economic and environmental benefits, as well as improvements in pricing, risk management, reliability, and market efficiency. According to energy demand forecasts, electricity demand is expected to continue to grow over the next decade, further emphasizing the impact of DR programs.

2.2 Overview of Demand Response Programs

Among DR solutions listed in the literature, one can mainly distinguish between incentive-based and time-based programs [3]. On the one hand, incentive-based DR consists in utilities offering financial compensations to consumers in order to modify their consumption pattern either through direct load control [4–8], or through more sophisticated programs, e.g. interruptible/curtailable rates, emergency demand response, capacity market, or demand bidding programs [9–11]. On the other hand, time-based DR relies on pricing schemes motivating consumers to adjust their consumption in response to changes in the electricity prices during a 24-hour period, e.g. time-of-use, real-time pricing, or critical-peak pricing programs [12–17]. For further reading, extended surveys on DR programs have been carried out in [18, 19].

Although individual consumers are encouraged to participate in such DR programs, it has been shown that uncoordinated participation will not achieve the benefits expected by utilities [20] and may result in undesirable issues such as rebound peaks. Coordination mechanisms are therefore necessary so that households can adjust their optimization strategies without negatively impacting the grid nor the whole power system. In [21], the authors propose an iterative approach to achieve a coordinated DR program in a smart neighborhood. First, households individually minimize their consumption cost. Afterwards, utilities sequentially ask households to adjust their power consumption in an attempt to flatten the overall pattern of the grid's load. This should be achieved without incurring any additional cost to consumers. Besides the sequential nature of the approach, last constraint considerably limits the expected benefit of coordination. A multi-objective particle swarm approach has been proposed in [22] to achieve a tradeoff between minimizing the consumption cost for consumers and minimizing power variability for utilities. As a result, consumers are provided with a set of solutions, corresponding to various tradeoffs between the aforementioned objectives. Households that participate in the DR program can then choose an appropriate solution out of this set.

In this chapter, we propose a new collaborative energy management solution in which neighbouring households can collaborate to achieve an overall goal in line with utilities' needs, while optimizing their own objectives. To this end, we formulate the robust collaborative neighbourhood problem as a linear mixed integer program, where robustness measures are applied to reduce the uncertainty associated with collaboration, i.e., whether households collaborate to achieve overall welfare or focus on their optimization goals. We recall that robust optimization aims at finding a solution that can be considered as optimal for any realization of uncertainties within a given set [23]. Multiple approaches dealt with the problem of energy resource/load scheduling considering uncertainties [24–27]. However, these approaches consider uncertainty brought by renewable resources, demand, or market prices. In the sequel, we refer to our approach as robust collaborative smart neighborhood.

3 Smart Neighborhood Model

In this work, we define a smart neighborhood as a group of $\mathcal{H}$ individual houses that are interconnected through an electrical and a communication network. Its size, $\mathcal{H}$, may range from a few houses to a few hundreds. A simplified illustration of the smart neighborhood architecture is given in Fig. 1.

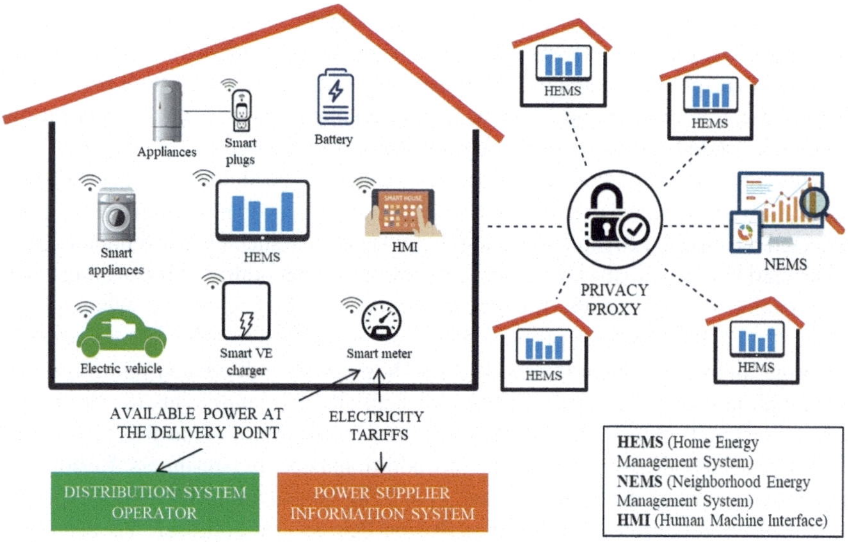

Fig. 1 A simplified illustration of the collaborative smart neighborhood

3.1 Smart Neighborhood Architecture

Each house is equipped with a smart meter that collects consumption and generation data, and communicates with power utilities. Furthermore, each house is equipped with a home energy management system (HEMS) that optimally schedules the household's day-ahead electric loads under time-variable pricing. Besides domestic appliances, some households have rechargeable battery packs and/or electric vehicles.

A neighborhood energy management system (NEMS) operates at the neighborhood level with the aim to minimize the maximum instantaneous power consumed by all the connected households. In a collaborative smart neighborhood, each household can choose either to reduce their power consumption costs by following the load schedule computed by their HEMS, or to collaborate with other households in order to achieve an overall welfare by following the load schedule computed by the NEMS. Consequently, the NEMS must take the collaboration uncertainty into account when performing its optimization.

In order to ensure the householder's privacy and prevent the NEMS from having access to their identity, an architecture similar to the one proposed in [28] can be used. In this architecture, a privacy proxy (PP) is inserted between the households' HEMS and the NEMS. Therefore, when submitting a load to the NEMS, a HEMS starts by encrypting the load using the public key of the NEMS and sends it to the PP along with the identifier of the household and a new symmetric encryption key. As such, the PP is unable to extract the characteristics of the load. Upon receiving a load, the PP stores the identifier of the household and forwards the load and the symmetric encryption key to the NEMS associated with an anonymous unique identifier. Upon receiving the anonymized load, the NEMS decrypts it and inserts it in the optimization process. Once the load has been scheduled, the NEMS encrypts the response with the consumer's symmetric encryption key and forwards the answer to the PP along with the associated anonymous unique identifier. Finally, the PP retrieves the identifier of the household and forwards the encrypted reply to their HEMS.

3.2 Load Modeling

Designing an efficient HEMS requires electric load modeling in order to estimate the impact of management strategies on household energy consumption. In the literature, two modeling approaches are followed, namely the top-down and the bottom-up approaches [29].

On the one hand, top-down approaches model the power consumption of an area (e.g. house) as a single unit. Therefore, only the total power consumption of the area is known [30]. Such approaches often rely on historical data to model the power consumption of an area, and are generally used to study the long-term (five years or more) effect of changes in load profiles. Bottom-up approaches, on the other hand, allow the energy consumption of individual appliances to be investigated separately.

By aggregating the consumption of each appliance, the load curve for a single household or for several households can easily be obtained [31]. In this work, we assume that households are equipped with HEMS such as the one we introduced in [32] which has been designed relying on a bottom-up approach.

Although the energy consumption of a household is determined by the electrical appliances they own and their lifestyle, we can classify the electric loads of a household into four main categories according to their operational specifications and the householder needs as follows:

(a) A *"primary load"* is an electric load that can neither be interrupted nor postponed to another time-slot. These loads include electrical appliances that operate continuously (e.g. refrigerator) as well as electrical appliances that the householder needs to operate instantaneously (e.g. vacuum cleaner).

(b) A *"deferrable load"* is a non-preemptive electric load of a predefined duration and power consumption profile. A deferrable load is to be executed within a time interval delimited by an arrival time and a delay. Moreover, a deferrable load cannot be interrupted once it has started its execution and must be completed before its deadline (e.g. a washing machine).

(c) A *"constant-power deferrable load"* is a particular type of deferrable loads that consumes a constant power in each time-slot wherein it is run. However, its duration is a parameter provided by the consumer himself (e.g. water heater/boiler).

(d) A *"modular load"* is an electric load that require neither a constant duration nor a constant power, but a given energy within a time interval delimited by an arrival time and a deadline (e.g. electric vehicle). During the charging process, instantaneous consumed power is determined by a power modulation coefficient that limits the power consumed to some predefined power levels. This charging process may be interrupted for some period of time if this is deemed necessary.

Bottom-up models enable energy management system designers to fine-tune management strategies. However, such models require data that is not readily available in a standard house. In order to compensate for the lack of detailed historical data, we use for numerical analysis data collected from the SMACH simulator [33]. SMACH[4] is a multi-agent model and simulator for the investigation of human behaviour regarding energy consumption in a household, at a fine-grain level (i.e. individual appliance's power consumption) based on weather, the household's activities and lifestyle.

4 Robust Collaborative Smart Neighborhood

We consider a neighborhood composed of $\mathcal{H}$ households where each household has different sets of primary and deferrable loads. Each household may have a rechargeable battery pack and/or an electric vehicle. Some of the households will respond positively to the collaborative signal sent by utilities, while others will prefer to perform their own scheduling optimization. Furthermore, we consider a time-variable

[4] SMACH is a French acronym for *Simulateur Multi-Agent du Comportement Humain*.

pricing scheme where the electricity tariffs are allowed to change every 10 minutes.[5] Under this configuration, the energy to power conversion ratio κ is equal to $1/6$.

4.1 Robust Optimization

In order to mitigate the impact of the collaboration uncertainty, we make use of robust optimization. Robust optimization is a technique that can be used to determine the best feasible solution for any realization of the uncertainty in a given set. For this purpose, we consider $\mathcal{S}$ potential scenarios and select, for each scenario, the households that are taking part in the collaborative smart neighborhood. The problem formulation is as follows:

4.1.1 Parameters

- The number of households $\mathcal{H}$ in the neighborhood.
- The number of potential scenarios $\mathcal{S}$ considered in the robust optimization.
- The ordered set Θ of all the 10-minute time-slots θ_t that span the simulation period $[0, \Delta]$.

$$\Theta = \{\theta_t,\ t = 1 \cdots T\} = \{\theta_1, \cdots, \theta_T\} \tag{1}$$
$$\text{such that } \theta_1 < \theta_2 < \cdots < \theta_T \text{ and } T = |\Theta|$$

- The vector $\mathcal{T} = \langle \tau_t, t = 1 \cdots T \rangle$ provides the electricity tariffs during the simulation period $[0, \Delta]$. Element τ_t is the electricity tariff for the power consumed during time-slot θ_t.
- The participation matrix $\mathcal{V} = \left[\upsilon^{\mathfrak{s},\mathfrak{h}}, \mathfrak{s} = 1 \cdots \mathcal{S}, \mathfrak{h} = 1 \cdots \mathcal{H} \right]$. Element $\upsilon^{\mathfrak{s},\mathfrak{h}}$ is a binary value specifying whether household $\mathfrak{h}$ accepts to collaborate in the potential scenario $\mathfrak{s}$ to achieve an overall welfare ($\upsilon^{\mathfrak{s},\mathfrak{h}} = 1$) or not ($\upsilon^{\mathfrak{s},\mathfrak{h}} = 0$).

Furthermore, for each household $\mathfrak{h}$ ($\mathfrak{h} = 1 \cdots \mathcal{H}$), we define the following additional parameters:

- The set $\mathcal{L}^{\mathfrak{h},\mathfrak{a}} = \{\lambda_i^{\mathfrak{h},\mathfrak{a}}, i = 1 \cdots L^{\mathfrak{h},\mathfrak{a}}\}$ composed of $L^{\mathfrak{h},\mathfrak{a}}$ primary loads needed by household $\mathfrak{h}$. A primary load $\lambda_i^{\mathfrak{h},\mathfrak{a}}$ is represented by a vector $\langle p_{i,t}^{\mathfrak{h},\mathfrak{a}}, t = 1 \cdots T \rangle$ where $p_{i,t}^{\mathfrak{h},\mathfrak{a}}$ is the power to be consumed by the load at time-slot θ_t.
- The set $\mathcal{L}^{\mathfrak{h},\mathfrak{b}} = \{\lambda_i^{\mathfrak{h},\mathfrak{b}}, i = 1 \cdots L^{\mathfrak{h},\mathfrak{b}}\}$ composed of $L^{\mathfrak{h},\mathfrak{b}}$ deferrable loads needed by household $\mathfrak{h}$. A deferrable load $\lambda_i^{\mathfrak{h},\mathfrak{b}}$ is represented by a tuple $\left(\alpha_i^{\mathfrak{h},\mathfrak{b}}, \beta_i^{\mathfrak{h},\mathfrak{b}}, \langle p_{i,t}^{\mathfrak{h},\mathfrak{b}}, t = 1 \cdots T_i^{\mathfrak{h},\mathfrak{b}} \rangle\right)$. The parameters $\alpha_i^{\mathfrak{h},\mathfrak{b}}$ and $\beta_i^{\mathfrak{h},\mathfrak{b}}$ denote the arrival time and the deadline of the load, respectively. The vector $\langle p_{i,t}^{\mathfrak{h},\mathfrak{b}}, t = 1 \cdots T_i^{\mathfrak{h},\mathfrak{b}} \rangle$ spans the duration

[5] This is in line with the time-variable tariff rates that can be retrieved from the French transmission system operator, RTE, website (http://clients.rte-france.com/).

$T_i^{\mathfrak{h},\mathfrak{b}}$ of the deferrable load where $p_{i,t}^{\mathfrak{h},\mathfrak{b}}$ represents the power to be consumed by the load at the tth time-slot where it is scheduled.

- The set $\mathcal{L}^{\mathfrak{h},\mathfrak{c}} = \{\lambda_i^{\mathfrak{h},\mathfrak{c}}, i = 1 \cdots L^{\mathfrak{h},\mathfrak{c}}\}$ composed of $L^{\mathfrak{h},\mathfrak{c}}$ constant-power deferrable loads needed by household $\mathfrak{h}$. A constant-power deferrable load $\lambda_i^{\mathfrak{h},\mathfrak{c}}$ is represented by a tuple $\left(\alpha_i^{\mathfrak{h},\mathfrak{c}}, \beta_i^{\mathfrak{h},\mathfrak{c}}, p_i^{\mathfrak{h},\mathfrak{c}}, T_i^{\mathfrak{h},\mathfrak{c}}\right)$ where $\alpha_i^{\mathfrak{h},\mathfrak{c}}$ is the arrival time of the load and $\beta_i^{\mathfrak{h},\mathfrak{c}}$ is its deadline. The parameter $p_i^{\mathfrak{h},\mathfrak{c}}$ represents the power to be consumed by the load at any time-slot where it is scheduled, while $T_i^{\mathfrak{h},\mathfrak{c}}$ represents the requested duration.
- The set $\mathcal{L}^{\mathfrak{h},\mathfrak{d}} = \{\lambda_i^{\mathfrak{h},\mathfrak{d}}, i = 1 \cdots L^{\mathfrak{h},\mathfrak{d}}\}$ composed of $L^{\mathfrak{h},\mathfrak{d}}$ modular loads needed by household $\mathfrak{h}$. A modular load $\lambda_i^{\mathfrak{h},\mathfrak{d}}$ is represented by a tuple $\left(\alpha_i^{\mathfrak{h},\mathfrak{d}}, \beta_i^{\mathfrak{h},\mathfrak{d}}, \varepsilon_i^{\mathfrak{h},\mathfrak{d}}, p_i^{\mathfrak{h},\mathfrak{d}}, \left\langle \rho_{i,j}^{\mathfrak{h},\mathfrak{d}}, j = 1 \cdots \varsigma_i^{\mathfrak{h},\mathfrak{d}} \right\rangle\right)$ where $\alpha_i^{\mathfrak{h},\mathfrak{d}}$ is the arrival time of the load and $\beta_i^{\mathfrak{h},\mathfrak{d}}$ is its deadline. The parameter $\varepsilon_i^{\mathfrak{h},\mathfrak{d}}$ represents the energy required by the load to reach its targeted state of charge, while $p_i^{\mathfrak{h},\mathfrak{d}}$ represents the maximum power that can be consumed by the load during a single time-slot. The vector $\left\langle \rho_{i,j}^{\mathfrak{h},\mathfrak{d}}, j = 1 \cdots \varsigma_i^{\mathfrak{h},\mathfrak{d}} \right\rangle$ provides the power modulation coefficients that can be used.
- The set $\mathcal{L}^{\mathfrak{h},\mathfrak{e}} = \{\lambda_i^{\mathfrak{h},\mathfrak{e}}, i = 1 \cdots L^{\mathfrak{h},\mathfrak{e}}\}$ composed of $L^{\mathfrak{h},\mathfrak{e}}$ rechargeable battery packs deployed in household $\mathfrak{h}$. A rechargeable battery pack $\lambda_i^{\mathfrak{h},\mathfrak{e}}$ is represented by a tuple $\left(\varepsilon_i^{\mathfrak{h},\mathfrak{e}}, p_i^{\mathfrak{h},\mathfrak{e},+}, p_i^{\mathfrak{h},\mathfrak{e},-}, \eta_i^{\mathfrak{h},\mathfrak{e},+}, \eta_i^{\mathfrak{h},\mathfrak{e},-}, \rho_i^{\mathfrak{h},\mathfrak{e},+}, \rho_i^{\mathfrak{h},\mathfrak{e},-}\right)$ where $\varepsilon_i^{\mathfrak{h},\mathfrak{e}}$ represents the full capacity of the battery. The parameters $p_i^{\mathfrak{h},\mathfrak{e},+}$ and $p_i^{\mathfrak{h},\mathfrak{e},-}$ correspond to the maximum charging/discharging power of the battery, while the parameters $\eta_i^{\mathfrak{h},\mathfrak{e},+}$ and $\eta_i^{\mathfrak{h},\mathfrak{e},-}$ are its charging/discharging efficiency coefficients. Finally, the parameters $\rho_i^{\mathfrak{h},\mathfrak{e},+}$ and $\rho_i^{\mathfrak{h},\mathfrak{e},-}$ are the maximum and minimum states of charge of the battery, respectively.
- The vector $\left\langle \pi_t^{\mathfrak{h}}, t = 1 \cdots T \right\rangle$ provides the available power at household $\mathfrak{h}$ during the simulation period $[0, \Delta]$. $\pi_t^{\mathfrak{h}}$ denotes the available power at time-slot θ_t.
- The vector $\left\langle \varpi_t^{\mathfrak{h}}, t = 1 \cdots T \right\rangle$ provides the instantaneous power to be consumed by household $\mathfrak{h}$ when it decides to perform its own scheduling optimization. The parameter $\varpi_t^{\mathfrak{h}}$ denotes the total power consumed by all the loads in household $\mathfrak{h}$ at time-slot θ_t, as computed by the HEMS.

4.1.2 Variables

Whenever a household is taking part in the collaborative optimization, the NEMS has to properly schedule each of the loads in order to achieve its objective. Moreover, the NEMS scheduling decision for a given household must be independent of the potential scenario and the decision of the remaining households to collaborate or not to the smart neighborhood. To this end, we introduce, for each household $\mathfrak{h}$ $(\mathfrak{h} = 1 \cdots \mathcal{H})$, the following decision variables:

- The binary variables $\varphi_{i,t}^{\mathfrak{h},\mathfrak{b}}$ $(i = 1 \cdots L^{\mathfrak{h},\mathfrak{b}}, t = 1 \cdots T)$ represent the starting time of the deferrable loads submitted by household $\mathfrak{h}$. $\varphi_{i,t}^{\mathfrak{h},\mathfrak{b}} = 1$, if the deferrable load $\lambda_i^{\mathfrak{h},\mathfrak{b}}$ of household $\mathfrak{h}$ is scheduled to start at time-slot θ_t. $\varphi_{i,t}^{\mathfrak{h},\mathfrak{b}} = 0$, otherwise.
- The positive integer variables $\phi_i^{\mathfrak{h},\mathfrak{c}}$ $(i = 1 \cdots L^{\mathfrak{h},\mathfrak{c}})$ represent the starting time of the constant-power deferrable loads submitted by household $\mathfrak{h}$. $\phi_i^{\mathfrak{h},\mathfrak{c}} = t$, if the constant-power deferrable load $\lambda_i^{\mathfrak{h},\mathfrak{c}}$ of household $\mathfrak{h}$ is scheduled to start at time-slot θ_t.
- The binary variables $\psi_{i,t}^{\mathfrak{h},\mathfrak{c}}$ $(i = 1 \cdots L^{\mathfrak{h},\mathfrak{c}}, t = 1 \cdots T)$ represent the time-slots where the constant-power deferrable loads submitted by household $\mathfrak{h}$ are executed. $\psi_{i,t}^{\mathfrak{h},\mathfrak{c}} = 1$, if the constant-power deferrable load $\lambda_i^{\mathfrak{h},\mathfrak{c}}$ of household $\mathfrak{h}$ is executed at time-slot θ_t. $\psi_{i,t}^{\mathfrak{h},\mathfrak{c}} = 0$, otherwise.
- The binary variables $\chi_{i,j,t}^{\mathfrak{h},\mathfrak{d}}$ $(i = 1 \cdots L^{\mathfrak{h},\mathfrak{d}}, j = 1 \cdots \varsigma_i^{\mathfrak{h},\mathfrak{d}}, t = 1 \cdots T)$ represent the power modulation coefficients that are selected for the execution of the modular loads submitted by household $\mathfrak{h}$ over time. $\chi_{i,j,t}^{\mathfrak{h},\mathfrak{d}} = 1$, if the modular load $\lambda_i^{\mathfrak{h},\mathfrak{d}}$ of household $\mathfrak{h}$ consumes $\rho_{i,j}^{\mathfrak{h},\mathfrak{d}} \times p_i^{\mathfrak{h},\mathfrak{d}}$ at time-slot θ_t where it is scheduled. $\chi_{i,j,t}^{\mathfrak{h},\mathfrak{d}} = 0$, otherwise.
- The binary variables $\delta_{i,t}^{\mathfrak{h},\mathfrak{e}}$ $(i = 1 \cdots L^{\mathfrak{h},\mathfrak{e}}, t = 1 \cdots T)$ represent the states of the battery packs deployed in household $\mathfrak{h}$ over time. The use of one single variable $\delta_{i,t}^{\mathfrak{h},\mathfrak{e}}$ at each time-slot θ_t ensures that the battery pack $\lambda_i^{\mathfrak{h},\mathfrak{e}}$ of household $\mathfrak{h}$ is either charging ($\delta_{i,t}^{\mathfrak{h},\mathfrak{e}} = 1$) or discharging ($\delta_{i,t}^{\mathfrak{h},\mathfrak{e}} = 0$).
- The binary variables $\zeta_{i,t}^{\mathfrak{h},\mathfrak{e},+}$ $(i = 1 \cdots L^{\mathfrak{h},\mathfrak{e}}, t = 1 \cdots T)$ represent the variation over time of the charging state of the battery packs deployed in household $\mathfrak{h}$. $\zeta_{i,t}^{\mathfrak{h},\mathfrak{e},+} = 1$ whenever the battery pack $\lambda_i^{\mathfrak{h},\mathfrak{e}}$ of household $\mathfrak{h}$ changes its state from discharging at time-slot θ_{t-1} to charging at θ_t. $\zeta_{i,t}^{\mathfrak{h},\mathfrak{e},+} = 0$, otherwise.
- The binary variables $\zeta_{i,t}^{\mathfrak{h},\mathfrak{e},-}$ $(i = 1 \cdots L^{\mathfrak{h},\mathfrak{e}}, t = 1 \cdots T)$ represent the variation over time of the discharging state of the battery packs deployed in household $\mathfrak{h}$. $\zeta_{i,t}^{\mathfrak{h},\mathfrak{e},-} = 1$ whenever the battery pack $\lambda_i^{\mathfrak{h},\mathfrak{e}}$ of household $\mathfrak{h}$ changes its state from charging at time-slot θ_{t-1} to discharging at θ_t. $\zeta_{i,t}^{\mathfrak{h},\mathfrak{e},-} = 0$, otherwise.
- The non-negative real variables $\vartheta_{i,t}^{\mathfrak{h},\mathfrak{e},+}$ $(i = 1 \cdots L^{\mathfrak{h},\mathfrak{e}}, t = 1 \cdots T)$. $\vartheta_{i,t}^{\mathfrak{h},\mathfrak{e},+}$ represents the power consumed at time-slot θ_t while charging the battery pack $\lambda_i^{\mathfrak{h},\mathfrak{e}}$ deployed in household $\mathfrak{h}$.
- The non-negative real variables $\vartheta_{i,t}^{\mathfrak{h},\mathfrak{e},-}$ $(i = 1 \cdots L^{\mathfrak{h},\mathfrak{e}}, t = 1 \cdots T)$. $\vartheta_{i,t}^{\mathfrak{h},\mathfrak{e},-}$ represents the power delivered at time-slot θ_t while discharging the battery pack $\lambda_i^{\mathfrak{h},\mathfrak{e}}$ deployed in household $\mathfrak{h}$.
- The non-negative real variables $\xi_{i,t}^{\mathfrak{h},\mathfrak{e}}$ $(i = 1 \cdots L^{\mathfrak{h},\mathfrak{e}}, t = 1 \cdots T)$. $\xi_{i,t}^{\mathfrak{h},\mathfrak{e}}$ is the energy available at time-slot θ_t in the battery pack $\lambda_i^{\mathfrak{h},\mathfrak{e}}$ deployed in household $\mathfrak{h}$.

Furthermore, we introduce the following decision variables:

- The non-negative real variables $\omega_t^{\mathfrak{s},\mathfrak{h}}$ ($\mathfrak{s} = 1 \cdots \mathcal{S}$, $\mathfrak{h} = 1 \cdots \mathcal{H}$, $t = 1 \cdots T$). $\omega_t^{\mathfrak{s},\mathfrak{h}}$ denotes the total power consumed in scenario $\mathfrak{s}$ by all the loads in household $\mathfrak{h}$ at time-slot θ_t.
- The non-negative real variable Ξ represents the maximum instantaneous power consumed by all the loads in all the scenarios at the NEMS' level.

4.1.3 Constraints

For each household $\mathfrak{h}$ ($\mathfrak{h} = 1 \cdots \mathcal{H}$), the solution of the robust collaborative smart neighborhood problem must satisfy:

- Each deferrable load $\lambda_i^{\mathfrak{h},\mathfrak{b}}$ of duration $T_i^{\mathfrak{h},\mathfrak{b}}$ must be scheduled exactly once between its arrival time $\alpha_i^{\mathfrak{h},\mathfrak{b}}$ and its deadline $\beta_i^{\mathfrak{h},\mathfrak{b}}$.

  $\forall i = 1 \cdots L^{\mathfrak{h},\mathfrak{b}}$,

$$\sum_{t=\alpha_i^{\mathfrak{h},\mathfrak{b}}}^{\beta_i^{\mathfrak{h},\mathfrak{b}}-T_i^{\mathfrak{h},\mathfrak{b}}} \varphi_{i,t}^{\mathfrak{h},\mathfrak{b}} = 1 \quad \text{and} \quad \sum_{t=1}^{\alpha_i^{\mathfrak{h},\mathfrak{b}}-1} \varphi_{i,t}^{\mathfrak{h},\mathfrak{b}} + \sum_{t=\beta_i^{\mathfrak{h},\mathfrak{b}}-T_i^{\mathfrak{h},\mathfrak{b}}+1}^{T} \varphi_{i,t}^{\mathfrak{h},\mathfrak{b}} = 0 \tag{2}$$

- Each constant-power deferrable load $\lambda_i^{\mathfrak{h},\mathfrak{c}}$ must be scheduled for $T_i^{\mathfrak{h},\mathfrak{c}}$ time-slots between its arrival time $\alpha_i^{\mathfrak{h},\mathfrak{c}}$ and its deadline $\beta_i^{\mathfrak{h},\mathfrak{c}}$.
  $\forall i = 1 \cdots L^{\mathfrak{h},\mathfrak{c}}$,

$$\sum_{t=\alpha_i^{\mathfrak{h},\mathfrak{c}}}^{\beta_i^{\mathfrak{h},\mathfrak{c}}-1} \psi_{i,t}^{\mathfrak{h},\mathfrak{c}} \geqslant T_i^{\mathfrak{h},\mathfrak{c}} \quad \text{and} \quad \sum_{t=1}^{\alpha_i^{\mathfrak{h},\mathfrak{c}}-1} \psi_{i,t}^{\mathfrak{h},\mathfrak{c}} + \sum_{t=\beta_i^{\mathfrak{h},\mathfrak{c}}}^{T} \psi_{i,t}^{\mathfrak{h},\mathfrak{c}} = 0 \tag{3}$$

- The time-slots assigned to a constant-power deferrable load $\lambda_i^{\mathfrak{h},\mathfrak{c}}$ must be contiguous. This is achieved by the following constraints:

  $\forall i = 1 \cdots L^{\mathfrak{h},\mathfrak{c}}, \forall t = 1 \cdots T$,

$$\psi_{i,t}^{\mathfrak{h},\mathfrak{c}} \leqslant 1 + \frac{t - \phi_i^{\mathfrak{h},\mathfrak{c}}}{1+T} \quad \text{and} \quad \psi_{i,t}^{\mathfrak{h},\mathfrak{c}} \leqslant 1 - \frac{t - \phi_i^{\mathfrak{h},\mathfrak{c}} - T_i^{\mathfrak{h},\mathfrak{c}} + 1}{1+T} \tag{4}$$

- Each modular load $\lambda_i^{\mathfrak{h},\mathfrak{d}}$ must be scheduled by selecting at most one appropriate power modulation coefficient for each time-slot between its arrival time $\alpha_i^{\mathfrak{h},\mathfrak{d}}$ and its deadline $\beta_i^{\mathfrak{h},\mathfrak{d}}$.
  $\forall i = 1 \cdots L^{\mathfrak{h},\mathfrak{d}}, \forall t = 1 \cdots T$,

$$\sum_{j=1}^{\varsigma_i^{\mathfrak{h},\mathfrak{d}}} \chi_{i,j,t}^{\mathfrak{h},\mathfrak{d}} \leqslant \begin{cases} 1 & \text{if } \alpha_i^{\mathfrak{h},\mathfrak{d}} \leqslant \theta_t < \beta_i^{\mathfrak{h},\mathfrak{d}}, \\ 0 & \text{otherwise.} \end{cases} \tag{5}$$

- The energy accumulated during the charging time of a modular load $\lambda_i^{\mathfrak{h},\mathfrak{d}}$ must be at least equal to the requested energy $\varepsilon_i^{\mathfrak{h},\mathfrak{d}}$ to reach its targeted state of charge.
$\forall i = 1 \cdots L^{\mathfrak{h},\mathfrak{d}}$,

$$\sum_{t=\alpha_i^{\mathfrak{h},\mathfrak{d}}}^{\beta_i^{\mathfrak{h},\mathfrak{d}}-1} \sum_{j=1}^{\varsigma_i^{\mathfrak{h},\mathfrak{d}}} \chi_{i,j,t}^{\mathfrak{h},\mathfrak{d}} \times \rho_{i,j}^{\mathfrak{h},\mathfrak{d}} \times p_i^{\mathfrak{h},\mathfrak{d}} \times \kappa \geqslant \varepsilon_i^{\mathfrak{h},\mathfrak{d}} \tag{6}$$

- The power consumed by the rechargeable battery pack $\lambda_i^{\mathfrak{h},\mathfrak{e}}$ if it is charging at time-slot θ_t, cannot exceed the maximum charging power of the battery.

$\forall i = 1 \cdots L^{\mathfrak{h},\mathfrak{e}}, \forall t = 1 \cdots T$,

$$\vartheta_{i,t}^{\mathfrak{h},\mathfrak{e},+} \leqslant \delta_{i,t}^{\mathfrak{h},\mathfrak{e}} \times p_i^{\mathfrak{h},\mathfrak{e},+} \tag{7}$$

- The power delivered by the rechargeable battery pack $\lambda_i^{\mathfrak{h},\mathfrak{e}}$ if it is discharging at time-slot θ_t cannot exceed the maximum discharging power of the battery.

$\forall i = 1 \cdots L^{\mathfrak{h},\mathfrak{e}}, \forall t = 1 \cdots T$,

$$\vartheta_{i,t}^{\mathfrak{h},\mathfrak{e},-} \leqslant (1 - \delta_{i,t}^{\mathfrak{h},\mathfrak{e}}) \times p_i^{\mathfrak{h},\mathfrak{e},-} \tag{8}$$

- Assuming that at the beginning of the simulation the battery packs were at their minimum state of charge, the energy available in the rechargeable battery pack $\lambda_i^{\mathfrak{h},\mathfrak{e}}$ at time-slot θ_t can be computed as follows:
$\forall i = 1 \cdots L^{\mathfrak{h},\mathfrak{e}}, \forall t = 1 \cdots T$,

$$\xi_{i,t}^{\mathfrak{h},\mathfrak{e}} = \begin{cases} \rho_i^{\mathfrak{h},\mathfrak{e},-} \times \varepsilon_i^{\mathfrak{h},\mathfrak{e}} + \kappa \times \eta_i^{\mathfrak{h},\mathfrak{e},+} \times \vartheta_{i,t}^{\mathfrak{h},\mathfrak{e},+} \\ \qquad -\kappa \times \frac{1}{\eta_i^{\mathfrak{h},\mathfrak{e},-}} \times \vartheta_{i,t}^{\mathfrak{h},\mathfrak{e},-} & \text{if } t = 1, \\ \xi_{i,t-1}^{\mathfrak{h},\mathfrak{e}} + \kappa \times \eta_i^{\mathfrak{h},\mathfrak{e},+} \times \vartheta_{i,t}^{\mathfrak{h},\mathfrak{e},+} \\ \qquad -\kappa \times \frac{1}{\eta_i^{\mathfrak{h},\mathfrak{e},-}} \times \vartheta_{i,t}^{\mathfrak{h},\mathfrak{e},-} & \text{otherwise.} \end{cases} \tag{9}$$

- Furthermore, the energy available in the rechargeable battery pack $\lambda_i^{\mathfrak{h},\mathfrak{e}}$ is always bounded by its minimum and maximum states of charge.

$\forall i = 1 \cdots L^{\mathfrak{h},\mathfrak{e}}, \forall t = 1 \cdots T$,

$$\rho_i^{\mathfrak{h},\mathfrak{e},-} \times \varepsilon_i^{\mathfrak{h},\mathfrak{e}} \leqslant \xi_{i,t}^{\mathfrak{h},\mathfrak{e}} \leqslant \rho_i^{\mathfrak{h},\mathfrak{e},+} \times \varepsilon_i^{\mathfrak{h},\mathfrak{e}} \tag{10}$$

- The binary variables $\zeta_{i,t}^{\mathfrak{h},\mathfrak{e},+}$ and $\zeta_{i,t}^{\mathfrak{h},\mathfrak{e},-}$ representing the variation over time of the charging and discharging states of the rechargeable battery pack $\lambda_i^{\mathfrak{h},\mathfrak{e}}$ are computed as follows:

$\forall i = 1 \cdots L^{\mathfrak{h},\mathfrak{e}}, \forall t = 1 \cdots T,$

$$\zeta_{i,1}^{\mathfrak{h},\mathfrak{e},+} = 0 \text{ and } \zeta_{i,1}^{\mathfrak{h},\mathfrak{e},-} = 0 \tag{11a}$$

$$\zeta_{i,t}^{\mathfrak{h},\mathfrak{e},+} \geqslant \delta_{i,t}^{\mathfrak{h},\mathfrak{e}} - \delta_{i,t-1}^{\mathfrak{h},\mathfrak{e}} \quad \forall t > 1 \tag{11b}$$

$$\zeta_{i,t}^{\mathfrak{h},\mathfrak{e},-} \geqslant \delta_{i,t-1}^{\mathfrak{h},\mathfrak{e}} - \delta_{i,t}^{\mathfrak{h},\mathfrak{e}} \quad \forall t > 1 \tag{11c}$$

- The number of charging/discharging cycles of the rechargeable battery pack $\lambda_i^{\mathfrak{h},\mathfrak{e}}$ is limited to one single cycle per day.
$\forall i = 1 \cdots L^{\mathfrak{h},\mathfrak{e}}, \forall \ell = 0 \cdots \Upsilon - 1,$

$$\sum_{k=1}^{24/\kappa} \zeta_{i,\ell\times 24/\kappa+k}^{\mathfrak{h},\mathfrak{e},+} \leqslant 1 \text{ and } \sum_{k=1}^{24/\kappa} \zeta_{i,\ell\times 24/\kappa+k}^{\mathfrak{h},\mathfrak{e},-} \leqslant 1 \tag{12}$$

where $\Upsilon = \lceil \frac{\kappa \times T}{24} \rceil$ is the number of days in the simulation period $[0, \Delta]$ and $24/\kappa$ is the number of time-slots within a single day.

Then, for each potential scenario $\mathfrak{s}$ ($\mathfrak{s} = 1 \cdots \mathcal{S}$) and each household $\mathfrak{h}$ ($\mathfrak{h} = 1 \cdots \mathcal{H}$), we enforce the following constraints:

- The available power at household $\mathfrak{h}$ and time-slot θ_t is imposed as follows:

$\forall t = 1 \cdots T,$

$$\omega_t^{\mathfrak{s},\mathfrak{h}} \leqslant \pi_t^{\mathfrak{h}} \tag{13}$$

- The maximum instantaneous power consumed at the NEMS' level can be deduced as follows:

$\forall t = 1 \cdots T,$

$$\sum_{\mathfrak{h}=1}^{\mathcal{H}} \omega_t^{\mathfrak{s},\mathfrak{h}} \leqslant \Xi \tag{14}$$

- Depending on the considered scenario $\mathfrak{s}$, the total power $\omega_t^{\mathfrak{s},\mathfrak{h}}$ consumed by all the loads in household $\mathfrak{h}$ at time-slot θ_t is either equal to $\varpi_t^{\mathfrak{h}}$ when the household decides to perform its own scheduling optimization ($\upsilon^{\mathfrak{s},\mathfrak{h}} = 0$) or is computed as the sum of the power consumed by all the loads in the household when it decides to take part in the collaborative smart neighborhood ($\upsilon^{\mathfrak{s},\mathfrak{h}} = 1$).
$\forall t = 1 \cdots T,$

$$
\begin{aligned}
\omega_t^{\mathfrak{s},\mathfrak{h}} = \left(1 - \upsilon^{\mathfrak{s},\mathfrak{h}}\right) \times \varpi_t^{\mathfrak{h}} + \upsilon^{\mathfrak{s},\mathfrak{h}} \times &\left(\sum_{i=1}^{L^{\mathfrak{h},\mathfrak{a}}} p_{i,t}^{\mathfrak{h},\mathfrak{a}} \right. \\
&+ \sum_{i=1}^{L^{\mathfrak{h},\mathfrak{b}}} \sum_{k=1}^{T_i^{\mathfrak{h},\mathfrak{b}}} \varphi_{i,t-k+1}^{\mathfrak{h},\mathfrak{b}} \times p_{i,k}^{\mathfrak{h},\mathfrak{b}} + \sum_{i=1}^{L^{\mathfrak{h},\mathfrak{c}}} \psi_{i,t}^{\mathfrak{h},\mathfrak{c}} \times p_i^{\mathfrak{h},\mathfrak{c}} \\
&\left. + \sum_{i=1}^{L^{\mathfrak{h},\mathfrak{d}}} \sum_{j=1}^{\varsigma_i^{\mathfrak{h},\mathfrak{d}}} \chi_{i,j,t}^{\mathfrak{h},\mathfrak{d}} \times \rho_{i,j}^{\mathfrak{h},\mathfrak{d}} \times p_i^{\mathfrak{h},\mathfrak{d}} + \sum_{i=1}^{L^{\mathfrak{h},\mathfrak{e}}} \left(\vartheta_{i,t}^{\mathfrak{h},\mathfrak{e},+} - \vartheta_{i,t}^{\mathfrak{h},\mathfrak{e},-} \right) \right)
\end{aligned}
\tag{15}
$$

It is worth noting that the term multiplying the parameter $\upsilon^{\mathfrak{s},\mathfrak{h}}$ is independent of $\mathfrak{s}$ and remains constant for a given household $\mathfrak{h}$ and a given time-slot θ_t in all scenarios where the household decides to collaborate to achieve an overall welfare.

4.1.4 Objective

From the utilities' standpoint, the objective is to minimize the maximum instantaneous power consumption at the neighborhood's level under uncertain collaboration. At the opposite, households prefer to reduce the total cost of their electricity. When the household is not collaborating to the overall welfare, the total cost of the electricity is a constant term determined by the HEMS and does not impact the optimization carried out by the NEMS. However, when participating in the smart neighborhood, the household would also require to minimize its total electricity cost. As stated earlier, this cost does not depend on the selected scenario.

In summary, when solving the robust collaborative smart neighborhood problem, the objective function is a tradeoff between minimizing the average cost over all the considered scenarios of the power consumed by all the households in the neighborhood (consumers' welfare) and minimizing the maximum instantaneous power consumption at the neighborhood level (utilities' welfare). This is mathematically expressed as:

$$
\min \quad \gamma_1 \times \Xi + \gamma_2 \times \frac{\mathcal{K}}{\mathcal{S}} \sum_{\mathfrak{s}=1}^{\mathcal{S}} \sum_{\mathfrak{h}=1}^{\mathcal{H}} \sum_{t=1}^{T} \omega_t^{\mathfrak{s},\mathfrak{h}} \times \tau_t \tag{16}
$$

where γ_1 and γ_2 are two non-negative real numbers.

4.2 *Case 1: Optimization at the HEMS' Level*

We consider one single scenario where we assume that all households are taking part in the smart neighborhood. We set as objective function the minimization of the average cost obtained by setting $\gamma_1 = 0$ and $\gamma_2 = 1$ in Equation (16). Solving the MILP will result in a solution that minimizes the cost for each household taken separately. Thus, we can easily mimic the case where each household decides to perform its own scheduling optimization carried out at the HEMS' level. The solution obtained at the end of this optimization problem will be used as the vectors $\langle \varpi_t^{\mathfrak{h}},\ \mathfrak{h} = 1 \cdots \mathcal{H},\ t = 1 \cdots T \rangle$.

4.3 Case 2: Overall Welfare as Expected by Utilities

The utilities' main objective is to achieve load flattening over time. For this purpose, we solve the MILP considering one single scenario. In this scenario, we assume that all the households are taking part in the smart neighborhood with the objective to first minimize the peak power, and then to minimize the power cost obtained by setting $\gamma_1 \ggg \gamma_2$ in Equation (16).

4.4 Case 3: Robust Optimization at the NEMS' Level

In order to solve the robust collaborative smart neighborhood optimization problem under uncertainty, we consider $\mathcal{S}$ different scenarios. In each potential scenario, a randomly selected subset of households will perform their own load scheduling optimization where they minimize their own power cost. Thus, the power consumed by these households is given by the vectors $\langle \varpi_t^{\mathfrak{h}},\ \mathfrak{h} = 1 \cdots \mathcal{H},\ t = 1 \cdots T \rangle$. The remaining households will collaborate to the smart neighborhood and will consequently align their load scheduling with the solution provided by the NEMS.

The main challenge of the NEMS is to find a resilient load scheduling solution that can still be judged as acceptable independently from which household is being collaborative. For this purpose, we solve the MILP while setting $\gamma_1 \ggg \gamma_2$ in the objective function.

5 Numerical Analysis

In order to evaluate the proposed robust collaborative smart neighborhood, we considered a two-day simulation period (January 15–16, 2015) and extracted the corresponding electricity tariffs from the day-ahead market in France.[6] We considered a neighborhood composed of $\mathcal{H} = 16$ households; among them, 8 randomly selected households are assumed to own an electric vehicle with the following power modulation coefficients $\langle \rho_{i,1}^{\mathfrak{d}} = 0.7, \rho_{i,2}^{\mathfrak{d}} = 1.0 \rangle$. We considered $\mathcal{S} = 100$ potential scenarios

[6]Electricity tariffs have been extracted from the French transmission system operator website (http://clients.rte-france.com/).

where, in each scenario, 3 or 4 randomly selected households were assumed to not participate in the collaborative neighborhood. This optimization problem has been solved using CPLEX solver.

When none of the households do have a rechargeable battery pack, solving case 1 results in an average tariff rate of $3.75 \times 10^{-5}€/Wh$ and a peak power equal to $26486.6W$. Similarly, the solution to case 2 is characterized by an average tariff rate of $3.78 \times 10^{-5}€/Wh$ and a peak power equal to $14314.7W$. The cost increase per user varies between 0.2% and 1.4% with an average increase of 0.8%. Finally, performing the robust optimization at the NEMS' level considering 100 potential scenarios results in a solution characterized by an average tariff rate of $3.79 \times 10^{-5}€/Wh$ and a peak power equal to $15359.9W$. Compared to a consumer performing its own optimization, collaborating to the smart neighborhood induces a cost increase varying between 0.5% and 1.5% with an average value of 1.06%. Without altering the obtained solution, checking it against all possible scenarios with 3 or 4 non-collaborating households pinpoints 320 scenarios where the peak power exceeds $15359.9W$. Indeed, the maximum observed peak power observed is equal to $16415.7W$. This represents a risk factor of 13.5% and a power excess of 7%.

Next, we assume that 8 randomly selected households do have a single rechargeable battery pack. Solving case 1 results in an average tariff rate of $3.16 \times 10^{-5}€/Wh$ and a peak power equal to $46187.7W$. Similarly, the solution to case 2 is characterized by an average tariff rate of $3.81 \times 10^{-5}€/Wh$ and a peak power equal to $9876.8W$. The cost increase per user varies between 2.1% and 39.4% with an average increase of 19.7%. Finally, solving the collaborative scenario under uncertainty results in a solution characterized by an average tariff rate of $3.25 \times 10^{-5}€/Wh$ and a peak power equal to $24713W$. Collaborating to the smart neighborhood will then induce a cost increase of 0.01% to 5.7% with an average value of 2.75%. Checking the obtained solution against all possible scenarios with 3 or 4 non-collaborating households pinpoints 461 scenarios where the peak power exceeds $24713W$. The maximum observed peak power is equal to $29146.2W$. This represents a risk factor of 19.3% and a power excess of 18%.

Figure 2 shows the power consumption over time for both configurations, with and without rechargeable battery packs. It should be noted that the solutions could be improved, both in terms of risk factor and power excess, by increasing the number of potential scenarios considered in our investigation.[7] Indeed, performing the robust optimization at the NEMS' level considering $\mathcal{S} = 250$ potential scenarios results in a solution characterized by an average tariff rate of $3.82 \times 10^{-5}€/Wh$ and a peak power equal to $24738W$. By checking the obtained solution against all possible scenarios with 3 or 4 non-collaborating households, we note that the risk factor decreased from 19.3% to 5.9% and the power excess decreased from 18% to 11.5%.

In summary, our proposed collaborative approach achieves a tradeoff between the utilities' welfare and the consumers' welfare. Indeed, the collaborative approach

[7] It is worth noting that the complexity of our approach increases with the size of the neighborhood, the covered simulation period, and the number of scenarios required for robust optimization. Metaheuristic approaches can be considered to further investigate the impact of these parameters.

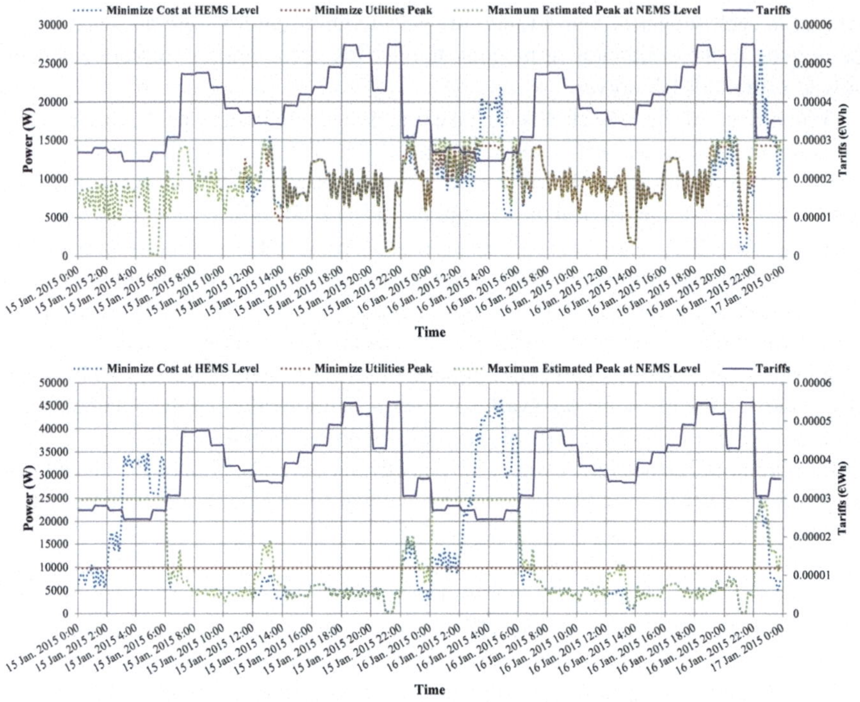

Fig. 2 Power consumption over time for the 3 compared cases: With (bottom figure) and without (top figure) rechargeable battery packs

reduces the power peak by almost 45% compared to the case where households only focus on reducing their power cost. Moreover, the additional cost induced for the consumers by the need to minimize the utilities' peak power was reduced from 39% to almost 6% in the worst case configuration. Finally, thanks to the robust optimization, the obtained result is resilient, with acceptable margins, to the number of households willing to participate in the collaborative smart neighborhood.

6 Conclusion

Although residential consumers are encouraged to participate in demand response programs, uncoordinated programs are likely to miss the expected benefits and result in undesirable issues such as rebound peaks. In this chapter, we proposed a novel collaborative energy management solution towards a smart neighborhood where households can choose either to collaborate in order to achieve an overall welfare or to only focus on their own optimization objective. In order to take into account the collaboration uncertainty, robustness measures have been applied. Numerical analysis highlights that our collaborative approach can achieve a significant tradeoff between utilities' and households' welfare. Indeed, compared to non-collaborative scenarios, collaborating to the smart neighborhood will increase the cost by about 1.06% to 2.75% while reducing the power peak by about 42% to 46%.

Acknowledgements Part of this work was carried out within SEIDO Lab, a joint research laboratory covering research topics in the field of smart grids, e.g.. distributed intelligence, service collaboration, cybersecurity, and privacy. It involves researchers from academia (Télécom Paris, Télécom SudParis, CNRS LAAS) and industry (EDF R&D).

References

1. Giffinger R, Pichler-Milanovi N (2007) Smart cities: Ranking of European medium-sized cities. Vienna University of Technology, Vienna
2. Liu C-C, McArthur S, Lee S-J (2016) Smart Grid Handbook. Wiley, New Jersey
3. Palensky P, Dietrich D (2011) Demand side management: Demand response, intelligenet energy systems, and smart loads. IEEE Transactions on Industrial Informatics 7(3):381–388
4. Shi W, Li N, Xie X, Chu CC, Gadh R (2014) Optimal residential demand response in distribution networks. IEEE Journal on Selected Areas in Communications 32(7):1441–1450
5. Vivekananthan C, Mishra Y, Ledwich G, Li F (2014) Demand response for residential appliances via customer reward scheme. IEEE Transactions on Smart Grid 5(2):809–820
6. Kaddah, R., Kofman, D., Mathieu, F., and M. Pioro. Advanced demand response solutions for capacity markets. In: IEEE International Conference on Innovations in Information Technology, pp. 23–28. IEEE (2015)
7. Fotouhi Ghazvini MA, Faria P, Ramos S, Morais H, Vale Z (2015) Incentive-based demand response programs designed by asset-light retail electricity providers for the day-ahead market. Elsevier Energy 82:786–799
8. Yi, Y., Zheng, A., Shao, X., Cui, G., Wu, G., Tong, G., and C. Gao. Joint adjustment of emergency demand response considering data center and air-conditioning load. In: IEEE International Conference on Energy Internet, pp. 72–76. IEEE (2018)
9. Aalami HA, Parsa Moghaddam M, Yousefi GR (2010) Demand response modeling considering interruptible/curtailable loads and capacity market programs. Elsevier Applied Energy 87(1):243–250
10. Horta, J., Kofman, D., Menga, D., Silva, S. Novel market approach for locally balancing renewable energy production and flexible demand. In: IEEE SmartGridComm Conference, pp. 1–6. IEEE (2017)
11. Horta, J., Kofman, D., Menga, D., Caujolle, M. Augmenting DER hosting capacity of distribution grids through local energy markets and dynamic phase switching. In: ACM Conference on Future Energy Systems, pp. 314–318. ACM (2018)
12. Mohsenian-Rad AH, Leon-Garcia A (2010) Optimal residential load control with price prediction in real-time electricity pricing environments. IEEE Transactions on Smart Grid 1(2):120–133
13. Pedrasa MAA, Spooner TD, MacGill IF (2010) Coordinated scheduling of residential distributed energy resources to optimize smart home energy services. IEEE Transactions on Smart Grid 1(2):134–143
14. Li, N., Chen, L., Low, S. H. Optimal demand response based on utility maximization in power networks. In: IEEE Power and Energy Society General Meeting, pp. 1–8. IEEE (2011)
15. Jacquot P, Beaude O, Gaubert S, Oudjane N (2019) Analysis and implementation of an hourly billing mechanism for demand response management. IEEE Transactions on Smart Grid 10(4):4265–4278

16. Kovács A (2018) On the computational complexity of tariff optimization for demand response management. IEEE Transactions on Power Systems 33(3):3204–3206
17. Luo F, Kong W, Ranzi G, Dong ZY (2020) Optimal home energy management system with demand charge tariff and appliance operational dependencies. IEEE Transactions on Smart Grid 11(1):4–14
18. Siano P (2014) Demand response and smart grids - a survey. Renewable and Sustainable Energy Reviews 30:461–478
19. Vardakas JS, Zorba N, Verikoukis CV (2015) A survey on demand response programs in smart grids: Pricing methods and optimization algorithms. IEEE Communication Surveys and Tutorials 17(1):152–178
20. Li Y, Ng BL, Trayer M, Liu L (2012) Automated residential demand response: Algorithmic implications of pricing models. IEEE Transactions on Smart Grid 3(4):1712–1721
21. Safdarian A, Fotuhi-Firuzabad M, Lehtonen M (2014) A distributed algorithm for managing residential demand response in smart grids. IEEE Transactions on Industrial Informatics 10(4):2385–2393
22. Herath, P., Venayagamoorthy, G. Multi-objective PSO for scheduling electricity consumption in a smart neighborhood. In: IEEE SSCI Symposium, pp. 1–6. IEEE (2017)
23. Ben-Tal A, Nemirovski A (1999) Robust solutions of uncertain linear programs. Operations Research Letters 25(1):1–13
24. Borges N, Soares J, Vale Z (2017) Multi-objective particle swarm optimization to solve energy scheduling with vehicle-to-grid in office buildings considering uncertainties. IFAC-Papers OnLine 50(1):3356–3361
25. Gerossier A, Girard R, Kariniotakis G, Michiorri A (2017) Probabilistic day-ahead forecasting of household electricity demand. CIRED - Open Access Proceedings Journal 2017(1):2500–2504
26. Ziel F, Steinert R (2018) Probabilistic mid- and long-term electricity price forecasting. Elsevier Renewable and Sustainable Energy Reviews 94:251–266
27. Yang X, Zhang Y, He H, Ren S, Weng G (2019) Real-time demand side management for a microgrid considering uncertainties. IEEE Transactions on Smart Grid 10(3):3401–3414
28. Petit, A., Cerqueus, T., Mokhtar, S. B., Brunie, L., Kosch, H. PEAS: Private, efficient and accurate web search. In: *IEEE Trustcom/BigDataSE/ISPA*, pp. 571–580. IEEE (2015)
29. Celik B, Roche R, Suryanarayanan S, Bouquain D, Miraoui A (2017) Electric energy management in residential areas through coordination of multiple smart homes. Elsevier Renewable and Sustainable Energy Reviews 80:260–275
30. Grandjean A, Adnot J, Binet G (2012) Review and an analysis of the residential electric load curve models. Elsevier Renewable and Sustainable Energy Reviews 16(9):6539–6565
31. Muratori M, Roberts MC, Sioshansi R, Marano V, Rizzoni G (2013) A highly resolved modeling technique to simulate residential power demand. Elsevier Applied Energy 107:465–473
32. Al Zahr, S., Doumith, E. A., Forestier, P. Advanced demand response considering modular and deferrable loads under time-variable rates. In: IEEE Globecom Conference, pp. 1–6. IEEE (2017)
33. Amouroux, E., Huraux, T., Sempé, F., Sabouret, N., Haradji, Y. Simulating human activities to investigate household energy consumption. In: ICAART Conference, pp. 71–80. Springer-Verlag (2013)

Governance in Smart City: An Approach Based on Social Network

Paramita Dey and Sarbani Roy

Abstract The governance for smart cities will be more citizen-centric and government policies will be based on the demand of the citizen. Social network has the potential of elevating the governance process to new levels. It enables government for instantaneous transmission of information to the targeted citizen, processing large scale data available through social media and can enable to take decisions based on that data in judicial way to increase transparency and accountability. For smart governance, it is required that within a limited budget government can propagate information to the maximum people. Social media analysis provides knowledge to ensure maximize the influence. Similarly for misinformation/rumour, influence can be minimized and SNA helps to identify communities and locations affected by this information. Moreover it can be used to identify those influential users who are able to spread information. This knowledge will empower Government agencies to take necessary precautionary measures such as targeted campaigning against the rumour, identified the source of information, block the source node or delayed rumour propagation. In this chapter we will study an conceptual framework for generating a governance system which will be able to identify the citizens those have maximum influence in the social network, campaigning of government policies through them for the maximum propagation to the citizen with minimum budget, detect rumour, found the source of misinformation, block the source node of rumour, detection of community, generating anti-spreading model for already propagated rumour.

Keywords Smart city · Online social network · Centrality measures · Influence maximization · Influence minimization · Community detection

P. Dey (✉)
Government College of Engineering & Ceramic Technology, Kolkata, India
e-mail: paramitadey@gcect.ac.in

S. Roy
Jadavpur University, Kolkata, India
e-mail: sarbani.roy@jadavpuruniversity.in

M. Khan et al. (eds.), *Smart Cities: A Data Analytics Perspective*,
Lecture Notes in Intelligent Transportation and Infrastructure,
https://doi.org/10.1007/978-3-030-60922-1_4

1 Introduction

In today's scenario, due to easy access to the user and faster reachability, social media is widely used for generating opinion in society. In recent past, these data play a pivotal role in Brexit and U.S. presidential elections for same reasons. Even in India, for prime minister Election, social media plays an important role. Different organizations use this huge user base for different purposes like advertising, campaigning, viral marketing etc. Smart city administration can use social network information for knowledge about public opinion on laws and policies and based on the feedback counter measures can be taken. In time of natural calamity like earthquakes, tsunamis can collect information in real time and planned appropriate strategy to handle the situation. Mainstream media, government and private agencies and other organizations can use the online social network data for information feed. They spend their money for advertising their policy/product in social media. Now an advertising through a famous person like footballer Leonel Messi in social media with huge fan-base can able to create a large promotion instantaneously but it will not be cost effective in terms of money. As social media is stored in the form of a graph, it is advantageous for some users due to their hierarchical positions to propagate the data in a more efficient manner. Moreover community detection over some criterion can identify groups of people who are more interested in that particular topic. Therefore a strategy to influence maximum number of people with minimum cost is important. A topic aware circulation of data can be more effective to reach intended users of the social network. Rumour spreading in social network is an important issue in social media. As there is no existing censorship for generating information in social network, it is important to resist misinformation in smart city framework which may cause harm to the society. For instance, in recent scenario of covid-19 pandemic, different rumours are circulating in social media like in hot and hu-mid weather the virus cannot persist, which was later proven false. It is important to track the flow of rumour and can take measurements to resist the propagation and find the source of rumour to make a control over it. The main constraints of social network data are the large volume and transient nature of information. Therefore it is required to generate the infrastructure which can use the paradigm like distributed computing, big data analytics and cloud computing.

Rest sections are organized in following manner. In Sect. 2 literature review and some critical issues about this topic are presented. An overview of system hierarchy for governance in smart city is discussed with the help of a flow diagram in Sect. 3. Data collection procedure is elaborated in Sect. 4. In Sect. 5, sentiment analysis is briefly discussed. In Sect. 6, information propagation techniques to reach maximum number of citizens are explained. Rumour detection and control strategies are discussed in Sect. 7. Conclusion and future research scopes are derived in Sect. 8.

2 Literature Review and Issues

2.1 Literature Review

The smart city is an idea or concept where citizen would be facilitated with all infrastructural amenities. The governance for smart cities must improve democratic processes in terms of transparency and citizen feedback centric governance. Development of the cities and political strategies will be based on the view of citizen. For this purpose, the government needs to collects information and feedback from the citizen.

Lots of research works have been performed to generate a framework for smart governance [1, 2]. Though a large number of infrastructure facilities are available in smart cities (already generated as pilot projects in different cities), there is no effective system in smart city infrastructure where government can interact with citizen in an interactive manner.

Online Social media platform empowers government to use common platform to share views with citizens of smart city and can reach to maximum number of peoples with minimum expenditure. Due to voluminous data of social network, it is trivial to collect and process online social network data. This chapter proposes a frame-work for smart governance through social media platform and uses some popular methodologies in social network analysis like influence maximization [3, 4], centrality measurements [5, 6], com-munity detection [7, 8], Rumour detection [9] and control strategies [10]. For smart city administration framework, all existing approaches are generally one directional, i.e. people can interact with government, but in proposed framework interaction is bidirectional and government would proactively participate in social media. Moreover it can reach to the maximum number of peoples with their policies in minimal budget of advertising.

2.2 Issues

For generating an infrastructure for smart governance, it is im-portant to concentrate on some issues which would be taken care of.

-Social network data are mostly unstructured data. Data collected using different crawlers are generally in the form of CSV (comma separated values). Therefore data has to store in NSQL (Not only SQL) database like Mongo DB.

-Generating a usage pattern from past history of the users is an important aspect. All users in the network are not participating in the communication in same way. Users, who is represented by the node of social network graph show different nature of participation like active (generating or sharing lots of post) or stiffler (does not participant in communication), ignorant (the information cannot reach to the user due to poor connectivity of the node in network). Moreover this nature can be topic specific. For example, who is very interested in politics cannot be interested in entertainment related news. There-fore an active node in one situation can be a stiffler node in other situation.

-As mentioned, social network can be used as a tool for controlling the influence of the information like maximization or minimization. Now it is not possible to choose a person with lots of follower to use for that purpose. Again it is not always effective as it generally passes through one or two hop from the source node. Finding influential nodes or links using graph properties can be more effective for this purpose. This nodes and links are generally denoted as centrality measures and it is important to identify them.

-Another issue of the analysis is large volume of social network data. Most of the computation complexity in social network is in the power of nodes count in the social graph. Therefore it is required to generate tools and algorithms for large scale data analytics. In the incubation period of social network analysis graph sampling was widely used to solve the problem but this methodology suffers with valuable information loss. Now with the paradigm like big data analytics, cloud computing it is become easier to process large scale data with large computation, time and storage complexity. Figure 1 de-scribes different methodologies for large scale data analysis of online social network.

3 System Hierarchies for Smart Governance

An overview of system hierarchy is presented through the flow diagram as Fig. 2. Online social network data will be extracted from social network through network crawling programmes. For generating feedback, identification of keywords plays an important role. To reach maximum number of people, influence maximization methodology can be used. Similarly, for misinformation techniques for influence minimization is important. Government policies can be modified depending on positive or negative feedback of the citizen of smart cities.

4 Collection of Data for Reviewing Government Policies

Government policies implemented for the people can have positive, negative or neutral effect. We can collect temporal and spatial online social network data from Twitter, Facebook about the governance of smart city and can find the impact of government policies. Online social networks generate a platform for human to share information at an unprecedented large scale. It provides access to vast amount of information which is of significant value to government agencies. Online social network data is characterized with 4 V: volume, variety, veracity and velocity. Volume denotes the large scale data, variety define different kind of data like text, audio, video etc., veracity indicates noise in data and velocity indicates transient nature of data.

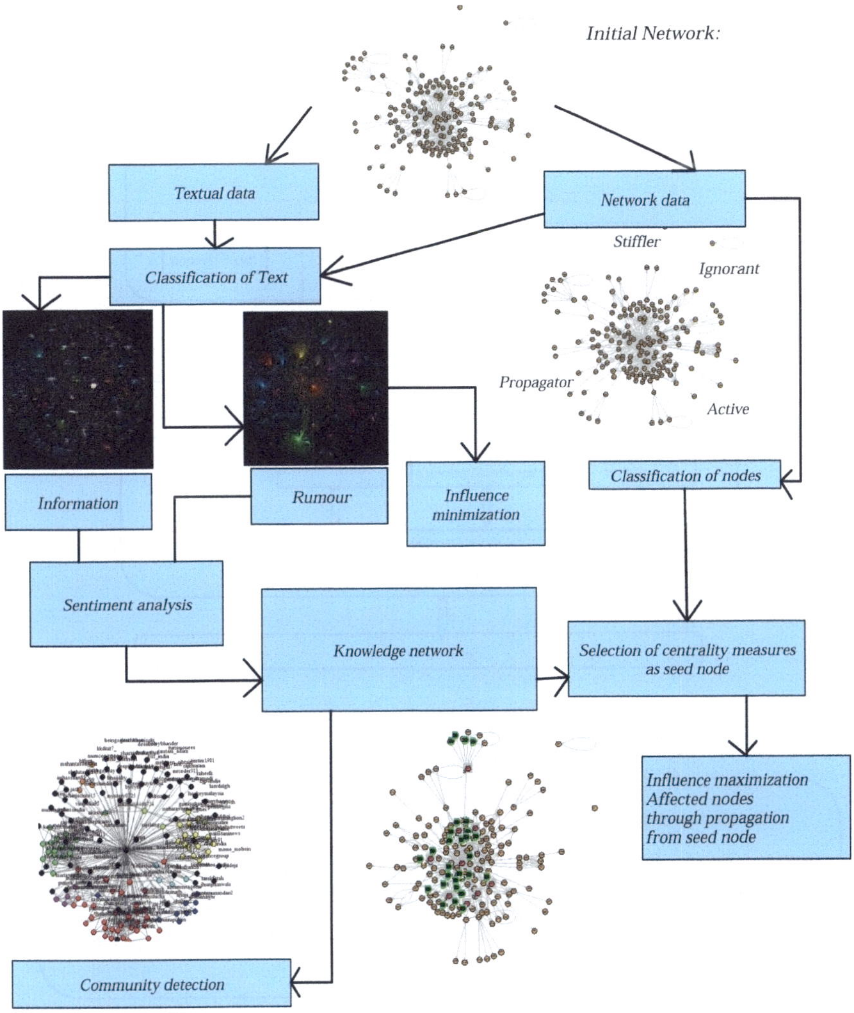

Fig. 1 Methodologies for large scale data analysis of online social network

Mining and analyzing the characteristics of information in online social networks provides valuable insights for decision making. Some positive feedback of government policy can be considered as incentive to work in that direction whereas negative feedback may cause reconsideration of government policies. Information spreading through the posts of users in the social media may cause harmful impact in the society also. It is important for news agencies and government organizations to verify the impact of in-formation and create necessary measures.

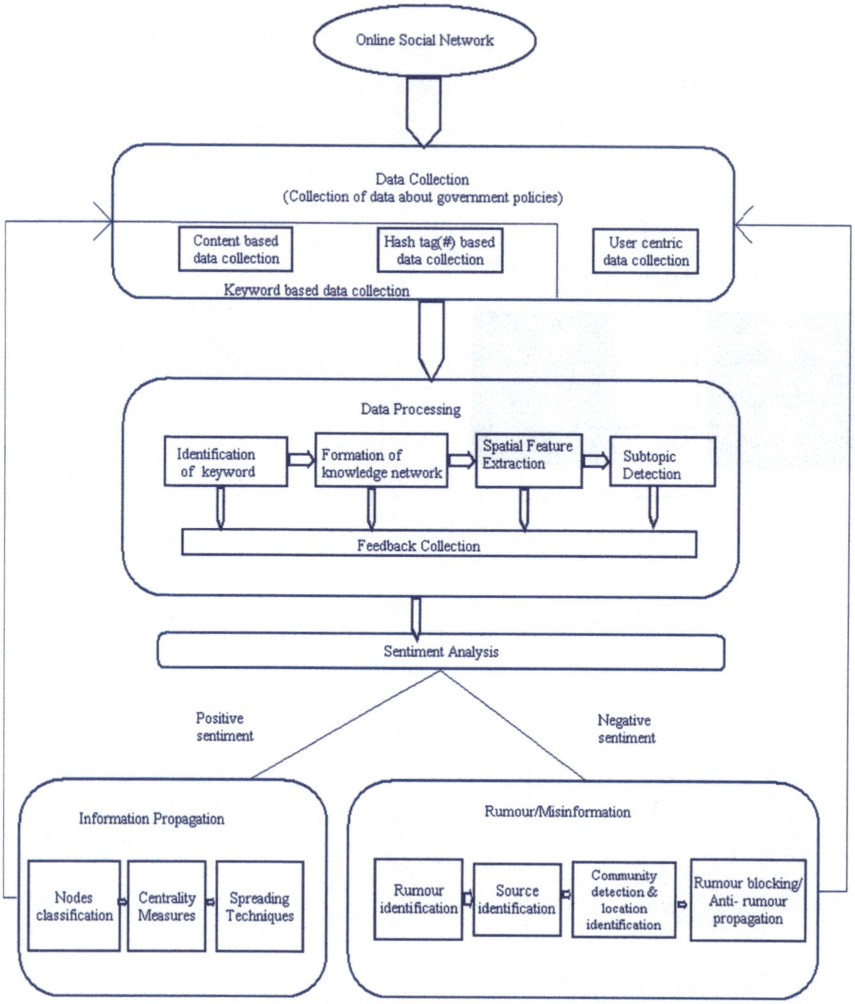

Fig. 2 SNA based system hierarchy for governance in smart city

4.1 Identification of Keywords

Social network data contains features such as textual, temporal, spatial and network information. Keyword detection, which are spread across the network and become popular, is of considerable interest to government agencies, for e.g. trending topics in Facebook, WhatsApp or Twitter data. These topics can represent a piece of information, meme or emoticon in a transient period of time. Generally temporal behaviour of trending topics in online social media show a burst nature of characteristics in short period of time. For example, when there was occurrence of some important sports event like world cup, it becomes more trending keyword.

Online social network (OSN) is a source of voluminous data with a lot of inbuilt meta-information such as time, location etc. in addition to the textual content. For instance, Twitter provides information on the user, time and location of the Tweet. Typically, topic models consider only textual information and to some extent the temporal information. But for a smart city application, topic models will consider spatial information related to that city in addition to the temporal and textual information. It is important to distinguish location and time at which an event happened along with the textual con-tent. For e.g. the posts of the people celebrating Christmas across the world have similar textual content but considering temporal and spatial information will help one to distinguish events happening in that particular region.

Traditional topic detection techniques used in network data may not work for social network data streams where temporal data show the nature of topic shifting and bursty data traffic. In [11, 12], the handling of bursty topic and tweet along with background topic modelling are considered respectively. For online infrastructure this topic detection modelling is rectified [13].

Detection of back topic or subtopic of a trending topic is far more difficult as it contain similar types of keywords. In [14, 15], A sub-topic detection methodology is presented where a graph of sequence of words is considered. All these algorithms are applicable if both topic and subtopics are equally trending. But for low tweeted sub-topic detection this algorithms cannot work efficiently. Hierarchical clustering algorithms can be used for social network topic detection related to their subtopics. As mentioned, online social network data are temporal in nature and for time dependent topic detection there exist topic detection model [16], where this temporal nature is taken care of. One disadvantage of this model is considering discrete time data whereas in online social network, continuous time data is generated. Therefore topic extraction along with subtopic detection requires more modelling.

The steps for trending topic detection from online social net-work can be described in following steps-

4.1.1 Data Extraction

OSN data are extracted through crawling of the social graph and derived data are formalized with their relative frequencies over a span of time. For example, Twitter data can be crawled over a period of time can be extracted in the form of metadata generally represented in CSV format.

4.1.2 Formation of Knowledge Network

Network can be generated through the active users based on their social relationships. Active users would be considered as the node of the network, similar kind of contents the relationship among the users of the network and relative frequency will define the edge of the network and thus knowledge network will be formed.

4.1.3 Temporal Features of Data

For each term, life cycle of the topic will be studied according to the relevance of that topic for specified time interval. After a certain time interval the features will be old and lost the significance. A set of emerging keywords would be generated based on their life span.

4.1.4 Spatial Features of Data

Generally online social network data provides us with spatial data. For example, in Twitter, it provides us with the latitude and longitude coordinate. For a smart city application of spatial features extraction will be twofold:

- From the citizen collecting information about the impacts of the government policies.
- The reaction from outside word.

Though reactions from the citizen have direct impact in the implementation of the policy but the outside reaction also should be counted.

4.1.5 Subtopic Detection

A topic graph that links the extracted emerging topics with their subtopics and background topic is required in order to generate a perspective about trending topic. Generally hierarchical clustering is a very good method for subtopic detection.

4.2 Feedback Collection

Data can be extracted from social network by crawling into network. Various softwares are available for that purpose. Data are generally extracted through two procedures:

- Using Keyword: Using identified keywords as hashtag or searching in the content, network information can be extracted from the network using the crawler like Tweepy. For example, from Twitter network data can be extracted through I-graph using # as shown in Fig. 3.
- Using user account: Networks are extracted based on some user accounts. For example, Facebook data can be extracted through Netvizz in the form of csv file [17] as shown in Fig. 4.

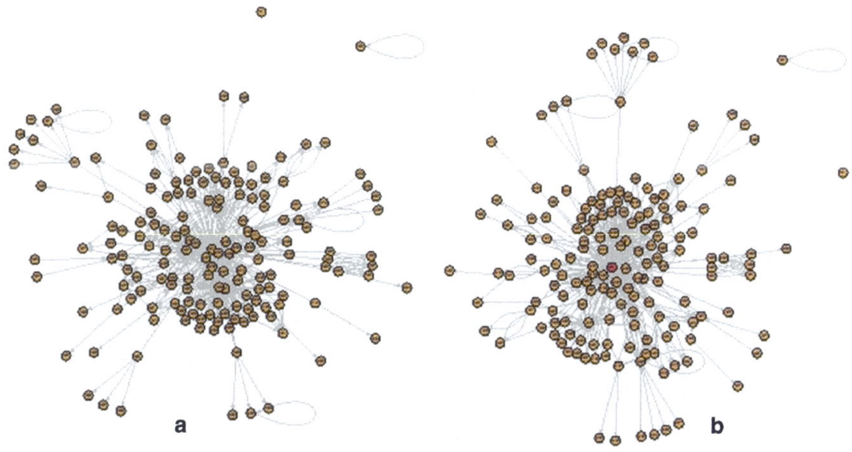

Fig. 3 Data extracted through hashtag(#) Scatter network B. Ego network

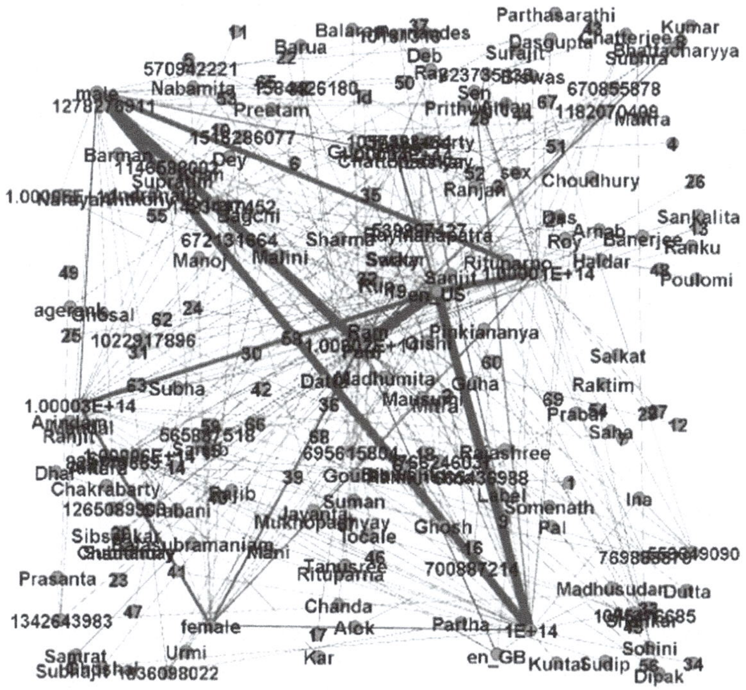

Fig. 4 Data extracted through user profile [18]

4.3 Classification of Information

The information extracted from online social network can be broadly categorized into five taxonomies.

4.3.1 Message-Based/content Based Features

This feature considers characteristics of the information content of the online post. Content can be used to find the following pattern from the online post:

- ***Lexical patterns and part-of-speech patterns***: Generally it provides some information about the origin of the post. This feature varies with place, profession, religion, country etc.
- ***Multimedia data***: used to check whether the post contains pictures, videos, or audios.
- ***URL***: This is used to check whether the post contains URL to an external source to support the content of the post.
- ***Time interval***: The time difference between the time of posting and user registration.
- ***Sentiment***: It analyzes the emotions of the content. Sentiment can be positive, negative or neutral.

4.3.2 User-Based Features

These features analyze the user characteristics whoever is generating the post or sharing it. These features can include

- ***Registration age***: As some post is more relevant to particular age group, it can be important features for classification.
- ***Number of followers***: The user who have more followers in the online network have greater influence in the network.
- ***Number of friends***: Number of friends connected to the user Number of user posts: Number of post signifies whether user is active or not in the social media.
- ***Number of Retweet***: Whether the post is shared by the followers of the user several times (measuring influence of the post).
- ***Number of sharing***: Measuring the count of shared information in the network (may not be directly connected with the user).
- ***Number of comments***: Measuring the count of comments on the post, may be positive, negative or neutral.
- ***User profile***: It includes gender, personal information like organization, real name etc.

- ***Spatial information***: Spatial information signifies the location where the event mentioned by posts occurred.
- ***Account credibility***: Whether the identity of the user is verified by the social network site.

4.3.3 Topic Based Features

Topic based features includes the features of the most discussed topics in the social network, and consider all content based and user based features for the analytics.

4.3.4 Propagation Based Features

Propagation based features consider network analysis if the propagated information. As the information is shares by several users, it generates several clusters within the shared network.

4.3.5 Spatial Features

Spatial features of the user can generate knowledge based community structures in the social network. It is obvious that impact of one incident can cause different impact in different places. So based on spatial features information can be classified.

5 Sentiment Analysis

Sentiment analysis is an important tool used for the different applications of extracted information. This analysis is generally generated using natural language processing and measures similarity of sentiment about a particular keyword. Sentiment would be positive when citizen are agreed with the government policies, neutral sentiment would be reflected by the zero value and negative value reflects disagreement. In Fig. 5, a sentiment analysis about GST (a tax system imposed in India in recent time) is shown as a scatter diagram. If the sentiment is negative about a policy, it is required to find whether there is some misinformation about this topic and take necessary actions about that. But if it is not a rumour, good governance can rethink or reconsider that policy.

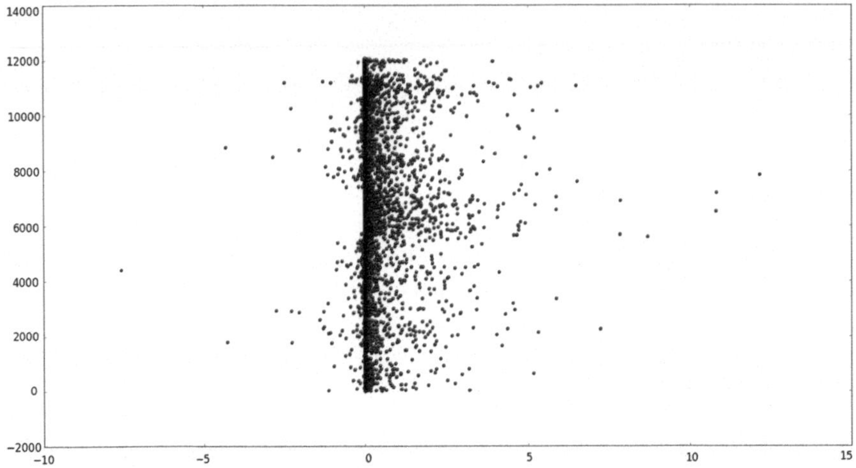

Fig. 5 Sentiment analysis for #GST in India

6 Influence Maximization

For influence maximization influential nodes can be used as the source node or propagator of the propagation. In online social network influential nodes can be identified through centrality measures.

6.1 Nodes Classification

Nodes, which signify the user in a network, can be classified according to their nature of participation in the network.

6.1.1 Daley-Kendal Model

According to Daley-Kendal (DK) model online social network users can be grouped into three categories:

- ***Ignorant***: the user who does not know about that particular information.
- ***Spreaders***: the user who know about the information.
- ***Stifler***: the users who know about the information but do not want to spread the information in the network.

6.1.2 Behavioural Model

According to the behaviour of the user present in online social network, nodes can be classified into two categories:

- ***Active***: The user who interacts with other users more often.
- ***Dormant***: The user with minimal interaction.

6.1.3 Flow Model

According to the flow of the information of the user present in online social network, nodes can be classified into three categories:

- ***Source***: the source node denotes the node who originates the data for particular flow.
- ***Sink***: The end node of the flow is indicated by sink node. This criterion can occur in two situations-

  The node is stifler in nature.
  The information lost its significance along with time.

- ***Participatory***: All other nodes besides source and sink node, participated in that information flow are considered as participatory node.

6.1.4 Case Study

A case study on classification of Twitter data are described in this subsection [19]. Information is extracted as the set of all Tweets containing a particular #hashtag.

Using #hashtag, a data filtering mechanism is used with algorithmic efficiency and statistical robustness at the processing of large metadata available in Twitter.

We define a directed Twitter User Graph. Its vertex set consists of every user on Twitter. Its edge set consists of all Following links: directed from every user, the followees, to each of their followers.

We distinguish users as Participants based on their role in rumour propagation through the network. We designate any user tweeting with a given #hashtag to be an active Participant in the rumour. Passive Participants include those not tweeting that particular #hashtag. To study the propagation of rumours on Twitter data, we define a directed Tweet Rumour Graph. Its vertex set corresponds to that of the Twitter User Graph. Stiflers and ignorant participants are designated as inactive nodes, while sources and propagators are designated active nodes, as shown in Fig. 6. We further label all active nodes with the timestamp of their 1st Tweet containing the specified #hashtag. The edge set denotes when both nodes are active. We get source node as user that has a lower timestamp than the other nodes in the in a connected component of the Tweet network. Thus, every connected component has its own source node in the Twitter graph.

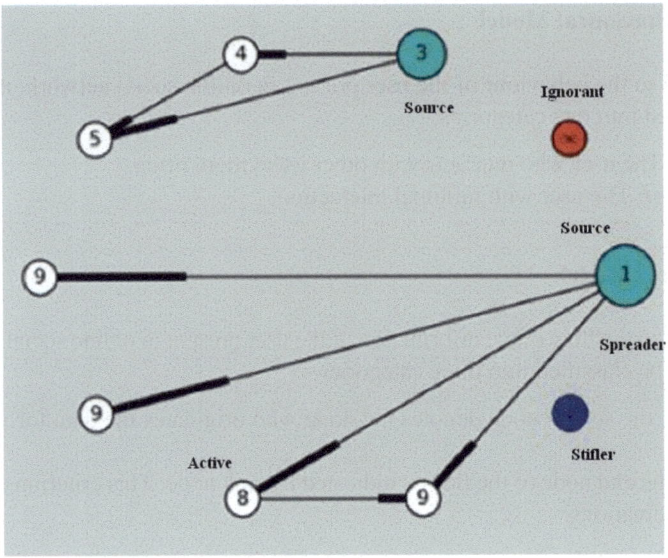

Fig. 6 Classification of nodes

6.2 *Centrality Measures*

For derivation of influential nodes in online social network, different centrality measures are widely used [20]. In a graph, centrality indicates the nodes that are mostly connected to other nodes in the network. Thus centrality measures in social graph indicate the most influential persons in the network.

6.2.1 Node Degree Centrality

Degree of a node v ∈ V can be defined as degree(v) = |{u: (v; u) ∈ E}|. This centrality measure [5] informs us of the number of connectivity among the vertices of the social graph. Nodes with a greater degree are connected to a large population within the graph, and are hence act as good seed candidates for information propagation.

6.2.2 Betweenness Centrality

Betweenness centrality of a node v ∈ V is defined by between(v)= $\sum_{s \neq v \neq t} \frac{\sigma_{st}(v)}{\sigma_{st}}$ where σ_{st} represents all possible shortest paths from node s to node t and $\sigma_{st}(v)$ denotes the number of shortest paths pass through node v. This centrality measure [18] indicates the capability of a node to quickly transfer information through the network.

6.2.3 Closeness Centrality

The closeness of a node v ∈ V can be defined as close(v) = $\frac{1}{\sum_{n \in V} \delta(n,v)}$, where $\delta(n, v)$ is the shortest distance from n to v. The closeness centrality [18] indicates how quickly information can spread from the node to the rest of the network.

We note here that if any node is v is unreachable from the node n, then close(v) = 0. This is a well-known problem with closeness centrality. As experts suggest, we thus redefine our implementations to use the mean harmonic distance: $\frac{1}{|V|-1} \sum_{v \neq n \in V} \frac{1}{\delta(n,v)}$.

6.2.4 Clustering Coefficient

Clustering coefficient for a node v ∈ V is defined as cluster(v) = $\frac{\lambda(v)}{\tau(v)}$, where $\lambda(v)$ denotes the number of triangles and $\tau(v)$ denotes the number of triplets containing v. It [5] reflects the property of the graph to cluster around some particular nodes.

6.2.5 K-core

K-core centrality of a network G denotes the maximal connected sub-graph of G where all nodes are assigned with minimum k number of other nodes. In other nodes after k-core decomposition, all nodes with node degree < k would be deleted from the graph [10].

6.2.6 PageRank

The PageRank of a node $v \in V$ is defined as $prank(v) = \frac{1-d}{|V|} + d \sum_{n:(v,n) \in E} \frac{prank(u)}{|\{n:(v,u) \in E\}|}$, where d is a dampening factor. It is a weighted centrality which reflects a node is connected to the nodes those who have more importance in the graph [21].

6.2.7 Case Study

A Twitter graph is extracted and three centrality values are derived. Distribution graph of three centrality measures are shown in Fig. 7.

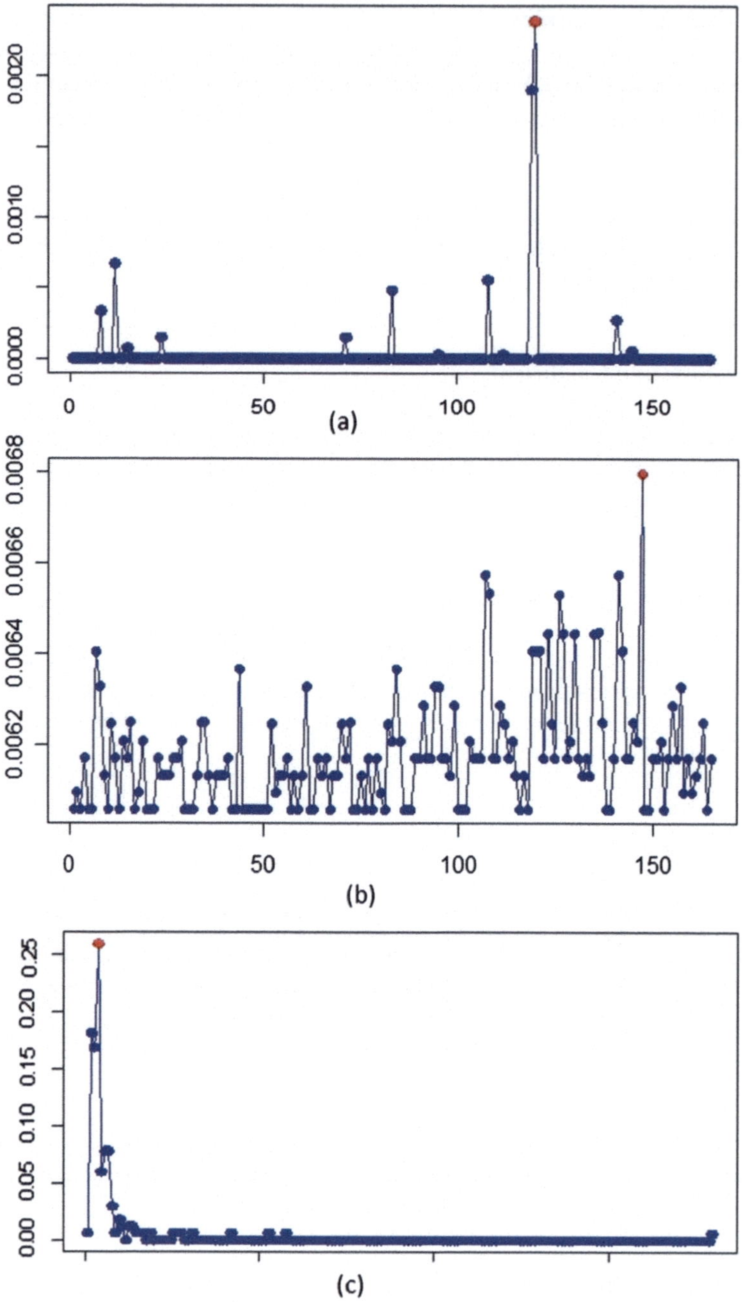

Fig. 7 Centrality measure from user distribution graph **a** Betweenness Centrality **b** Closeness Centrality **c** Node degree Centrality

6.3 Spreading Techniques

The maximum information propagation will be beneficial for smart governance as it ensures to reach maximum number of citizens. Some information propagation simulator is discussed in this subsection. A case study based on random walk is also presented.

6.3.1 Breadth First Search

In 1958 E. F. Moore propose Breadth-first search (BFS) for finding the shortest path among a grid network, similarly at almost same time in 1961 same algorithm was independently proposed by C. Y. Lee for wire routing algorithm [18]. BFS is used for propagation or searching in data structures like tree or graph. BFS starts from the seed node selected randomly from the network and propagate to the immediate neighbour with a probability. The neighbours then hopped to their next level neighbours and the process is continued. One of the disadvantage of BFS is that, nodes with a higher degree would be traversed more frequently which in term causes more biasing to the nodes with higher degree and results in local maximization with higher degree node. BFS is most studied and applied in online social network for finding user behaviour pattern, measurement and topological characterization of social network.

6.3.2 Forest Fire

This model is based on the concept of cellular automata. For a grid with dimension d and length l, the propagation follows in ld in the network using the following criterion:

- The burning node turns will be converted into empty node
- A node will be effected if at least one neighbor of the node is burning
- A node will ignites with the probability p even if no neighbor node is burning
- An empty space will fills with a node with probability s

6.3.3 RandomWalk

This propagation is based on a path that consists of a continuation of randomly chosen advancement following some mathematical space such as the integers. Random walk is generally associated with Markov chains or Markov processes or Markov model or other variants of the model. Random walk started with the seed node and propagates to other nodes randomly with a probability p and traversed through the network.

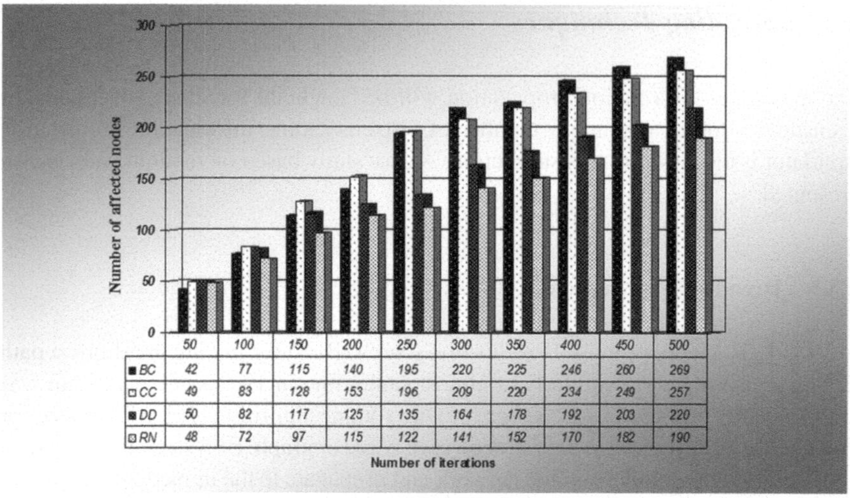

Fig. 8 Compare chart of propagation based on different centrality measures as seed node. *Abbreviations: BC: Betweenness centrality CC: Closeness centrality DD: Degree Distribution RN: Random Node*

6.3.4 Susceptible-Infectious-Recovered (SIR)

It is a probabilistic method with three states: Susceptible: no knowledge about the information, Infectious: have knowledge and want to spread and Recovered: Aware but not interested in spreading. Three states are more analogous to practical scenario and widely used to simulate the propagation.

6.3.5 Case Study

Using three highest centrality measures of previous result, information propagation is measured using random walk. Figure 8 represent the affected value in the network. This simulation is a small case study to reflect the utilization of centrality measure as source node of information propagation.

7 Rumour Detection & Influence Minimization

As discussed in earlier section the spread of malicious or rumour in online social network, can have negative effects in society. In that situation smart governance should provide a mechanism which can identify rumour and take proper steps to minimize the influence in social media.

7.1 *Taxonomy of Rumour*

Depending on the nature of rumours, it can be classified into following sections:

- ***Misinformation***: Post that provides false information. It also includes posts that misrepresents the facts or represent information out of context. This type of information may cause harm to the society. For example, the propaganda against polio vaccination in India restricts the Indian government policies of removing polio from India.
- ***Controversial information***: A post that can create dispute in another post, article, video or image unsubstantial information: A post that contains information that is either unsubstantiated or made some inference depending on the viewpoint of user.
- ***Reporting misinformation***: A source that reports some rumour and supplies a secondary source such as newspaper reporting or some hyperlink.
- ***Linked dispute***: A post that oppose some social network posts.
- ***User belief***: A post that reflects opinion of that user.

7.2 *Identification of Rumour*

We can identify rumours or misinformation as unverified statement which starts from some sources without authentication and truthfulness and spreads in the network. A rumor can result in three directions: true (factual), false (nonfactual) or without conclusion. For the detection of rumours following analysis can be used.

7.2.1 Lexical Analyzer

Both semantic and syntactic analyses are required for lexical analysis of the data nodes involved in rumour propagation:

- ***Semantic Analysis:***

Semantic features include uses of opinion word/vulgar word/emoticons. It is common practice to identify commonly used opinion word and is a part of classification of sentiments which is used for identification of words reflecting strong negative or positive opinion (like trustworthy, terrible, traitor etc.). These types of words signify strong emotions and reflect personal view and indicate rumour. Similarly, the vulgar words tend to signal personal viewpoint. Emoticon, often used in online social network is also a good indicator of rumour.

Speech Act word, such as expressions and recommendations are also a good indicator of user view point. It comprises of speech act verb like ask, promise, report and speech act phrase like I belief, we demand etc.

The uses of strong adjectives to qualify some incidents, events, persons are also an indicator of rumour.

- ***Syntactic Analysis:***

Punctuations like ?(note of interrogation) and !(note of exclamation) often used for expressing the emotions and thus reflect personal view point.

Part-of-speech tags, generated from the parser tree of dependency, can be used to identify the use of adjectives and interjections. In the sentence, where Interjections are used, mostly reflects emotion and thus can indicate personal expressions in the information. In the same way adjectives are used to indicate recommendations or personal expression of the user. Both of them are indicator of personal belief and thus indicator of misinformation.

7.2.2 Bursty Term Analysis

A keyword with bursty frequency generally appears only during a small time span. Generally it suddenly breaks out and end within a very short time span and show a tendency of rapid increase and decrease within the social network. The frequency of a general keyword generally continued as low steady level whereas frequent term occurrences always remain at high. But a sudden up and down of information frequency generally indicates a rumour. Periodic terms always occur once in a while, which implies it is less probable that periodic terms appears as bursty terms. Periodicity score of the topic can be scored and measured to find out the characteristics of the nature of occurrence.

7.2.3 Skewness of Information Distribution

For a periodic event, generally information generates a Gaussian distribution. For example, when there is football world cup, the information about the event grows in a regular pattern, reaches its peak at the time of event and then gradually goes down. But for a rumour, distribution of the keyword is generally show skewness (asymmetric distribution). Skewness score is a good measure for rumour detection in real-time.

7.2.4 User Identities

User's identity is used for identification of the rumour. It includes following criterion for identification:

- ***Controversiality***: it indicates the follower of the user
- ***Originality:*** it indicates whether the statement is original or not

- ***Credibility***: Indicates whether online social network account is verified or not through the appropriate authority
- ***Influence***: the information shared by the persons can create impact (positive or negative) or not
- ***Role***: the kind of roles played by that user like active or stiffler user
- ***Engagement***: It measures how active the user in online social network till joining

7.2.5 Source of Information

Source identification is an important criterion for rumour detection. Generally rumours are generated from fewer numbers of sources and spread in an exponential manner. But for general information the sources are evenly distributed.

7.2.6 Sentence Clustering

After the mostly used terms have been extracted from the network through knowledge community detection, clustering of sentence is the next iteration to be followed. Sentence weight is defined as the average of all the weights of all the terms it comprises of. Sentences rank are determined and top k (number of k can be predefined or learned) sentences are considered as the centers of initial clusters and clusters are generated through clustering algorithm like K-means clustering.

7.3 Source Identification of Rumour

Source node can be identified from a social graph through verification of timestamp. As mentioned in earlier section, the two-phase algorithm, developed by Google, can convert a social graph into the union of stars where each star represents the connected components. Each vertex is assigned with a unique id. Searching and connecting all connected neighbours of the graph, it generates the star which is the union of all large and small of the network which includes that user as the vertex. From that star node with lowest time stamp can be considered as the vertex/apex node and signifies the source node of information.

7.4 Community Detection and Location Identification

For mapping the contagion of a rumour, it is important to find communities and where the nodes are affected by the information.

7.4.1 Community Detection

Communities are the structural property of the network indicating users of same community are interacting with each other as compare to other users of the network. Individuals of same community share more common and similar properties.

Community detection algorithms play crucial roles because characterizations of some clusters help us for the characterization of the network as a whole. This feature is very useful for scale free real network like online social network. For example we can identify the group of people who are interested in some particular topic. Similarly identification of group of users generating misinformation may results in proper measures against them. Identification of cyber-crime, terrorism can lead to resistance against the crime. Community detection has several application areas in smart governance.

7.4.2 Case Study

In the case study, Twitter account data are extracted and analyzed.

Using Twitter login the user should enter into a Twitter account and after authentication with Node XL, it will import Twitter data in form of adjacency matrix. For the data processing the downloaded data is stored as CSV file. Now this exported CSV file is used to plot the graph and apply different community detection algorithm to it. I-graph is used for graph visualization and R embedded in Hadoop platform is used for the analysis of community. For our case study we are using Edge betweenness algorithm which is a hierarchical algorithm. Edge betweenness score can be expressed as the ratio of shortest paths that include that particular edge and all possible shortest paths. In edge betweenness algorithm based on the edge betweenness value edges are removed in the decreasing order until the graph breaks into all disjoint components as shown in Fig. 9.

7.4.3 Location Identification

After the detection of community we can filter the information based on the location of data. In online spatial information are available: the location of the user from where it is registered and the location of the incidence, which is the topic of the rumour. For a smart city governance data will be collected based on this location of smart city.

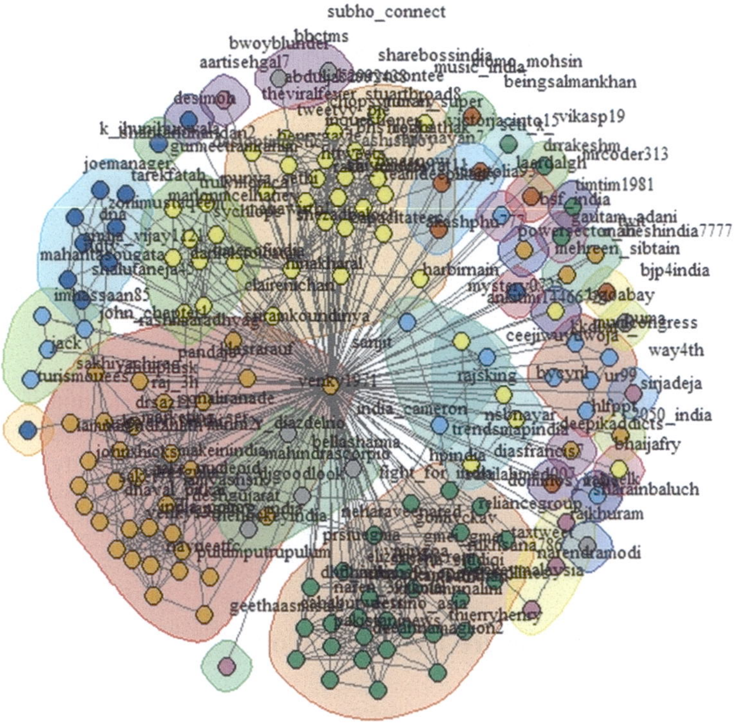

Fig. 9 Community detection using edge betweenness algorithm [22]

7.5 *Influence Minimization*

Propagation of rumour can be controlled or restricted when there is an adverse effect in the society due to that propagation. Different rumour controlled strategy for rumour controlling can be adopted for that purposes.

7.5.1 Blocking the Source Node

From the previous subsection it is observed that source nodes of the rumour are generally small in number and can be identified from the network. Therefore it is possible that source node can be identified and blocked.

7.5.2 Blocking Centrality Nodes and Edges

For online social network, identification of rumour is associated with some time delay. At the time when information is recognized as rumour, information is already spread into the medium. So it is required to restrict some influential nodes or edges from the network. Centrality nodes are somehow representing most influential node in the social graph. Blocking most influential node in the network, it is possible to restrict inoculation of the rumour in the network. It is observed that instead of blocking the centrality nodes, which costs the restriction of all connected edges of that node, it is less costly and more effective to block important edges. Centrality edges can be derived using some graph topological properties like edge betweenness, graph cut etc.

7.5.3 Beacon Model

In this model, Beacon nodes start anti-rumour campaigning after detection of rumour in the network and influence the user. In [10], random nodes are used as beacon node but centrality nodes will be more efficient for that purpose as centrality nodes are more efficient for spreading information [23]. The efficiency of Beacon model will increase with less delay time for the detection of rumour.

8 Conclusion and Future Research Scope

In this chapter, a framework for smart city governance based on online social network is proposed. For governance in a smart city, the prime criterion is that the governance should be citizen centric. As citizen has no means to communicate their view with the government directly, it will be the responsibility of the government to interact with the people proactively. Lot of research work is going on in that area, especially in information propagation and rumour detection. The fields that involve in this area are network analysis, big data analysis, natural language processing, machine learning and other interrelated disciplines. The main research challenges lies to its large volume of data and real time analysis. Lots of research scopes are present in this direction.

References

1. Alawadhi S, Scholl HJ (2016) Smart governance: a cross-case analysis of smart city initiatives. In: 49th Hawaii international conference on system sciences (HICSS), Koloa, HI, pp 2953–2963
2. Chowdhury C, Roy S, Song H, Srinivasan R, Sookoor T, Jeschke S (2017) Mobile crowdsensing for smart cities. In: Smart cities foundations principles and applications. Wiley, Hoboken, NJ, USA
3. Wang Q, Jin Y, Lin Z, Cheng S, Yang T (2016) Influence maximization in social networks under an independent cascade-based model. Physica A: Statistical Mech Appl 444:20–34. ISSN 0378-4371

4. Dey P, Chatterjee A, Roy S (2019) Influence maximization in online social network using different centrality measures as seed node of information propagation. Sadhana, Springer
5. Guzman JD, Deckro RF, Robbins MJ, Morris JF, Ballester NA (2014) An analytical comparison of social network measures. IEEE Trans Comput Soc Syst 1(1):35–45
6. Rodrigues FA (2019) Network centrality: an introduction. In: From nonlinear dynamics to complex systems: A Mathematical modeling approach, Macau E. (eds), Chapter-10, Springer
7. Fortunato S (2010) Community detection in graphs. Phys Rep 486:75–174
8. Raghavan UN, Albert R, Kumara S (2007) Near linear time algorithm to detect community structures in large-scale networks. Phys Rev E 76, 036106
9. Zhang Q, Zhang S, Dong J, Xiong J, Cheng X (2015) Automatic detection of rumor on social network. In: Li J., Ji H., Zhao D., Feng Y. (eds) Natural language processing and chinese computing. Lecture Notes in Computer Science, vol 9362. Springer, Cham
10. Rudra A, Tripathy M, Bagchi A, Mehta S (2010) A study of rumor control strategies on social networks. In: Proceedings of the 19th ACM international conference on Information and knowledge management, Toronto, ON, Canada, pp 1817–1820
11. Diao Q, Jiang J, Zhu F, Lim EP (2012) Finding bursty topics from micro blogs. In: Proceedings of the association for computational linguistics (ACL). pp 536–544
12. Zhao WX, Jiang J, Weng J, He J, Lim EP, Yan H, Li X (2011) Comparing twitter and traditional media using topic models. In: Proceedings of the 33rd European conference on advances in information retrieval, pp 338–349
13. Kentaro S, Tomohiro Y, Takeshi F (2014) Online topic model for twitter considering dynamics of user interests and topic trends. In: Proceedings of the 2014 conference on empirical methods in natural language processing, pp 1977–1985
14. Guille A, Hacid H, Favre C, Zighed DA (2013) Information diffusion in online social networks: a survey. SIGMOD Records 42(2):17–28
15. Polykarpos M, Giannis N, Francois R, Yannis S, Michalis V (2015) Degeneracy-based real-time sub-event detection in twitter stream. In: International AAAI conference on weblogs and social media
16. Diao Q, Jiang J (2014) Recurrent chinese restaurant process with a duration-based discount for event identification from twitter. In: SIAM international conference on data mining, pp. 388–397
17. Dey P, Roy S (2016) A comparative analysis of different social network parameters derived from Facebook profiles. In: Satapathy SC et al. (eds) Proceedings of the second international conference on computer and communication technologies, Advances in Intelligent Systems and Computing, vol 379
18. Dey P, Roy S (2016) Social network analysis. In: Meghanathan N (eds) Advanced methods for complex network analysis. Hershey, PA: IGI Global
19. Sarkar A, Chattopadhyay S, Dey P, Roy S (2017) The importance of seed nodes in spreading information in social networks: a case study. In: Communication systems and networks (COMSNETS), IEEE Xplore Digital Library
20. Arruda GFd, Barbieri AL, Rodriguez PM, Moreno Y, Costa LdF, Rodrigues FA (2014) The role of centrality for the identification of influential spreaders in complex networks. In: Social and information networks, physics and society
21. Page L, Brin S, Motwani R,Winograd T (1999) The PageRank citation ranking: bringing order to the web. In: Technical Report, Stanford InfoLab, pp 1–17
22. Dey P (2019) Characterization of online social network: a case study on Twitter data. In: Enabling technologies and architectures for next-generation networking capabilities, pp 272–299. IGI Global, USA
23. Dey P, Pyne S, Roy S (2017) Information spreading in online social networks: a case study on Twitter network. In: Proceedings of the 18th ACM international symposium on mobile Ad Hoc networking and computing

Algorithms in Smart Cities

An Overview of the Machine Learning Applied in Smart Cities

Reinaldo Padilha França, Ana Carolina Borges Monteiro, Rangel Arthur, and Yuzo Iano

Abstract Smart cities are a rapidly evolving reality that is emerging as a path to be followed in the development of urban centers outline at the sustainability and quality of life of its inhabitants worldwide. They surprise us with the creativity of the solutions for more efficient use of available resources, reducing the impact of our lives on the environment through digital transformation. Machine Learning is a technology that allows models to be trained on data sets before they are implemented, it is a type of algorithm that improves automatically and gradually with the number of experiments in which it is placed to train. Where computers can learn according to the expected responses through associations of different data, which can be images, numbers, and everything that this technology can identify. This artificial intelligence can also stimulate changes in utility business models, which means that users can benefit from better services resulting in greater mobility and comfort. Solving connected problems related to the optimization of urban planning and integrating city services for personalized results, concerning the use of specific services by the inhabitants. In this context, this chapter is motivated to provide a scientific contribution related to the discussion and overview of Smart Cities and Machine Learning, addressing their key points and their importance, their interconnection, and use, with a precise bibliographic background, singularizing and stereotyping the competence of technologies.

R. P. França (✉) · A. C. B. Monteiro · R. Arthur · Y. Iano
School of Electrical and Computer Engineering (FEEC), University of Campinas, UNICAMP, Av. Albert Einstein – 400, Barão Geraldo, Campinas – SP, Brazil
e-mail: padilha@decom.fee.unicamp.br

A. C. B. Monteiro
e-mail: monteiro@decom.fee.unicamp.br

R. Arthur
e-mail: rangel@ft.unicamp.br

Y. Iano
e-mail: yuzo@decom.fee.unicamp.br

M. Khan et al. (eds.), *Smart Cities: A Data Analytics Perspective*,
Lecture Notes in Intelligent Transportation and Infrastructure,
https://doi.org/10.1007/978-3-030-60922-1_5

Keywords Machine learning · Smart cities · Smart architecture · Data analysis · Artificial intelligence

1 Introduction

The idea of Smart Cities implies sustainability, energy, loss of resources, waste management, traffic improvements, assimilation and aggregation between public systems, civil service offerings, and many other aspects. The concept of "smart city" is to allow the evolution of the utilization of available resources, interpreting data and making that data become useful information for those who are in that city, i.e., in which the citizen is the focus [1].

Machine Learning can contribute in several ways to the construction of smart cities, since this technique allows the creation of intelligent and efficient services for the population, with, for example, monitoring data on transport, control over the use of public services and monitoring in real-time from municipal security cameras. It is a process of data analysis that motorizes and technicalizes the construction of analytical models. It is a strand of artificial intelligence (AI) that consists of the concept that systems can learn and understand from data, identifying patterns, and having conditions to make decisions with small human intervention. The iterative aspect of machine learning (ML) is significant due when models are presented to new data, in the sense that they can adapt independently and autonomously. Still evaluating that they have the ability to learn from previous computations producing reliable and repeatable decisions, and even results [2].

In Transport, an example of the use of technology can be seen in public transport. Companies will be able to optimize services according to user demand. And passengers will be more accurate about bus arrival times. That's because artificial intelligence contributes to data monitoring and analysis. Thus, public managers can control everything that occurs at stations, checking the traffic of people, and identifying the hours with the largest audience. Citizens, on the other hand, can follow information about the exact arrival time of the bus through smartphone applications [3].

In Safety, machine learning also contributes to the safety of cities. The technology allows for the quick identification of people moving through the regions monitored by cameras. Artificial networks will be able to mimic the functions of the human brain to interpret data and perform facial recognition, even if the environment has many people. The systems used are also capable of associating information recorded in the name of people, such as the vehicle they own, family members, and frequent visits [4].

In Health, artificial intelligence will also be of great use. Like robotic surgery in patients, or robots used to monitor the operation, through a camera that is inserted into the patient. He can also handle scalpels, tweezers, and scissors. The doctor does the surgery on a control table. The camera image can be enlarged during the operation to give the doctor much more accuracy during the procedure [5].

And technology can further improve treatments. By using the concept of artificial intelligence, the machines will be able to suggest possible diseases that could affect individuals over the years. The diagnosis will be made based on the patient's history and the indicators identified in his exams. Consequently, doctors will be able to carry out preventive work or initiate certain treatments earlier [6].

In Education, classes can be more dynamic and students can count on virtual platforms to talk to teachers and perform exercises. Using this platform top, students can more easily monitor their performance and identify what they need to improve to pass the year. The intelligent system used by the institution can identify, through algorithms, the ability of each student to understand the subjects and indicate which classes he needs to attend again to clear his doubts [7].

Many cities are evolving in a similar way beyond connected digital infrastructure and smarter devices. In some cities, citizens (population), visitors (tourists), and businesses (investors among others) act as "mobile sensors", carrying smartphones that are an essential part of a smart connected digital ecosystem. Requiring smart cities to integrate and compose capabilities and features, and even coordinate applications and services, through massive volumes of data, that at one or more levels in the operational digital domains for multiple stakeholders and participants. In order to generate a network of sensors, with common digital standards, connected and interconnected to a computer platform governed by the city itself. Allowing that the network, relating real-time digital infrastructure is based on artificial intelligence and machine learning, has the capacity to change easily as the priorities of the city and its citizen's change [8].

Since with the use of machine learning and artificial intelligence (AI), it is possible to predict network problems as accurately as possible and anticipate trends concerning the captured data promoting actionable intelligent insights. Leveraging these insights manages to help municipalities developing smarter and more resourceful, data-driven business and government policies that provide them to adapt to constituent and component needs in a secure and real-time manner [9].

In the same sense as to meet the needs of citizens and for the development of new smart cities, technologies such as chatbots (service robots) and machine learning are essential to understand the needs of citizens and even engage them in this process. With the analysis of the collected and accumulated data, it is possible to create algorithms that improve the interaction of residents with smart cities [10].

In light of the above, this chapter outline to afford an updated overview of Machine Learning towards Smart Cities, addressing its evolution and fundamental concepts, showing its relationship as well as approaching its triumph, with a precise bibliographic background, singularizing and stereotyping the competence of technologies.

2 Methodology

This research study was developed based on the analysis of scientific papers and scientific journal sources referring to Machine Learning towards Smart Cities, aiming to gather pertinent information regarding thematic concerning evolution and fundamental concepts of technology. Thus, it is also possible to boost more academic research through the background provided through this study.

3 Smart Cities Concepts

Smart cities exist in the most varied ways, after all, each place has unique needs and specificities, but one feature is recurrent: they are the future of urban life. Smart cities are automated and more sustainable, those that use technology to generate efficiency in urban operations and create sustainable solutions, in such a way that it maintains its economic development while improving the quality and attribute of life of the population. The generation of smart cities is a subject that increasingly arouses the interest of the government and the population. They are those that optimize the use of resources to better serve citizens, with a great emphasis on mobility. Using technology to improve the quality and style of life of the city population is one of the key points of Smart Cities, which seeks to accelerate development without giving up sustainability [1, 3].

Technology is essential in a smart city, but it is only a way to solve a series of urban problems and achieve goals that are becoming increasingly important for large cities. Provided they use technology to promote the well-being of residents, economic growth, and, at the same time, improve sustainability. From the data point of view, they make optimal use of interconnected information to understand and better control operations and resources [4].

Technological evolution, especially the development of very recent resources, such as the Internet of Things (IoT), has been an extremely essential point in the development of solutions to problems in the contemporary world. Smart cities can also be considered as those that utilization information and communication technologies (ICT) in order to make public transport networks and lighting, heating, and water supply systems in urban space more efficient and sustainable, as well as to ensure safer public spaces that are better suited to the needs of the population [11].

Some essential factors in relation to smart cities are that they are targeted at increasing the quality and style of life of citizens; they need the involvement of stakeholders (universities, companies, NGOs, and the like) to be successful; they are built based on planning; optimize the use of resources; have clean and accessible mobility, with the help of technology [12].

Six key areas are prioritized for the development of actions with respect to smart cities, such as Economy, Mobility, People, Governance, Environment, Quality of Life. And according to the Cities in Motion Index, from the IESE Business

School in Spain, certain variables can indicate the level of intelligence in a city. As human capital, social cohesion, economy, environment, urban planning, international reach and connection, mobility and transport, governance, public administration, technology [8].

Two types of smart cities can be adopted, where one is more focused on the use of technology, and the other is centered on the sustainable economy. They are not mutually exclusive, they just change the position of the protagonist, given the way they are applied. Smart cities that focus on the use of technology, what happens is a series of solutions that optimize the time, access, and security of the population. This happens from the integration, processing, and monitoring of information. These are improvements made for citizens. In these processes, there is a concern with security, in relation to the confidentiality of personal data, and with excessive exposure and privacy. Smart cities that focus on the economy, where education, health, and entrepreneurship, digital and creative economy, are fostered in order to generate value for the city, increasing the quality of life, and strengthening the economy with new businesses and attracting investors. These are improvements made by citizens [10].

The human capital assessment includes indicators such as the proportion of the population in secondary and even considering higher education, as well as the number of universities, museums, schools, and theaters. Governance, in turn, takes into account factors such as the number of embassies, consulates, and research and technology centers, as well as respect for the rights of the population and perception of corruption [13].

Among the technologies implemented are the reprogramming of traffic lights automatically if there is traffic intensity on certain roads, which is done by the application of AI/ML. Traffic density can be controlled by underground sensors, where these obtained data are processed through AI/ML interfaces and algorithms. In the city, there may also be innovative systems for collecting garbage, which practically eliminates the need for collection, regarding the use of AI/ML technology in the sorting and management of solid waste, through data collection by intelligent sensors and vision systems using machine learning software in recycling facilities, which assist in the classification and identification of items contaminated with food and other substances. In the same sense as reflecting on several examples that include combining drone delivery services, using ML system to solve logistical obstacles, delivering products in less and less time, from address identification (AI). Still considering the identification (ML algorithm) of any obstacle in the middle of the path, whether static or mobile, since there is an infinity of objects of the most diverse characteristics, in the drone route. Or even city health services with pharmacies, regarding the use of AI to diagnose patients, considering ML algorithms to detect skin cancer, or those AI software that "listen" to emergency calls, and by the tone of voice and the background noise of the person can define the best relationship approach, or ML algorithms that examine medical records, habits and genetic information gathered in health institutions, nourishing doctors with more information to better serve the patient. Or even private vehicles, in what derives the transport by an application that recognizes the route (ML algorithm) and the choice of the driver through the app interface (AI), still evaluating the existence of possible particular rules, either by the driver or passenger. However, another context is the transportation of garbage on demand, based on digital monitoring by residents, it performs intelligent management of domestic waste, through AI interfaces, the residues are transported from the residences to the processing centers, where the sorting and treatment are carried out (through the ML algorithm), instead of dumps based on sensors. Still evaluating the use of real-time data (ML and predictive algorithms), to improve vehicle traffic, at the same time that the results of these algorithms are analyzed by means of AI, evaluating the traffic points that had reached their capacity [14].

In certain smart cities around the globe, the mouths of the brass are connected to a piping system buried a few meters from the surface. Dealing with a large sucker that sucks in the garbage, and so the bags travel through the pipe, arriving at the final destination which is a collection center (performing the sorting through ML algorithm). There, they are deposited in containers, which are then filled to a sorting plant, where plastic, paper, and metal are recycled and organic waste is transformed into fuel to move turbines that generate electricity, this resulting selectivity is based on the results of ML algorithms analyzed using AI. Making fewer collection trucks stop circulating in the city on a daily basis, representing a highly sustainable system that encompasses the entire waste production chain [15].

Smart cities are still considered to compose competence and technique, which have the property to coordinate large volumes of data, services, and even applications, in a process considering one or more levels in the operational domains for multiple participants. With the objective of creating a network of sensors, with common standards, connected to a platform, with digital infrastructure in real-time, it consists of machine learning and artificial intelligence techniques, managed by the city itself. This concept act on the infrastructure of bicycle paths, encouraging the population to use bicycles instead of cars, contributing considerably to reducing the intensity of traffic and pollution, and carbon emissions. However, in addition to bike lanes, bicycle parking, rental, and drop-off locations, equipped with GPS and sensors that, considering that through these sensor data (air humidity, temperature, and lighting) directed to ML algorithms, through AI analysis it is possible to estimate air quality, and users still receive information about congestion in real-time, as well as public transport devices such as trains, buses, and subways that are capable of transporting them. Evaluating that through mobile applications (ML) they allow passengers to buy tickets, check schedules and status, and even find nearby stations. And combined with AI features, they provide users with location tracking, notification, and suggestion resources, supported by processed data sets (ML) that adjust and continuously improve over time [16].

We can also classify smart cities through its approach: bottom-up, where cities work with data from sensors installed by the city, monitoring cameras, social networks, among others. Through the data collected by these devices, ML solutions allow automated counting of people through facial analysis, differentiating faces, and identifying them among several others. Allied to AI systems that perform the crossing of data and the identification of situations (ML and predictive algorithms), or even the instant identification of dangerous situations (behavioral and predictive ML analysis), optimizing the work of public and private security agents, contributing to the construction of increasingly intelligent cities. Everything is integrated on a single platform, which allows efficient management of services. In other words, decisions are made according to users' attitudes and thinking. This format is widely used in large cities that seek to make their services intelligent because it is a matter of adaptation. Taking a more democratic approach, since citizens can give their opinion and move so that the actions work. Or top-down, which recreates cities from scratch, making the initial structure work in their favor. In this example, the entire structure is assembled so that the city is intelligent [17].

The smart cities that are created can come in two forms: the first is to invest in planned cities and include sustainable technologies and actions in their previous planning. And the second is to reassess the processes of existing cities and identify the improvements that can be made according to the needs of residents and the location. In an environment where everything is instantaneous, identification technologies using ML algorithms (people, vehicles, or objects), considering those intelligent systems (AI) offering greater efficiency, allowing suspects to be identified (ML) or through evidence of recorded theft safety equipment. In other words, through learning people's behavior (ML), with information crossings (Big Data and AI), they can be used together to improve security solutions. Cities that invest in IoT in the management of urban services achieve cost reduction, since the technology offers better visibility for management processes, making it easier for new investments to be made until reaching the level of smart cities. In this way, Smart Cities can be pointed out and seen as systems of people that interact, use materials, consume energy, use services, and still finance to catalyze economic development through digital resources that impact on improving quality of life. Reflecting on these interaction flows, it is appreciated as intelligent due to actions that making strategic use of digital infrastructure, considering services, information, and communication linked with urban planning and management, which respond to the economic and social needs of modern society [18].

As cities grow, making them smart is essential. Universities, companies, and public institutions can join forces so that solutions are applied and serve citizens more and more effectively. The incorporation of millions of mobile "sensor citizens" and connected devices that interact with users, machines of the most diverse types, and cloud structures that require the essential reconsideration of the network. It is necessary that, this digital assimilation be able to scale quickly to move large amounts of data in real-time, which meets the dynamic needs of new smart city platforms, with the respective networks that must be more predictive and agile [19].

All of this implies the need to include technologies for trend analysis or application of AI and machine learning techniques in addition to a centralized environment that offers access and communication with all these devices. With so many people living concentrated in big cities, problems arise related to urban mobility, the environment, access to housing, and public services. However, this environment also provides innovation, and solutions are increasingly using technologies. Cities having an appropriate governance environment for a smart city, where data from the city's various sensors collected and sent to the cloud, on a cloud platform, where all devices are configured and can be tracked and managed remotely. In this way, they are getting smart unveiling to deal with the proliferation of sensors (IoT), people (mobile or wearable), automobiles, and a variety of intelligent devices that require access to the network and generate an impressive volume of data. However, the condition of "smart" is not an instance in time and a "smart city" relates to not being static [21].

As much as there are many concepts, this is the main objective. What changes is how this level is reached, whether through advanced technologies, whether through simple projects, but which completely changes the lives of the population.

4 Artificial Intelligence Concepts

Artificial intelligence (AI) in its essence, allows systems to make decisions independently, accurately and supported by digital data, is a strand of computer science research that, through computational symbols, seeks to build mechanisms and develop devices that simulate the human capacity to reason, learn, perceive, and decide which paths to follow, in a rational way, in the face of certain situations, i.e., making decisions and solving problems, in short, the ability to be intelligent [21].

Before, computers needed good data models for classification, processing, and analysis; access large amounts of raw data; and powerful computing at an affordable cost to make this processing fast and efficient, forming three major pillars to evolve from simple computing to artificial intelligence. With the computational evolution, artificial intelligence has made a great advance in computational analysis, and the machine can even make analysis and synthesis of the human voice. With the evolution of these three segments, i.e., big data + cloud computing + good data models, artificial intelligence has finally become possible [3].

We can say that AI learns like a child. Gradually, the system (depending on the purpose for which it was developed) absorbs, analyzes, and organizes the data in order to understand and identify what are objects, people, patterns, and reactions of all kinds. AI impacts and affects several other areas, such as automated planning and scheduling, autonomous control, games, medical diagnostic programs, computer programs, security applications for information systems, robotics (auxiliary robots), devices for handwriting recognition hand and voice recognition, among many others [4, 5].

Through applications and even in large cities, in projects that involve the concept of Smart City, in which certain urban spaces are the stage of experiences of intensive use of communication technologies and information of urban management. The more integrated AI resources, the more intelligent the city is, which is largely based on video resources. Besides, a Smart City is based on several pillars, such as security, education, sustainability, mobility, among others, and AI must integrate each of them [22].

Concerning Security, systems, and cameras that use AI and image processing algorithms provide efficiency for urban management. The more automation, the more managers will be able to focus on other issues that need human intelligence to solve. The use of AI is possible to perform access control to cities, monitoring, facial recognition in the search for missing persons, integration with the Police, and other authorities. Considering the use of intelligent vehicles with vehicular DVR (Digital Video Register) has also been a highlight, as well as the analysis behavior of individuals in order to prevent kidnappings and robberies [23].

Other solutions with respect to urban mobility are cameras for vehicle counting and traffic statistics, since there are systems that use intelligence to recover stolen or irregular cars. Making some smart cities already have this siege or digital wall, an extremely smart and effective way for the police. Since with facial recognition technology, it is also possible to assist in the search for wanted criminals, making the use of these technologies contribute to the efficient control of entry and exit of criminals, weapons, drugs, and illegal goods [24].

With respect to city management, the use of AI (Artificial intelligence) consists of the principle of optimization, dynamization, and expansion of the scope of the most several operations carried out within the city, where its systems are programmed to identify patterns carrying out predictions and actions with speed and accuracy. It is possible to manage garbage collection, police patrols, detect fires, fire and smoke, vandalism, monitor events, and even earthquakes and floods, avoiding various tragedies and improving the quality and efficiency of public services. Efficiency can be obtained by applications, sensors, among other apps and devices, relating the quantity and quality of data. Which is the urban context, intelligent technology based on the use of AI is evaluated as a way to sophisticate and polish the management of these territories, especially those that are more populous and of greater extension [20, 25].

With regard to urban mobility, systems integration is essential. Since a smart city needs interconnection. In terms of traffic, there must be a synchronization of traffic lights, intelligent cameras that identify the events that occur. Bus stops that add safety, accessibility, and comfort to users of public transport, using technology tools, such as the camera system for monitoring and their connection with the police and the public transport control center [26].

As well as monitoring traffic in real-time, detecting accidents by issuing the necessary alerts to the police, supervisory, and health agencies, if victim assistance is required, and even before they happen and issuing warnings and alerts to the authorities. Facilitate the real-time monitoring of large cities with the use of AI (artificial intelligence), through the application of machine learning to process videos and images in real-time. Recognizing people, accident locations, license plates, and a multitude of possibilities that can be extracted from videos and images. Making cities safer and smarter. However, for all of this to be possible, it is necessary to have quality data. This data can come from public transport, sensors, cameras, police reports, and many others. In addition, video and image processing is a great source of a lot of data. However, it is a complex task and it can require a lot of computational resources, which analyzes real-time difficulty. Before, and even in many modern cities, this image monitoring is done by a human who is by nature unable to monitor multiple images in real-time and recognize patterns automatically. What through AI this task can be performed with greater efficiency [19, 26].

About Health, the application of AI is extremely comprehensive. As in the creation of systems in which the citizen can know if his doctor is available for consultations or even if the medicine he needs is available in the public network, making the service more efficient. As for Sustainability, technologies such as AI, ML Big Data, Internet of Things (IoT), are associated with these management strategies, projects, and technologies aimed at increasing the quality and style of life of the population and greater efficiency in resources and services of various types [27].

The utilization of new technologies, such as solar panels to capture energy, also makes cities smarter and promotes the use of clean energy, supplying electric cars, thus giving rise to more solutions to the issue of sustainability in cities. And with AI, improving urban systems, urban engineering, and even city management can be closely bonded to each other. The management of smart cities can be increasingly efficient through the use of data collected in real-time combined with the skills of AI. Making government officials able to learn more and more about the managed city, applying this knowledge to improve infrastructure, security, and resource allocation [28].

It is important to remember that before a city becomes smart it needs to be digital, because to use AI tools, there needs to be connectivity and energy. These areas are all interconnected, as the integration of equipment, data, and information depends on connectivity. Much of the public infrastructure of large smart cities is overused, inefficient, or even unused. And with the use of real-time information shared between people, and with the government itself, it can be useful in a variety of situations. The smarter the city, the greater the number of cameras, and other smart devices, creates a greater demand for monitoring all videos produced in real-time. Consequently, the blindness paradox is noticed, which is associated with the smaller number of people involved in the visual monitoring process, i.e., the more cameras installed, the lower the human attention to events [29].

However, it is currently possible to know how long the desired bus will pass, but there is still a lack of intelligent solutions that determine the number of people on each bus, which could make the user wait a little longer and travel more comfortably. It is in this scenario that AI enters, because it takes care of this task, generating alerts of what needs to be monitored by human eyes, facilitating the work, and optimizing the processes. In this way, it is perceived that the use of AI will be applied more and more frequently, not only in Smart Cities, but also in the routine of the citizen as a whole, allowing the quality of life of people to improve and live in cities to be improved. make it easy. Smart Cities use technology to generate efficiency in urban activities, while promoting economic development and ameliorating the quality of life of the population, in a sustainable way [30].

Smart cities foster technological development, think of strategic solutions for the city, and the lives of the people who live in it, supported by innovative and sustainable planning that contributes to urban mobility. And with IoT, AI and Big Data, make smart cities use interconnected and interoperable systems for the strategic use of services, infrastructure, and communication. Reducing infrastructure, operation, and maintenance costs for municipalities and enabling public management to get the social and economic requirements of society [20].

4.1 Machine Learning

Solutions that use computational intelligence have already managed to solve several historical problems of humanity, from autonomous machines to automatic translation. With the advent of AI, machines are being able to achieve even better performance than humans, in some applications, such as object detection and recognition. Machine Learning is a technology where computers have the ability to learn according to the expected responses through associations of different data, which can be images, numbers, and everything that this technology can identify. It allows models to be trained on data sets before they are implemented. An application or software with Machine Learning is a type of program that improves automatically and gradually with the number of experiences in which it is placed to train. It evolved from the study of pattern recognition and the theory of computational learning in AI [31].

This technology is considered a subfield of AI, which works on the ground that machines can learn on their own when they have access to large volumes of data. The simplest definition is that machines can detect patterns and create connections between data, using Big Data and sophisticated algorithms, to learn how to perform a task on their own. Making a parallel, when developing a machine learning system, the structure used in programming is different from traditional software programming [32].

In the traditional method, a set of rules is created to generate a response from the processing of the data entered. Machine Learning algorithms are created from the data that will be analyzed and the answers/results expected from this analysis, at the end of the process, the system creates its own rules or questions. Basically, the algorithms use improved statistical analysis of the data they receive, resulting in more accurate responses and predictions. Machine Learning allows software applications to become more "intelligent" and predictive, without having to program them frequently so that they "learn" [33].

In the first stages, training is assisted. The iterative process leads to an improvement in the types of associations made between data and elements, which are presented in large numbers. Due to this large amount of data that will be analyzed, the patterns and associations made only by human observation could be inefficient, in case they are made without support from Machine Learning. After the initial training of an ML algorithm, it can be employed in real-time to learn alone with the data, presenting greater precision in the results over time. To work with the machine learning system, it is necessary to use a certain set of data. Big Data technology allows data to be virtualized so that it can be stored most efficiently and economically, either on-premises or in the cloud. In addition to the efficiency that Big Data also helps in improving the speed and reliability of the network, removing other physical limitations associated with managing large amounts of data [31, 34].

Machine learning technology can be categorized as supervised which is when the algorithms are supervised, this means that a human being controls the input and output of desired data and provides comments on the accuracy and rigor of predictions during the realization of training. When complete, the trained algorithm applies what has been learned to new data. It is mainly used when the system already knows which inputs are associated with which outputs and needs to learn a way to understand this association. It uses pattern detection to establish predictions, such as the categorization of emails, separating what is relevant and what is SPAM [35].

In the unsupervised category, the algorithms there is no need for training in order to obtain desired outcome data. Since in general, this class uses an iterative approach known as deep learning. Unsupervised algorithms are employed for more complex processing tasks. His approach is to discover the relationships implicit in a set of unlabeled data. In this case, it identifies patterns for labeling the data. The online recommendation system is an example of unsupervised learning. There is also the Reinforcement Learning category, where the computer is encouraged to learn based on trial and error. The process is optimized through direct practice, teaching the system to prioritize certain habits, such as systems that play chess and autonomous vehicles [36].

Despite the advantages offered in the process, a company does not need to have Big Data to work with Machine Learning. But it must seek to offer knowledge to computers (machines) through observations, data, insights, and interactions with the external world. This obtained knowledge grants these computers to precisely generalize new events and configurations. A common fact for any learning algorithm is the need for data. Data is essential for Machine Learning, since one does not work without the other. The use of previously available data, commonly called training data, to train a model that is able to generalize the phenomenon well and predict a result when presented with new data, generally called test data [32, 34].

Every learning algorithm is composed of only 3 components. (1) Representation, which refers to any phenomenon must be represented in some formal language that a computer can interpret. (2) Similarly, choosing a representation for a phenomenon means choosing a set of models that can be learned from that representation. This set is called the hypothesis space. Besides, it is also important to take into account which characteristics of the phenomenon are important for a given representation. (3) Evaluation as an evaluation function is necessary to distinguish good models from bad models. There are internal and external functions. An internal function is used by the model in its optimization task, which normally, learning algorithms aim to minimize error. External functions are used to evaluate the final model. E Optimization is related to the need for a method to find, among the various models available, one that maximizes performance [35].

The fundamental purpose of Machine Learning (ML) is to generalize and abstract beyond the examples in the training set. That's because, no matter how much data there is for model training, since it is very unlikely that these exact examples will be seen again in the forecast time. A good model must be able to observe input data and model the phenomenon in a way that is neither too generic nor too specific, because if too generic, the predictions will be as good as a random guess, and if too specific, the model will have decorated input data and will not know how to handle new input data, this is considered generalization [31, 38].

Learning Machine is profitable for human beings because, with all their computational and processing power, they are able to highlight accurately or even find patterns more quickly in Big Data (huge volume of data) that would otherwise be impractical to be done by humans. Computational intelligence is a tool that can be used to improve human beings' ability to resolve problems and make inferences and deduction about a broad range of problems, from diagnosing diseases (in the medical field) to coming up with solutions to global climate change (from an ecological point of view). So, ML (Machine learning) contributes in several ways to building smart cities. Since it makes it possible to create smart and efficient services for the population, such as monitoring data on transportation, controlling the use of public services, and real-time monitoring of municipal security cameras [39].

Whether in Transportation with the use of technology in public transportation, where companies can optimize services according to the demand of users. And passengers will be more accurate about the bus arrival time. This is because technology contributes to data monitoring and analysis. Thus, public managers can control everything that occurs at stations, checking the traffic of people, and identifying the hours with the largest audience. Citizens, on the other hand, can follow the information on the exact arrival time of the bus through smartphone applications. Security, allowing the quick identification of people who travel in the regions monitored by cameras, interpreting data, and doing facial recognition, even if the environment has many people. Assisting the identification of citizens and the consequent arrest of criminals [40].

5 Discussion

Smart cities use advanced technology solutions to promote improvements in the quality of life, sustainable development, and greater economic growth. In the context of a smart city, everything is connected at high speed, which makes it possible to interconnect lighting systems, traffic, public transport, and much more. What makes the subject gain so much prominence today, as the need to extract value from the amount of data available.

Related that from the historical point of view, never in the history of mankind has so much data been produced as today. And considering the increased processing speed of computers, the cheaper technologies, and the evolution of GPUs. And a huge amount of data available is what fuels and strengthens Machine Learning.

These data can be collected today on a large scale, at a low cost, through intelligent sensors, which can be installed in movement devices throughout the city to collect data on the various desired variables. In the case of an application for smart cities, sensors can help determine (through ML data processing), in the transport system, the fluidity of traffic, and the speed of travel (through AI analysis), among many other aspects.

Viewing this scenario, ML is one of the evolutions of functional analysis, which is the branch of mathematics focused on the study of function spaces. It is a data analysis method that automates and mechanizes digitally the development of analytical models. Through this technology, it is possible to use algorithms for data collection and learning from the data, taking into account the entire history and then making a determination, resolution, or even prediction regarding something or situation in the real world. So, the machine is 'trained' to acquire data skills. Enabling 'learn and predict situations' based on data from a sensor or a given database.

What happens is that the computer/machine learns through techniques that use computational algorithms. These algorithms work with a certain database, so they can learn through training and predict situations with the experience gained from that data. The algorithms work in a somewhat autonomous way, returning results that were not even programmed, that is, they begin to have insights about something specific.

Data is essential, so the first step is to choose which data will be worked on and made available for the Machine Learning process. And depending on the algorithm used, the technology can act autonomously, returning consistent results for which they were not even trained.

Whether with Supervised Learning where the machine is trained through a data set where the output is known for each input. Since the data for this type of method must have labels. The challenge is when you want to predict future situations. With data entry, what are the expected results for that situation. In this type of scenario, the algorithm must adjust to arrive at the correct results and with the maximum accuracy. For that, learning can be constant, thus increasing the experience with that problem.

Whether with unsupervised learning that unlike the previous one, the data has no label, that is, the correct output is not informed. In this type of scenario, the algorithm must discover the database and what is being shown, exploring the data, and trying to find some structure in it. Depending on the technique used, the algorithm will find groupings between these data, approximating the data that has some similarity to each other.

However, not all data sets, whether acquired in open databases, collected in real-time or from a database, will be structured and ready to be passed on to the machine to learn. The machine needs a reasonable data set to be able to learn too. Many times, the data is completely unstructured and out of a standard, and for that, it is necessary to perform techniques such as Data Cleaning.

Data Cleaning, is a technique used to work with data. Doing a process in which data normalization, structuring, standardization, and contextualization are carried out. It may be that not all databases need to go through this process, however, it is good to validate depending on the problem and what the desired result is. And so, as a result of having a structured data set, the next step in this learning process is the application of a learning method together with a Machine Learning technique.

In this context, it is notable that the improvement of urban systems linked to AI technologies and the management of cities can be closely linked to each other. With regard to stations and transportation points, and considering the provision of information to passengers (through AI apps), it is necessary to operate systems that optimize bus networks with real-time timesheets, evaluating the use of data (through ML data processing) such as the number of daily trips per line, and consequently the volume of passengers. In what derives these analyses, can be obtained through user experiences being collected, stored, and combined through data by applications derived on the use of ML, enabling possible adaptations to user paths and even expanding the adhesion to the public transport network, using multimodal transport.

In a city with thousands of sensors considering their heterogeneity, whether from different manufacturers, brands, and models, creating sensor networks with the ability to capture air samples, perform analysis of them (ML data processing), and warn (notify, clarify or indicate through AI interface) the population of a given city (large or small) about their quality. Besides, to pollution indications, the measurement and mensuration of fossil fuel particles or fires can be made through data capture and ML processing, being able to indicate in real-time if there is burning near a forest that represents a danger of devastation, through analysis using AI techniques.

Smart applications across the city, it is possible to receive data (information) from the users themselves (collected), which through an AI interface can be indicated (made available) whether on the way (route) is evaluated safe for women, with respect to the presence of movement in the streets, commercial establishments, policing, public lighting, or even harassment (ML data processing), for example.

Reflecting on facial recognition, and even behavioral patterns (ML data processing), linked to the agencies that are responsible for safety in cities, is an approach that can be employed as a strategy to a reduction of violence. In addition, AI monitoring systems can provide the counting of numbers of people at determining events, with respect to the identification of individuals by Machine Learning.

Traffic systems based on Machine Learning give cities the power to improve the monitoring, follow up, and analysis of this data (AI) through traffic light control, intelligent traffic management, camera monitoring, among others. Through video systems, it is possible to recognize different modes of transport, identify accidents, and differentiate between pedestrians and vehicles (by applying ML), using such data to trigger flow control intelligent devices and analyze strategies and approaches for the future (by applying AI). Since those responsible for public administration will thus be able to guarantee the normal operation of transport systems and see the existence of bottlenecks such as congestion.

The use of Machine Learning and AI predicts network problems as accurately as possible and is able to anticipate trends by a huge volume of performance data into valuable insights. Leveraging these insights helps municipalities employ smarter data-driven policies which provide them to adapt to urban requirements securely and in real-time, as needed by the city.

One of the ways utilized to discover water leaks can be through capture flow frequencies in pipes (through sound sensors). And through an application that analyzes data reading (by applying ML) correlating through a database stored in the cloud (by applying AI), it has the properties of differentiating sounds from pipes with a normal flow from those with leaks using ML/AI techniques.

It is possible to use Big Data in conjunction with ML/AI application to correlate with unstructured data sources such as weather reports, newspapers, and magazines, among other sources of information useful to manage well-being in the city.

All of this implies the need to include technologies for trend analysis or application of AI and Machine Learning techniques in addition to a centralized environment that offers access and communication with all these devices. Making this integration fully possible, either with legacy systems in on-premise environments or with systems and applications residing in private or public clouds.

In Health, when using the concept of Machine Learning, machines will be able to suggest possible diseases that may affect individuals over the years. The diagnosis can be made based on the patient's history and the indicators identified in their exams. Consequently, doctors will be able to carry out preventive work or initiate certain treatments earlier, operating together with AI applications.

As in Education, it is possible to change teaching strategies through more dynamic classes and students will be able to count on virtual platforms (AI application) for dialogue with teachers and carrying out exercises. And so, by identifying, through ML algorithms, the ability of each student to understand certain subjects and indicate which classes he needs to attend again to clear his doubts. And teachers can monitor students' performance, through the AI interface of this type, take complimentary classes, and correct exercises automatically.

One of the challenges to be faced is the integration of all the data collected to achieve the best answers and find the most appropriate solutions to any problems detected. They need to operate in an integrated manner, solutions need to overcome the inherent challenges of interoperability, interchangeability, response time, and cybersecurity. Considering that these pioneering premises can become great examples for municipalities around the world, which could transform the way of living and using technology.

6 Conclusions

Smart Cities are increasingly in evidence, their concept is related to cities that are able to adopt, in a decentralized way, the set of new applications from various providers focused on urban services that are increasingly present in the daily lives of the population. In the same way as the management and monitoring of all urban services in the city.

The implementation of smart cities that use AI, IoT, and big data technology to monitor their citizens, must be aware that demographic density is not something that is generated, it is something that is allowed to happen from an existing demand for territory, and that cities are not made of technologies or buildings, but of people with very simple needs. From a social point of view, it is seen that these simpler needs are related to adequate housing, basic sanitation, adequate means of transport, accessibility to the labor market, and services. It is in these aspects, it is possible to achieve this without fiber optics, cell phone applications, or command centers.

Before discussing technology, it is worth noting that many underdeveloped countries still have hundreds of thousands of people living in slums, especially Brazil. These citizens, live in precarious conditions, without access to the minimum infrastructure and services, considering also hundreds of thousands more in semi-formal housing, without following the rules of technology that affect the city. It is worth saying that this is the result of a lack of planning, a lack of regulation. However, certain regulations affect and affect the decision of where enterprises and businesses will be located since the very foundation of certain cities with intelligent planning.

In fact, the objective must be just the opposite, to make a city more attractive and more accessible, sustainable growth must be managed. Making it evolve beyond connected digital infrastructure and intelligent devices. In some cities, the population (citizens), visitors (tourists), and businesses (domestic or foreign economy) act as "mobile sensors", carrying smartphones that are a piece of a smart connected digital ecosystem. In this sense, a smart city can be divided into some regions, each being monitored considering specific problems that can be solved or mitigated from constant monitoring using disruptive technologies such as IoT, AI, and Machine Learning.

With respect to Machine Learning, it explores the construction of algorithms that can understand and learn from their mistakes and perform predictions about data from two learning approaches, be supervised, unsupervised, and reinforcement. This allows you to produce reliable and repeatable decisions and results. Such algorithms can make predictions from samples or make decisions guided solely by data, without any type of programming. Although similar in some respects to computational statistics, which makes predictions with the use of computers, machine learning is used in computational tasks where the creation and programming of explicit algorithms are impractical.

The notion of intelligence is structuring for the definition of the term Smart Cities. Whether from the perspective of the citizens, responsible for shaping the city from their interactions, or from the perspective of the territory, whose role is not limited to the stage where social relations take place.

A complex ecosystem for handling large volumes of data, mission-critical and fully connected systems in real-time, mixing IoT, AI, machine learning, deep learning make all of these things work together. Computer vision with machine learning models trained to detect and recognize things to be avoided (holes) and not collide (with other cars, people, objects, among others).

Making sure that there are no limitations for creativity, for innovation; about the insights provided by all the data involved, considering new business models, about opportunities for entrepreneurs to develop new products and new services, about supplying governments with data generated to make correct and quick decisions, and about opportunities for product R&D and also for engineers, scientists and application developers.

Finally, there are new opportunities, including new careers, for the entire ecosystem, the different IoT equipment becomes intelligent and generates data and machine to machine (M2M) communication, optimizing processes and guaranteeing real-time information to all those involved, citizens to managers. In other words, this is a new lifestyle for users and for new businesses that are emerging as fast as we can imagine.

7 Trends

ML generally get data generated by various applications, such as healthcare apps, cars with Internet access, among many others, and uses its tools are able to customize the smart city experience, to identify patterns and learn how to optimize the set of services, apply aggregating information about the most used roads, in the context of a transport system, for example [31].

In relation to state-of-the-art and 5G computing technologies can allow for better connectivity, new services, and lower latency, while machine learning and AI allow real-time analysis so that users can make decisions. A strong trend that will bring significant changes to Smart Cities, with faster speeds with less energy consumption, definitely enabling the consolidation of IoT. The data will be retrieved from various sources, such as traffic cameras, GPS/vehicle sensors, among others, and will be used to examine and understand traffic patterns [41].

Cities will have numerous sources of possible data, such as ticket sales on public transport, police reports, intelligent sensors on the roads, local tax information, and even local weather stations. A vast source of raw data that employed to AI pattern recognition technology will make expressively more manageable like videos and photos. ML and AI can be useful in garbage collection still reflecting on proper disposal and management, responding effectively to the maintenance of vital municipal activity. Intense urbanization and industrialization made it mandatory for cities to adopt smart techniques for waste management. Thus, the implementation of AI and Machine Learning for smart waste management and smart recycling can offer a sustainable smart waste management system. AI-enabled sensors and devices installed in garbage containers that deliver notifications to dispatch garbage collection vehicles as soon as they are about to fill up [4].

Smart cities need smart policing, where law enforcement organizations organize evidence-based data tactics that are effective and economical. Like AI computer vision systems, they can allow computers (machines) to locate thousands of elements of urban life in a given space, encompassing people, civil servants, cars, disasters, trash, accidents, fires, and many others. It helps to identify, for example, people who are smoking in prohibited areas or who are lingering in multi-story accommodation. The cameras allow authorities to observe the density of crowds and the cleanliness of public spaces [42].

The Internet of Medical Things (IoMT) or even known as Healthcare IoT refers to the entire set of IoT-enabled medical devices, medical sensors, and applications that can be conducting communicate autonomously (machine to machine) and linked to a cloud platform structure. Related to the increase in connected smart devices playing a vital role in assisting the health domain. They will act as a support system for a range of e-health services, improve access to critical data, and allow monitoring of chronic diseases. The technologies AI and ML are transforming the way cities operate, provide, and preserve public amenities, today's smart city programs using these technologies improve city services and citizens' lives, including lighting, power, transportation, security, health services, recycling connectivity, among others [43].

The future of urban mobility is designing the next steps for autonomous vehicles. In addition to saving time, the idea is that autonomous cars will be smaller in the future and take up less space. With automated traffic, they could move very close to each other, making the area of cars even smaller. Using IoT it is possible to predict preventive maintenance and even corrective cars, and provide insights and inputs for smart cities. As well as providing insights and information to the local government, which can create more fuel stations; create the type of spine capable of transferring energy from the mechanical impact of car tires to connected gears, to generate electrical energy; Ditto many other insights and information that may arise from these initiatives [44].

The main concept is the car inserted in a city for people. Autonomous cars can collect, process, send, and receive data to make decisions and take action. A complex ecosystem to handle large volumes of data, mission-critical, and fully connected systems in real-time, mixing IoT, AI, Deep Learning, ML, making all of these technologies work together. As a consequence, an increase in the area for pedestrians, for cyclists, for parks and other living spaces. All this without a major investment in the infrastructure for the cars, since the change would be basically all computerized [1, 4].

Following the concept of intelligence when thinking of a city that takes into account the various modes and creates a solution so that cars do not occupy the city too much and leave it unhuman. The great long-term vision and panorama of smart cities is total interconnectivity, with autonomous cars, automotive sensors, trucks, smart bikes and scooters, and even buses, all talking to each other, still taking into consideration smart roads, traffic lights, and even smart garages. And smart city spending can impact multiple businesses. Since reduced urban traffic would mean cheaper shipments and people having the possibility to spend more time at work and less time-shifting (in transit) between them. In the same sense that fewer accidents can result in lower insurance financial costs for everyone, from the general point of view [3].

The entire system will work towards moving people with an optimal degree of efficiency, efficacy, and safety. A tightly connected system with capabilities to save lives (medical angle), saving time (urban angle), resources (ecological angle), and fuel (transit angle). A smart reality that will become more feasible as governments move towards requiring vehicle-to-vehicle communication to become a reality. As a consequence, the growth potential of these intelligent systems and the companies focusing on the concept of digital intelligence make them meaningful. In this bias, therefore, that innumerable large companies have focused more on smart city technology [8].

References

1. Yigitcanlar T, Han H, Kamruzzaman ML (Eds) (2020) Approaches, advances and applications in sustainable development of smart cities. MDPI
2. Muller AC, Guido S (2017) Introduction to machine learning with Python: a guide for data scientists. O'Reilly Media
3. Abduljabbar R, Dia H, Liyanage S, Bagloee SA (2019) Applications of artificial intelligence in transport: an overview. Sustainability 11(1):189
4. Ullah Z, Al-Turjman F, Mostarda L, Gagliardi R (2020) Applications of artificial intelligence and machine learning in smart cities. Comput Commun
5. Monteiro ACB, Iano Y, França RP, Arthur R (2020). Development of a laboratory medical algorithm for simultaneous detection and counting of erythrocytes and leukocytes in digital images of a blood smear. In: Deep learning techniques for biomedical and health informatics, pp 165–186. Academic Press
6. Lau AY, Staccini P (2019) Artificial intelligence in health: new opportunities, challenges, and practical implications. Yearb Med Inf 28(01):174–178
7. Kučak D, Juričić V, Đambić G (2018) Machine learning in education-a survey of current research trends. Annals of DAAAM & Proceedings 29
8. Song H, Srinivasan R, Sookoor T, Jeschke S (2017). Smart cities: foundations, principles, and applications. John Wiley & Sons
9. Wang W, Siau K (2019) Artificial intelligence, machine learning, automation, robotics, future of work and future of humanity: a review and research agenda. J Datab Manag (JDM) 30(1):61–79
10. Cavada M, Hunt D, Rogers C (2017) Smart cities
11. França RP, Iano Y, Monteiro ACB, Arthur R (2020). Improvement of the transmission of information for ICT techniques through CBEDE methodology. In Utilizing educational data mining techniques for improved learning: emerging research and opportunities, pp 13–34. IGI Global
12. Campbell T (2013) Beyond smart cities: how cities network, learn and innovate. Routledge
13. Bolívar MPR (2015) Smart cities: Big cities, complex governance?. In Transforming city governments for successful smart cities, pp 1–7. Springer, Cham
14. Bacco M, Delmastro F, Ferro E, Gotta A (2017) Environmental monitoring for smart cities. IEEE Sens J 17(23):7767–7774
15. Pellicer S, Santa G, Bleda AL, Maestre R, Jara AJ, Skarmeta AG (2013, July) A global perspective of smart cities: a survey. In 2013 Seventh International conference on innovative mobile and internet services in ubiquitous computing, pp 439–444. IEEE
16. Halegoua G (2020) Smart cities. MIT Press
17. Ismagilova E, Hughes L, Dwivedi YK, Raman KR (2019) Smart cities: advances in research—An information systems perspective. Int J Inf Manage 47:88–100
18. Saravanan K, Julie EG, Robinson YH (2019). Smart cities & IoT: evolution of applications, architectures & technologies, present scenarios & future dream. In: Internet of things and big data analytics for smart generation, pp 135–151. Springer, Cham
19. Crainic TG, Perboli G, Rosano M, Wei Q (2019) Transportation for smart cities: a systematic review. CIRRELT
20. Allam Z, Dhunny ZA (2019) On big data, artificial intelligence and smart cities. Cities 89:80–91
21. Jackson PC (2019) Introduction to artificial intelligence. Courier Dover Publications
22. Arrieta AB, Díaz-Rodríguez N, Del Ser J, Bennetot A, Tabik S, Barbado A, Chatila R (2020) Explainable Artificial Intelligence (XAI): concepts, taxonomies, opportunities and challenges toward responsible AI. Information Fusion 58:82–115
23. Bao H, He H, Liu Z, Liu Z (2019, June) Research on information security situation awareness system based on big data and artificial intelligence technology. In: 2019 International conference on robots & intelligent system (ICRIS), pp 318–322. IEEE
24. Lemayian JP, Al-Turjman F (2019) Intelligent IoT communication in smart environments: an overview. In: Artificial intelligence in IoT, pp 207–221. Springer, Cham

25. Impedovo D, Pirlo G (2020) Artificial Intelligence applications to smart city and smart enterprise
26. Alcantara MN, Gonzaga ADS, Kneib EC (2019) Regenerative mobility: disruption and urban evolution. Int J Environ Sci Sustain Dev 4(3):41–55
27. Natarajan K, Prasath B, Kokila P (2016) Smart health care system using internet of things. J Netw Commun Emerg Technol (JNCET) 6(3)
28. Xu Y, Li J, Tan Q, Peters AL, Yang C (2018) Global status of recycling waste solar panels: a review. Waste Manage 75:450–458
29. Anand, P. B., & Navío-Marco, J. (2018). Governance and economics of smart cities: opportunities and challenges.
30. Cunha MA, Przeybilovicz E, Macaya JFM, Santos FBPD (2016) Smart cities: transformação digital de cidades
31. Alpaydin E (2020) Introduction to machine learning. MIT press
32. L'heureux A, Grolinger K, Elyamany HF, Capretz MA (2017) Machine learning with big data: challenges and approaches. IEEE Access 5:7776-7797
33. Molnar C (2019) Interpretable machine learning. Lulu. com
34. Qiu J, Wu Q, Ding G, Xu Y, Feng S (2016) A survey of machine learning for big data processing. EURASIP J Adv Signal Process 2016(1):67
35. Sethi P, Bhandari V, Kohli B (2017, October) SMS spam detection and comparison of various machine learning algorithms. In 2017 international conference on computing and communication technologies for smart nation (IC3TSN), pp 28–31. IEEE
36. Xie S, Zheng Z, Chen L, Chen C (2018, July) Learning semantic representations for unsupervised domain adaptation. In: International conference on machine learning, pp 5423–5432
37. Burrell J (2016) How the machine 'thinks': understanding opacity in machine learning algorithms. Big Data Soc 3(1):2053951715622512
38. Mullainathan S, Spiess J (2017) Machine learning: an applied econometric approach. J Econ Perspect 31(2):87–106
39. Kurakin A, Goodfellow I, Bengio S (2016) Adversarial machine learning at scale. arXiv preprint arXiv:1611.01236
40. Mohri M, Rostamizadeh A, Talwalkar A (2018) Foundations of machine learning. MIT press
41. Li S, Da Xu L, Zhao S (2018) 5G Internet of things: a survey. J Ind Inf Integr 10:1–9
42. Stone KE (2018) Smart policing and the use of body camera technology: unpacking south africa's tenuous commitment to transparency. Policing: A J Policy Pract 12(1):109–115
43. Joyia GJ, Liaqat RM, Farooq A, Rehman S (2017) Internet of Medical Things (IOMT): applications, benefits and future challenges in healthcare domain. J Commun 12(4):240–247
44. Litman T (2017) Autonomous vehicle implementation predictions. Victoria Transport Policy Institute, Victoria, Canada, p 28

Clustering Techniques for Smart Cities: An Artificial Intelligence Perspective

Mohamed Tharwat and Ahmed Khattab

Abstract The recent exponential growth of the number of connected devices in smart cities is expected to further increase over the next decade. Smart cities exploit the Internet of Things (IoT) technology to create cyber-physical systems. Smart city evolution through the IoT technology brings a new era of smart healthcare, industrial, automotive, security, precision agriculture and military systems. Such smart cyber-physical systems face many challenges due to their design and operation characteristics as well as their energy, processing, and memory constraints. Another major challenge faced by smart cities is the dynamic changes in the topology due to the mobility of the devices and the damaged devices which frequently change the system characteristics. Furthermore, the scalability of such systems requires robust design techniques to handle the large number of connected devices in a limited area. Clustering is a prominent technique used to solve the scalability problems and provide a robust operational network in highly dynamic environments such as IoT-based smart systems. Recently, artificial intelligence and machine learning have been exploited to revolutionize the way clustering is performed. In this chapter, we present a comprehensive survey of the artificial intelligence-based clustering techniques for cyber-physical systems in smart cities.

Keywords Smart city · Internet of Things (IoT) · Clustering · Artificial intelligence · Machine learning

M. Tharwat · A. Khattab (✉)
EECE Department, Cairo University, Giza, Egypt
e-mail: akhattab@ieee.org

M. Tharwat
e-mail: mohammed.thrwat@eng-st.cu.edu.eg

M. Khan et al. (eds.), *Smart Cities: A Data Analytics Perspective*,
Lecture Notes in Intelligent Transportation and Infrastructure,
https://doi.org/10.1007/978-3-030-60922-1_6

1 Introduction

Smart cities have been recently receiving increasing interest from academic, industry, and governmental bodies because of the tremendous impact it brings to the well-being of the residents. In addition, smart cities are proven to positively impact the global economy growth by introducing new possibilities, techniques and sustainability to solve problems and optimize the use of available resources [42]. Several big cities are transforming into smart models of resources optimization and operation such as Paris, Stockholm, Rio de Janeiro, Hong Kong and Tokyo [6]. The smart city infrastructure is composed of several interacting smart environments such as: e-healthcare, remote assisted living, precision agriculture, smart civil structure health (e., bridges' vibration, buildings' tension) monitoring, smart air pollution control, early forest fire detection, smart sportsmen care, quality tracking of shipment conditions, water quality monitoring, autonomous waste management control, smart parking, vehicles auto-diagnoses, autonomous traffic congestion resolution, smart lighting, smart shopping, electromagnetic levels monitoring, home automation. However, the economic and societal potential of such smart systems are extremely higher than what has so far been realized. Thus, further research and development are still needed to get the full potential of such smart systems.

The smart environments within smart cities are Cyber-Physical Systems (CPSs) that integrate embedded computing, networking capabilities and physical processes. Embedded computing exploits Wireless Sensor Networks (WSNs) to monitor target physical processes (e.g. agriculture). Feedback loops are also available to receive actuation decisions to implement sets of control actions to the physical processes (e.g. spray certain amounts of pesticides). WSN technology builds on the disciplines of device miniaturization, embedded controllers and embedded software to transform devices which were not primarily intended to perform computation—such as cars, home appliances, medical instruments and factory machines—to become smart.

The networking capability of CPS components of smart cities is enabled through the Internet of Things (IoT). The IoT technology connects the WSN embedded sensor and actuator devices to the Internet to allow the exchange of information to/from the WSN devices and any authorized Internet-enabled device. Furthermore, IoT empowers CPSs and smart systems by exploiting the Internet's new capabilities and applications such as cloud computing, artificial intelligence, distributed processing and data visualization.

CPSs provide abstractions of the target physical processes through modeling, design, and analysis techniques that capture the dynamics of the physical processes. Such CPS abstraction models are integrated with the embedded computing system through the IoT to result in integrated systems that perform the target autonomous smart purposes.

Therefore, the WSN and IoT technologies are the key enablers of smart cities. Smart WSN and IoT devices are designed to operate in diverse harsh and/or hard to reach environments that are unfeasible for human to reach in certain applications such as volcano monitoring. Smart Cyber-Physical Systems applications in smart cities are generally classified according to the nature of sensing into [39]:

Terrestrial Systems: Consist of hundreds to thousands of embedded devices that are deployed either in structured (factory or power plant) or unstructured (forest or smart farm) manner.

Mobile Systems: Comprised of smart devices that move while interacting with the surrounding physical environment. Examples include vehicular networks and Unmanned Arial Vehicle (UAV) systems.

Multimedia Systems: Consist of several embedded devices equipped with cameras and microphones to enable the tracking and monitoring of people, things, or events in the form of multimedia such as imaging, video, and audio. Such systems heavily rely on computer vision applications.

Underground Systems: Usually consist of embedded devices buried underneath the surface of the ground to measure underground conditions such as soil moisture.

Underwater Systems: Consist of a number of sensor devices and vehicles placed under water surfaces. Underwater vehicles collect data from the sensor devices and relay it to the IoT system.

All such systems face similar challenges—despite the differences—because of the resource constraints of the embedded devices (limited power, processing and storage capabilities) and the complex design and operation nature of such CPSs. However, the most important challenge comes from the fact that usually the number of embedded devices might be in the order of tens to hundreds, or even thousands, of devices in a confined area. Controlling and handling a large number of devices requires scalable architectures and robust strategies which are quite challenging to design [1].

To design scalable protocols for CPS and IoT-based smart systems, complete centralization is a poor option as it introduces a single point of failure in the system and requires high transmission power to enable far devices to connect to the central entity. This contradicts with the low-energy constraint and the limited radio communication nature of the devices. To overcome these problems, the devices should cooperate to form and operate the network. Several distributed algorithms and protocols have already been presented in the literature for scalable WSN and IoT networks [23]. However, distributed algorithms solve some of the complete centralization problems but at the expense of introducing other problem such as causing high packet exchange overhead. Therefore, combined techniques which mix centralized and distributed strategies were introduced [23]. Locally grouping the network devices in a certain area—in what is known as "clustering"—divides the coverage areas into small clusters. In each cluster, a device is either nominated or elected to act as a controller for this group and is called the Cluster Head (CH). The CH coordinates the medium access control, collects data, processes the data to some extent and communicates with the IoT gateway or other centralized entities. The CH role is dynamically rotated among the devices within the cluster to avoid consuming the energy of one device and to balance the energy consumption between all devices.

Several legacy clustering algorithms have been presented in the literature [1]. However, such legacy clustering techniques does not fully suit the highly dynamic nature specific to CPSs within smart cities. Thus, new clustering techniques have recently been introduced which exploit artificial intelligence (AI) and machine learning (ML) techniques to produce clustering techniques that autonomously adapt to the changes in CPSs. Artificial intelligence techniques revolutionized the way to deal with optimization problems such as those encountered in clustering problems. In this chapter, we present an extensive study of the recent AI-based clustering techniques that solve the scalability problems in IoT-based CPSs and smart cities.

This chapter is organized as follows. In Sect. 2, we discuss the clustering objectives, properties and the clustering process in smart cities. In Sect. 3, we present an extensive survey of the literature of AI-based clustering techniques. Section 4 concludes the chapter and discusses the potential future works that can further enhance AI-based clustering techniques.

2 Clustering for Smart Cities

WSN and IoT technologies are the pillars on top of which smart cities and CPSs are built. To leverage the benefits of both technologies in the architecture of smart environments, cooperation between different devices is mandated to share and exchange information. However, the size of smart city WSN and IoT networks is very large and it is difficult to organize and operate such size of networks. Robust and scalable techniques are needed to handle such architectures given the dynamic nature of the devices within many smart city applications.

Clustering is a widely explored technique to enable efficient system operation by downscaling flat networks with large node population. Clustering divides the large-sized network into smaller logical groups based on the characteristics of the network and smart city application requirements as shown in Fig. 1. A cluster-based architecture is composed of cluster heads and cluster members. A Cluster Head (CH) is the main orchestrator of the cluster as it is the node which collects, processes and sends the data to the main controller and also controls the medium access decisions. Cluster member nodes join the CH based on predefined algorithms to form the cluster. CHs may also collaborate with each other to form a backbone network to connect multiple clusters forming the overall network [48].

Each cluster nominates or elects a node either randomly or by selecting the node that is most capable of performing the functionalities of the cluster head. In some applications, the cluster head might be preassigned by the network designer. The nodes' relations within clusters might be dynamic or fixed according to the mobility or the type of the application. Clustering also opens a new era of in-network processing strategies to best utilize the nodes' resources as in many applications to prolong the network lifetime by suppressing redundant data before sending the information to the main controller. In particular applications, data aggregation functions are used by the CH to just send the minimum, maximum, or the average of the data instead of sending all data, which significantly reduces the used transmission energy. The value of in-network processing can be further increased by preassigning the CH role to a

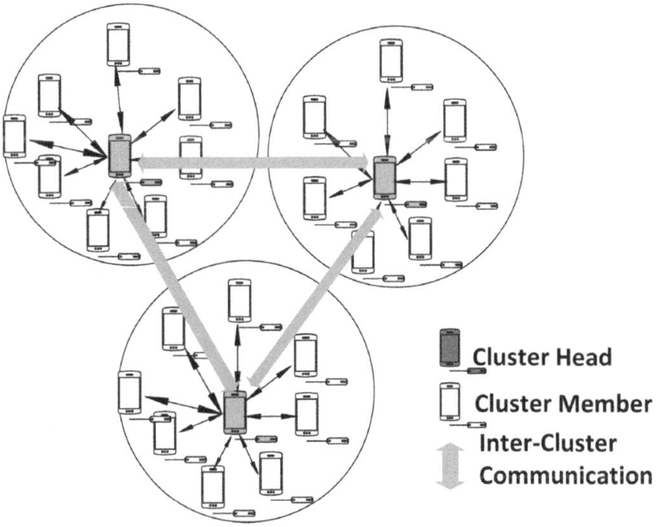

Fig. 1 Architecture of cluster-based smart systems

special and powerful node with advanced signal processing capabilities and extra battery. In smart cities, many devices are typically located near the power sources such as lights, traffic signals, air conditioners, ..., etc. Such devices could be used as cluster heads for clusters.

Clustering, a key feature of self-X (self-organizing, self-healing and self-configuring) networking, is more prominent in smart city applications with random node deployments compared to applications with planned node deployments. In planned deployments, sensor nodes are placed in predefined places and routing paths are also predefined, hence, clustering is unnecessary. In random node deployments, the distribution and type of nodes are important factors in the clustering operation and whether or not special feature nodes exist. The normal distribution of nodes facilitates the efficient operation of the network, even if no special nodes are available to serve as CHs. In this case, the basic sensing features of some nodes can be reduced to allow them to operate as CHs.

2.1 Clustering Objectives

The objectives of clustering vary according to the target smart city application of the sensor network. For instance, applications such as vehicular information for self driving cars necessitates realtime services and delay sensitive information, unlike other common applications in which the most important objective is to prolong the system lifetime given the energy constraints (e.g., healthcare applications). The main objectives of the clustering process in different smart city applications can be classified as follows:

Scalability: Scalability is defined as the ability to maintain certain performance characteristics regardless the size of the network. WSN and IoT networks in smart cities are intended to use tens to hundreds or maybe thousands of nodes. Clustering attributes should be carefully optimized with scalability optimization goals in mind for optimal handling of such large numbers of nodes, especially in smart city architectures that combine several applications.

Load Balancing: Clusters should be approximately of the same size and have almost the same characteristics to have a balanced network in terms of resources. To balance the energy consumption, the CH responsibilities should rotate over all the nodes in the cluster to avoid draining a specific node's energy which might cause a hole in the network especially if the CH is not a special powerful node. In this case, it becomes crucial to have balanced clusters for prolonging the network lifetime. In-network processing should also be carefully taken into consideration when designing a balanced clustering technique [17, 46].

Prolonging System Lifetime: Prolonging the system lifetime is a crucial objective of the clustering process. Instead of having all the nodes communicating frequently and directly to the centralized controller, nodes just communicate with low power with their immediate CH. Thereby, limiting the communication range and the transmission frequency which significantly reduce the energy consumption.

Delay Bound: Delay is a critical parameter in some smart city applications like forest fire detection, military applications and healthcare monitoring. It is important to consider the delay as a constraint when designing the clustering algorithm [41].

Fault Tolerance: WSN and IoT nodes might experience hardware failures, energy drainage, interference from various sources and excessive delays. Therefore, WSN and IoT systems in smart cities should be capable of reconfiguring themselves without human intervention. This is curial with the large number of nodes and particularly when the nodes are located in inaccessible spots or when the system operates in harsh environmental conditions. Fault tolerance should be accounted for while designing a clustering protocol to ensure the accuracy of the aggregated data. Using a backup CH or routinely performing cluster maintenance procedures have been proposed to enable network reconstruction upon the malfunctioning of CHs [4, 5, 47].

Reduce Network Holes: Network holes result when a part of the network becomes uncovered with nodes due to random deployment or because of node failures. WSN and IoT nodes are resource-constrained in terms of battery and computational resources, especially since such resources drain with time. Clustering algorithms should adapt to node failures to overcome any potential network hole problems.

Minimal Cluster Count: In case of using special powerful nodes to serve as CHs, network designers prefer to reduce the number of clusters to reduce the number of used powerful nodes which scales down the overall system cost.

2.2 Cluster Properties

For efficient and optimized clustering process, the following characteristic should be carefully considered while forming the clusters within the smart system.

Cluster Count: This is the number of clusters in the system. Typically, a small cluster count is preferred to ensure energy efficiency. In some scenarios, the cluster count is pre-defined by the network designer [23].

Cluster Size: The cluster size measures the largest path length between a cluster member and a cluster head. It is preferred to be small to limit the needed power for communication with the CH.

Cluster Density: Is defined as the number of the cluster member nodes.

Message Count: The total number of messages needed for electing the CH.

Stability: A cluster is said to be stable if its nodes do not vary often with time as this increases the stability of the network.

Intra-Cluster Topology: Identifies whether the way the communication inside the cluster is single-hop or multi-hop.

Inter-Cluster Head Connectivity: Identifies whether the way the cluster will communicate with the main controller is either direct or through relaying nodes.

2.3 Cluster-Head Capabilities

The characteristics of the cluster heads drive the clustering process in terms of the number of clusters, network lifetime and network stability. The most important CH characteristics are:

Type: The network designer can add special powerful nodes with rich resources in terms of memory and energy to act as cluster heads.

Mobility: Mobility of the CH adds extra constraints on how to communicate with the member nodes and the main controller.

Role: In WSN and IoT-based CPSs, CHs can either perform data aggregation and fusion tasks or just relay the cluster members' information without any sort of processing.

As clustering techniques solve the scalability problem in high density networks as the case in smart cities, the CH selection remains a challenging task. Several objectives are to be met by the clustering process while determining whether a node should be a CH or just a cluster member. It is proven that the clustering process is a Non-deterministic Polynomial (NP)-hard optimization problem [2]. Solutions of NP optimization problems require searches through large spaces of possible solutions. Artificial intelligence techniques have been recently applied to many NP optimization problems and resulted in promising results. In the next section, we discuss how the main AI techniques were applied to the WSN/IoT clustering process in CPSs and smart cities.

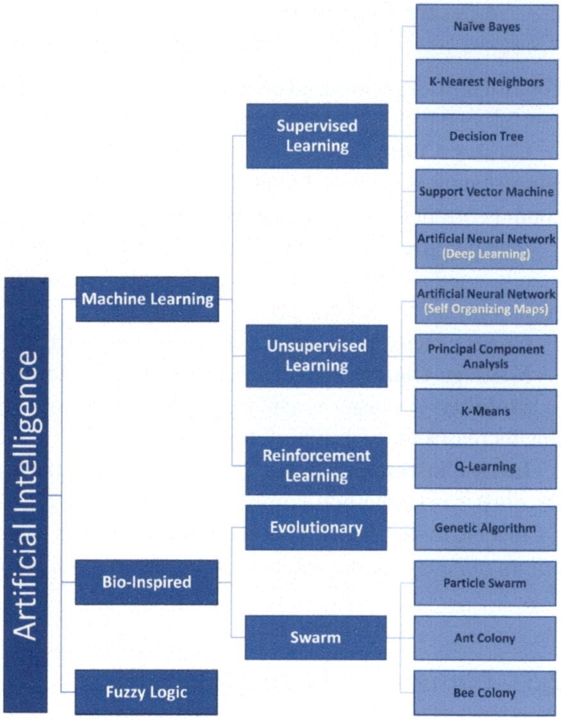

Fig. 2 Classification of artificial intelligence techniques

3 Artificial Intelligence Based Clustering

The term "Artificial Intelligence" consists of two terms: "Artificial" which implies something made or produced by humans rather than naturally occurring but typically imitating something natural; and "Intelligence" which is the ability to acquire and apply knowledge and skills. Combining the two terms results in the definition of AI that is computer-based smart systems that are developed to mimic the human brain power to perform tasks such as perception, decision-making, pattern/image/speech recognition, language translation and other tasks that are normally carried out through human intelligence [36]. In this section, we discuss smart city clustering approaches that are designed based on the main artificial intelligence techniques categorized as shown in Fig. 2.

3.1 Machine Learning Based Clustering

Machine learning is a data analysis technique that autonomously results in analytical models. It is an AI approach in which the smart system learns from the available

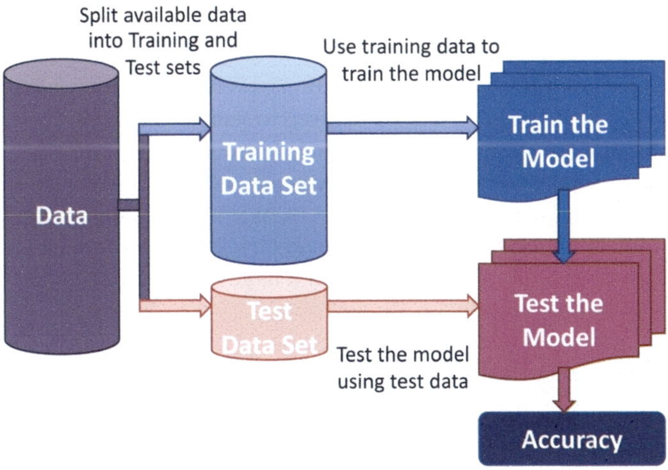

Fig. 3 Supervised learning process flow

data and makes decisions based on the identified patterns in the data without human intervention. The relationship between the inputs and outputs categorizes machine learning techniques into three main categories:

- **Supervised Machine Learning**: A function that maps an input to an output is learnt based on already known input-output pairs that serve as examples. The available data set is divided into two parts: a training set that is used to infer the model and a test set that is used to assess its accuracy as shown in Fig. 3.
- **Unsupervised Machine Learning**: The model trains itself without labeled training data. The model gradually discovers its characteristics from the interaction with an unlabeled data set.
- **Reinforcement Learning**: The computerized system is programmed to just have a goal to achieve. The computerized system then learns how to achieve that goal through a trial and error approach in which actions are tried within the environment and according to the received rewards further actions are taken.

3.1.1 Naïve Bayes Based Clustering

In Naïve Bayes supervised learning, the classification methodology is based on the Bayesian theorem assuming an independent feature set. Naïve Bayes techniques do not only easily and rapidly predict the class of test data set, but also perform well in multi-class prediction.

A cluster head selection methodology was presented in [20] using the node's residual energy and the sum of local distances of member nodes to the CH node. The proposed CH selection approach creates a labeled data set from a random iteration, then applies the naïve Bayes mechanism on the network based on this data set.

Simulations showed enhancements over the legacy Low-Energy Adaptive Clustering Hierarchy (LEACH) protocol [19].

Naïve Bayes based clustering results in simple protocols with low computational complexity without the need for labeled data sets to setup the clustering process. Such protocols are scalable and do not need a main controller, and hence, they best fit in smart city applications with random deployments without main controllers targeting moderate system permanence. Examples include wildlife monitoring, forest fire detection, military applications and other smart city scenarios in which inexpensive sensor nodes are randomly deployed to cover a large area.

3.1.2 Decision Trees Based Clustering

Decision-tree-based supervised learning represents a type of classification techniques that is based on combining several binary questions [16]. The simplest goal of decision trees is to find the optimal tree structure based on the training data.

The authors of [3] proposed a decision tree algorithm for selecting the cluster head based on four factors: the node's distance from the cluster centroid, the node's remaining energy, its degree of mobility and a vulnerability index which reflects the impact of the failure of a specific node on the whole network (i.e., will cause a network disconnection or partitioning) and is calculated based on graph theory. The nodes are grouped into two categories of CH candidates and cannot-be-CH nodes. The decision tree is built based on training data provided by experts (rather than being from a real network) for the four factors for different nodes. The decision tree algorithm runs on the main network controller after the broadcast messages are sent from the sensor nodes. The algorithm execution is carried out in two phases. A startup phase in which the main controller runs the algorithm from the broadcast data and produces the candidate CHs based on a predefined number of the needed CHs. System nodes connect to the CH according to the received signal level. In the steady state phase, the CH role is rotated after the main controller runs the algorithm on the updated data from the CHs about the status of all nodes on a periodic manner with a period that might be predefined.

This algorithm is considered semi-centralized as the algorithm runs on the main controller. The algorithm has regular update messages from the nodes and CHs to the main controller which will impact the energy levels of the nodes. The algorithm has a limited scalability and requires a large and highly accurate training data to get the candidate CH nodes correctly.

Clustering using decision trees is suitable for critical smart city applications in which an event is to be detected in a particular coverage area based on answering a set of YES/NO questions. Smart cities have many critical applications within limited areas such as intruder detection for buildings or time restricted areas like museums and banks. In such applications, the sensor nodes are deployed and decision tree based clustering can be used to collect the information and send alerts to the main control room, or in some scenarios to the police office, based on pre-collected labeled data once an event is triggered. It is worth mentioning that the performance of decision tree based clustering protocols is improved if labeled data sets are available from system expertise [16]. The same applies to border tracking application such as monitoring the border of a compound or a residential area to keep the main operation room updated with all up-normal activities.

3.1.3 Support Vector Machine Based Clustering

Support Vector Machines (SVMs) represent a type of supervised machine learning based on hyperplane classification or clustering. The optimal plan that divides the data into different plans is learnt from the training data set. The number of hyperplanes relies on the independent features of the training data which might be a multidimensional hyperplane.

A SVM-based routing algorithm was proposed in [24] by searching for the smallest sphere that encompasses a data set after transforming the data space to a high dimensional space using Gaussian kernel. After identifying this sphere, it re-transforms back to the data set space producing the cluster boundaries contour. The proposed algorithm calculates the cluster boundaries from the training data and a node joins a cluster if either the actual distance or the minimum distance is less than a predefined threshold value.

Support vector machine based clustering is suitable for smart city applications in which training data can be gathered. It is typically applicable in monitoring smart city activity regions such as hospitals in which labeled training data can be easily obtained. It is worth noting that as the size of the available data set increases, the performance of SVM-based clustering significantly improves. Sensor nodes are deployed in the activity regions inside the smart city to continuously capture data to produce accurate monitoring of stealth incidents and to track and monitor kids. Particular beneficiaries of the SVM-based clustering approach are the emerging medical IoT and remote assisted living in which medical sensor nodes are deployed to monitor the health conditions of patients.

3.1.4 Neural Networks Based Clustering

Artificial Neural Networks (ANNs) consist of a set of nodes called neurons to resemble the neural system of the human brain. The neurons of an ANN are connected in several ways. In supervised learning ANNs, every neuron adds a non-linear function to its input signal producing the output signal. When artificial neural networks are designed with many hierarchical levels to perform the process of machine learning, it is referred to as deep learning. Artificial neural networks can also be used for unsupervised learning if configured as Self Organizing Maps (SOMs). A SOM is a type of an unsupervised artificial neural network used for feature detection by producing a space of the training data with a much lower dimensions.

The authors of [31] proposed a coverage-aware neural network based algorithm for electing the CH using a neural network that is composed of three layers: input, competition and output layers. All the candidate nodes to be elected as CHs are the input to the input layer and are fully connected to the competition layer. The neurons in the competition layer compete with each other producing the elected CH with the smallest energy associated with the delivery ratio of the packet originating from source nodes and correctly received at destination nodes. A learning rate is controlling the speed of the convergence of the competition layer. The algorithm shows an enhancement over the Power-Efficient and Adaptive Clustering Hierarchy (PEACH) protocol presented in [45].

Whereas [8] proposed a clustering process that consists of four phases: (1) Initialization where a base station collects information from the nodes using broadcast messages; (2) Cluster setup phase using four parameters which are the residual energy, importance of the node, concentration of a group of nodes in a specific location and CH frequency using a SOM-based neural network that consists of 4 inputs and 16 outputs through which the CH nodes are elected; (3) Data transmission phase which first checks if the residual energy level is above the node's critical energy level below which the node is considered dead; (4) CH rotation is initialized if the residual energy of any serving CH becomes less than a certain threshold. Simulations depict that such an algorithm outperforms the legacy LEACH [19] in terms of increased network lifetime.

The authors of [29] proposed a clustering algorithm to maximize the confidentiality rate (security) of the Industrial IoT (IIoT). Deep Feed-forward Neural Network (DFNN) was initially used to solve the power allocation problem with maximizing the confidentiality rate as an objective. The deep learning network was used to increase the security between nodes during their communications. The resulting power needed to satisfy certain security measures is then compared to a threshold. If the resultant needed transmit power exceeds the threshold, this implies that the resulting confidential capacity is maximized. Consequently, the master node begins to form node pairs based on their power requirements forming the cluster. The proposed algorithm outperforms the adaptive particle swarm optimization scheme presented in [40] at the expense of a significantly higher computational complexity.

The authors of [30] proposed an algorithm to recognize CHs in Multiple-In Multiple-Output (MIMO) based sensor networks. The proposed algorithm applies Back Propagation Neural Network (BPNN) jointly with a distributed gradient drop technique to calculate the position of the unknown CH. By collecting the location information from several randomly deployed anchors, the distance between the anchor node and the cluster head is measured, and hence, the position relationship is obtained. Particle swarm optimization is applied to initialize the weight of the NN. The training step of the BPNN is started to minimize the network cost function. The results show a significantly lower error rate for CH positioning.

NN-based clustering provides high system accuracy at the expense of high computational requirements. Hence, NN-based clustering is suitable for smart city dense and urban area applications such as IIoT and traffic monitoring systems. In the emerging IIoT applications, thousands of nodes are deployed to monitor and control different devices in the industrial zones. Clustering techniques are used to collect data and trigger actions sent to/from the main dispatch room. Since there is no power constraints in IIoT systems, the computation power is not an issue. Hence, NN-based clustering techniques are preferred to support the optimization of the application parameters such as delay and high accuracy. Similar concepts applies to traffic monitoring systems that use thousands of nodes deployed on the street level to collect the traffic conditions.

3.1.5 K-Means Based Clustering

K-means unsupervised learning algorithms are iterative algorithms that try to partition the data-set into K non-overlapping groups. A data sample only belongs to one group. In the partitioning algorithm, K-means learning forms a cluster by grouping similar data points together while trying to make the clusters as much different. A data point is assigned to the cluster with the minimum sum of the squared distances between the data point and the cluster's centroid.

A clustering algorithm using the K-means technique based on the Euclidean distances between nodes was proposed in [32]. An initial clustering step randomly selects CHs from a random formation of clusters. A second step of re-clustering the nodes by calculating the cluster centroid and assigning the nearest node to the centroid as a new CH.

The authors of [35] proposed a K-means technique for clustering static sensor nodes with a known location from the main controller. A midpoint algorithm was used for selecting the initial CH nodes based on the known information. In the next iteration, the residual energy is taken into consideration for selecting the CH.

K-means based clustering protocols are simple, and hence, can be used in different smart city applications especially the noncritical ones. Examples includes the sensor networks deployed to monitor the weather conditions in the city where thousands of sensor nodes are deployed to collect field information. K-means based clustering is typically suitable for dense smart city applications such as street-level sensors.

3.1.6 Q-Learning Based Clustering

Reinforcement learning enables programmed machines to autonomously define the ideal action that results in the best performance in a specific situation. A feedback reflected by a corresponding reward is sent to the agent to learn its behavior. This feedback is defined as the reinforcement signal [21]. Q-Learning represents a model-free reinforcement learning technique that allows agents to figure out how to ideally act in Markovian domains by experiencing the consequences of the taken actions [44].

An algorithm for CH identification based on Q-Leaning was proposed in [15]. The CH is the node with the minimum routing cost to all sinks. The proposed Q-learning algorithm utilizes the information in the data request packets to identify the next hop and to determine whether a specific node can act as a CH or not. The used information is the hop count to all the neighboring nodes, the geographic location and the battery status. Simulation results show that such approach prolongs the network lifetime measured by the time to the first node death.

The authors of [28] proposed a Q-learning algorithm through which non-CH nodes adaptively choose the CH. First, a restriction on the minimum energy for candidate nodes to be CH is applied to the distributed energy efficient clustering proposed in [33] to identify the CH nodes. Then, the member nodes select the CH based on a Q-Learning algorithm for data transmission. Simulations prove the effectiveness of the proposed algorithm on large data sets.

Q-learning based clustering is suitable for smart cities with mobile devices in which a goal is targeted without any training data available whatsoever. Therefore, the trial and error approach of Q-learning is preferred not only to reach the goal but also to react to the changes in the system. Example applications include object tracking, vehicular networks and unmanned aerial vehicle networks. Smart city futuristic plans will significantly depend on the aerial vehicle networks and drones to monitor and collect information from the field and send it directly to a dedicated control room according to the application. Hence, the autonomous and continues learning features of Q-learning based clustering techniques will play a vital role in such scenarios to organize and orchestrate the operation of these drones.

3.2 Bio-Inspired Clustering

Biological systems have attractive properties and functions that can be used in many fields by analogy. The most important property is the ability of biological systems to efficiently adapt to dynamic environments [14]. For example, immune systems, evolutionary computation, particle swarm optimization, ant colony optimization and behavioral systems inspire the smart city node clustering research community to derive such analogy to develop bio-inspired AI algorithms as discussed in this section.

3.2.1 Genetic Algorithms for Clustering

Genetic Algorithms (GAs) are evolutionary algorithms that apply adaptive heuristic search based on natural selection and genetics principles. Such principles direct the search process to the region resulting in a better performance in the solution space by intelligently exploiting random searches combined with historical data. GAs are widely used to solve optimization problems and search problems such as the search for the CH nodes.

The authors of [7] focused on ultra reliable low latency WSNs and proposed a multi-objective optimization function for longest network lifetime, highest network connectivity and reliability. A static centralized algorithm that pre-learns a near-optimal clustering network while not overloading the CHs was proposed. The energy consumed in transmissions and the distance are added to the clustering objective function. Principal Component Analysis (PCA) supervised learning is used to get rid of any dependencies between the multiple objectives of the optimization problem. Then, its ranks the objectives according to their significance levels to define the fitness function that is used to evaluate the different chromosomes.

The authors of [5] proposed an evolutionary clustering algorithm based on the Non-dominated Sorting Genetic Algorithm III (NSGA-III) [10] for optimizing clustering parameters to best fit Vehicular Ad-hoc Networks (VANETs). The proposed algorithm imposes a many-objective optimization problem to optimize the Cluster Head Lifetime (CHL), Cluster Member Lifetime (CML) and Control Packet Overhead (CPO) of the Double Head Clustering (DHC) algorithm presented in [4]. The authors proposed a three-step process to solve this optimization problem: metaheuristic optimization stage, solution evaluation stage and a solution selection stage. The resulting clustering protocol after this optimization shows outstanding results compared to the non-optimized DHC.

GA-based clustering has a remarkable system accuracy that comes at the expense of high computational complexity. As the case with NN-based clustering, GA-based clustering is also suitable in smart city applications with high node density and in urban environments where accurate and realtime data is to be collected for further analysis. Examples include IIoT, e-healthcare, traffic monitoring and vehicular networks.

3.2.2 PSO Based Clustering

With the objective of simulating the behaviors of bird flocking and fishing schooling, Particle Swarm Optimization (PSO) was introduced in [12]. For example, a group of birds searching for a piece of bread in a particular area. Even tough all the birds do not initially know exactly the coordinates of the piece of bread, they learn the distance to it in each iteration. Therefore, the best way to find the bread is to track the bird with the minimum distance to it. In the optimization procedure, each bird (also called particle) is considered as a solution in the search area or search space. Each particle is assigned a fitness value. A fitness function—that should be optimized—is then used to evaluate the fitness values of all particles [37].

A hierarchical particle swarm optimization technique that works on the two hops between the member node and the CH was proposed in [13]. The clusters are formed by the main controller by applying the PSO algorithm on the nodes after collecting broadcast messages from the nodes. The algorithm identifies the candidate nodes to be CHs then attaches the nodes to the CH with the minimum received signal level, thereby, forming the first tier of the cluster. The main controller next assigns the nodes that are not attached yet to a CH to first tier nodes also according to the received signal strength indicator with all neighbor nodes. The selection of CHs targets the minimization of the total number of CHs to maximize the energy efficiency of the system and to maximize the quality of the links between the cluster members and their attached nodes. The nodes use time division multiplexing to communicate with the CH when the system reaches the steady state. After sending its data, each node goes to the sleep mode for energy saving. The results show that such a technique outperforms LEACH [19] and its variant LEACH-C as well as and PSO-C [27] in terms of the energy saving per node.

In [34], a cluster head selection based on PSO and a cluster formation technique were introduced. The CH selection strategy exploits information such as the nodes' residual energy and various computed distances such as the distance from the main controller and the distance to sink nodes. Using the broadcast information sent from the nodes to the main controller in the setup phase, the main controller uses the PSO algorithm to identify the locations of CHs based on a proposed multi-objective fitness function. This fitness function takes into account the average intra-cluster distance, the average sink distance, the CHs current energy levels and the velocity and position updates. The PSO algorithm iterates for a while to increase the accuracy in identifying the CH locations. The node closest to an optimal location is elected to serve as a CH. Then, the cluster formation phase allows nodes to join a CH using a weighted function which depends on the CH residual energy, distance between the node and the CH, distance from CH to the base station and the CH node degree. Simulation results show enhancements over LEACH [19] and PSO-C [27] for a small number of CHs. However, it was shown that such enhancements decrease with increasing the number of CHs.

Also, [38] proposed a semi-distributed PSO-based CH selection algorithm referred to as PSO-SD. PSO-SD exploits the average distance from the nodes in a cluster, the residual energy, the node's degree and head count in the proposed fitness function. The used energy model is based on the successful transmissions, receptions and re-transmissions of the collided packets. The simulation results of the PSO-SD approaches indicate its remarkable performance compared to the legacy LEACH [19] algorithm in terms of total nodes alive after a certain time.

One of the main drivers for smart cities is enhancing the wellbeing of the citizens. Therefore, several dedicated applications have been developed to measure the quality of the environment such as water quality and air pollution. The sensor nodes in such applications typically have limited power sources and are usually hard to reach to change their batteries. Hence, prolonging the system lifetime is one of the major design objectives of such smart city systems. PSO based clustering techniques are suitable for such application as they cluster the nodes in a way that mainly minimizes the energy consumption of the system.

3.2.3 Ant Colony Based Clustering

Ant Colony Optimization (ACO) was presented in [11] as a bio-inspired metaheuristic that can be used to solve hard combinatorial optimization problems. ACO is inspired by the ant behavior in searching for the shortest route to the food, and how an ant shares its route information with all other ants in the search space. A similar methodology is used by ACO metaheuristics to solve optimization problems [11], including smart city node clustering. The key idea is to construct a path-graph that models the states of the problem at hand and search for best path [26].

The authors of [43] proposed a two-step clustering algorithm that exploits ant colony optimization and multi-hop communication. In the cluster formation Step, a CH is elected using random probability locally produced by every node using the node's residual energy and CH count. Member nodes join the cluster closest to them. In the cluster stability step, ACO is applied for each CH to identify the shortest path to reach all nodes inside the cluster taking into consideration the residual energy as a metric for choosing the paths. Simulations show the enhancements compared to LEACH [19].

Ant colony based clustering can be used in smart city applications such as temperature monitoring in a forest or military applications where maximizing the nodes' lifetime is a main optimization goal as the case in PSO-based clustering. Recall that the sensor nodes in such applications are designed to be in the field for a long time without human accessibility.

3.2.4 Artificial Bee Colony Based Clustering

Artificial bee colony optimization algorithms are motivated by how honeybees search for the food. Such an approach was introduced for optimizing numerical problems. An analogous flow is followed to solve optimization problems in which an initialization step is used, followed by defining a fitness function, iterating over all possible solutions, and memorizing the best possible solution which maximizes the fitness function.

A clustering technique similar to LEACH algorithm [19] but selects the CHs based on the artificial bee colony algorithm was proposed in [22]. The main controller runs an artificial bee colony algorithm with a predefined set of clusters. Initially, random CHs are selected. Then, the fitness function is calculated taking into account the distances between the nodes and the energy using the artificial bee colony algorithm.

Artificial bee colony based clustering can be used wherever ant colony based clustering is used. Furthermore, it can be used when a predefined set of clusters are needed during the engineering phase of the smart city application. Examples include precision agriculture and smart building applications in which the number of sensor nodes to be deployed needs to be well planned during the engineering phase.

3.3 Fuzzy Logic Based Clustering

Unlike probabilistic logic which deals with the degree of belief as is either true or false, fuzzy logic deals with facts with a degree of truth that can take any value between 0 and 1. Fuzzy logic learning systems are rule-based systems that exploit the practical experience of an operator [9].

A CH election algorithm in which the centralized controller runs a fuzzy logic algorithm was presented in [18]. Initially, the controller collects the broadcast messages from nodes that include the residual energy and other parameters to identify the centrality and concentration of nodes. The controller then identifies the potential CHs by a produced chance obtained by applying fuzzy logic IF-THEN rules. The achieved network lifetime is higher than that achieved by LEACH [19].

The authors of [25] proposed a localized version of a prior algorithm [18]. Every node computes a chance value using its residual energy and the sum of distances between the node and other nodes within a certain range. Every node then computes a chance value based on IF-THEN rules. A CH is elected by exchanging messages with neighbors. Simulation results showed enhancement in the time to first node failure with respect to LEACH [19].

Clustering using fuzzy logic results in scalable protocols that do not rely on a main controller, and hence, best fit smart city applications in which nodes are randomly deployed. Such protocols also have low computational complexity that comes at the expense of moderate system performance. Hence, they can used in wild life monitoring and forest fire detection.

4 Concluding Discussion

Smart cities have attracted significant attention over the past few years. Their proven economic benefits and the tremendous impact they bring to the wellbeing of the residents. The core base of smart cities and their cyber-physical systems are wireless sensor networks and IoT systems. The number of devices in a typical smart city infrastructure is in the range of thousands with many different types to support different applications creating many challenges to deal with. Clustering is a well known technique to support the scalability of such infrastructures. Clustering is an NP-hard optimization problem due to its many objectives that need to be optimized at the same time. Artificial intelligence techniques have been recently exploited to change the way we deal with optimization problems.

In this chapter, we have presented a state-of-the-art survey of clustering techniques based on artificial intelligence. To the best of our knowledge, this chapter has presented the first detailed survey of AI-based clustering summarized in Table 1. The performance benchmark for most of these work is the pioneer LEACH algorithm. The comparison illustrated in Table 1 is based on different parameters: (1) The smart city applications the clustering approach best fits; (2) The overhead which indicates the messages to be exchanged to setup the clustering process; (3) The scalability as an indication of the extent to which the system can accommodate new nodes and if there is a need to re-run the algorithm with each node joining the system; (4) The algorithm complexity as a measure of the computations required by the clustering process; and (5) The need for the main controller for orchestrating the clustering process or the nodes themselves can operate the clustering can process locally.

The surveyed AI-based clustering techniques fit different smart city applications as discussed in the chapter. Depending on the objective, requirements and constraints of the application, the best technique can be chosen. The existence of a main controller is also a key aspect in choosing the most suitable technique.

Table 1 AI based clustering algorithm comparison

Category	Technique	Smart city applications	Paper	Overhead	Scalability	Complexity	Main controller
Machine learning	Naive Biase	Forest and Wildlife Monitoring	[20]	Low	Good	Low	No
	Decision Trees	Border Tracking and Smart Building Security	[3]	High	Limitted	High	Yes
	SVM	Medical IoT Systems and Monitoring Activity Regions	[24]	Low	Moderate	Moderate	Yes
	NN	IIoT Systems, Realtime Traffic Monitoring and Vehicular Networks	[31]	High	Moderate	Moderate	Yes
			[8]	High	Moderate	High	Yes
			[29]	High	Moderate	High	Yes
			[30]	High	Moderate	High	Yes
	K- Means	City Weather Condition Monitoring	[32]	Low	Good	Low	No
			[35]	Low	Good	Low	Yes
	Q-learning	Unmanned Aerial Vehicular Networks	[15]	Low	Good	Low	Yes
			[28]	Moderate	Moderate	Moderate	Yes
Bio-Inspired	Genetic	IIoT, e-healthcare, Traffic Monitoring, and Vehicular Networks.	[7]	High	Moderate	High	Yes
			[5]	High	Low	Moderate	Yes
	PSO	City Environment Monitoring (e.g., Water Quality and Air Pollution)	[13]	High	Low	High	Yes
			[34]	Moderate	Low	High	Yes
			[38]	High	Moderate	High	Yes
	ANT Colony	Temperature Monitoring and Military Applications	[43]	Moderate	Low	High	Yes
	Bee Colony	Precision Agriculture and IIoT	[22]	Moderate	Low	High	Yes
Fuzzy Logic	Fuzzy Logic	Wildlife Monitoring and Forest Fire Detection	[18]	Low	Moderate	Low	Yes
			[25]	Low	Moderate	Low	No

As neural networks and particle swarm optimization are very powerful tools to solve complex optimization problems, they have been widely used in AI-based clustering for smart cities. As a future research direction, combining many parameters as optimization constraints will increase the overall system performance of the existing techniques. Even thought this will increase the complexity and computational cost of the system, the fixed resource-rich nodes available in the smart city infrastructure can be exploited to provide such computational requirements. Furthermore, adding different and dedicated controllers within the smart city infrastructure should be considered to facilitate the clustering process. Similar improvements can be applied to genetic algorithms based clustering.

Currently, the most used optimization parameter is the residual energy in the nodes. Given the new advances in the battery technologies and energy harvesting techniques, the residual energy will have a little impact on the overall system performance. In smart city infrastructures, new challenges have been identified resulting from the heterogeneity of the applications, the high density of devices in small areas, time-critical applications and the high computation requirements. Hence, optimizing the parameters related to these challenges should be considered in future AI-based clustering techniques for smart cities. Another potential future research direction is the development of new distributed algorithms. Such techniques are needed to overcome the dynamic and evolving nature of the smart city infrastructure rather than relying on centralized approaches as the case with the existing clustering techniques.

References

1. Abbasi AA, Younis M (2007) A survey on clustering algorithms for wireless sensor networks. Comput Commun 30(14–15):2826–2841
2. Agarwal PK, Procopiuc CM (2002) Exact and approximation algorithms for clustering. Algorithmica 33(2):201–226
3. Ahmed G, Khan NM, Khalid Z, Ramer R (2008) Cluster head selection using decision trees for wireless sensor networks. In: IEEE international conference on intelligent sensors, sensor networks and information processing, pp 173–178. IEEE (2008)
4. Alsuhli GH, Khattab A, Fahmy YA (2019) Double-head clustering for resilient VANETs. Wireless communications and mobile computing 2019
5. Alsuhli GH, Khattab A, Fahmy YA (2019) An evolutionary approach for optimized VANET clustering. In: IEEE international conference on microelectronics (ICM), pp 70–73. IEEE (2019)
6. Cathelat B (2019) Smart cities: shaping the society of 2030. United Nations Educational, Scientific and Cultural Organization (UNESCO)
7. Chang Y, Yuan X, Li B, Niyato D, Al-Dhahir N (2018) Machine-learning-based parallel genetic algorithms for multi-objective optimization in ultra-reliable low-latency WSNs. IEEE Access 7:4913–4926
8. Cordina M, Debono CJ (2008) Increasing wireless sensor network lifetime through the application of som neural networks. In: IEEE international symposium on communications, control and signal processing, pp 467–471. IEEE (2008)
9. Cox E (1992) Fuzzy fundamentals. IEEE Spect 29(10)

10. Deb K, Jain H (2013) An evolutionary many-objective optimization algorithm using reference-point-based nondominated sorting approach, part I: solving problems with box constraints. IEEE Trans Evol Comput 18(4):577–601
11. Dorigo M, Blum C (2005) Ant colony optimization theory: A survey. Theoret Comput Sci 344(2–3):243–278
12. Eberhart R, Kennedy J (1995) A new optimizer using particle swarm theory. In: IEEE international symposium on micro machine and human science, pp 39–43. IEEE (1995)
13. Elhabyan RS, Yagoub MC (2014) PSO-HC: Particle swarm optimization protocol for hierarchical clustering in wireless sensor networks. In: IEEE international conference on collaborative computing: networking, applications and worksharing, pp 417–424. IEEE (2014)
14. Floreano D, Mattiussi C (2008) Bio-inspired artificial intelligence: theories, methods, and technologies. MIT Press (2008)
15. Forster A, Murphy AL (2009) Clique: role-free clustering with q-learning for wireless sensor networks. In: IEEE international conference on distributed computing systems, pp 441–449. IEEE (2009)
16. Geurts P, Irrthum A, Wehenkel L (2009) Supervised learning with decision tree-based methods in computational and systems biology. Mol BioSyst 5(12):1593–1605
17. Gupta G, Younis M (2003) Load-balanced clustering of wireless sensor networks. In: IEEE international conference on communications (ICC), vol 3, pp 1848–1852. IEEE (2003)
18. Gupta I, Riordan D, Sampalli S (2005) Cluster-head election using fuzzy logic for wireless sensor networks. In: CNSR, pp 255–260 (2005)
19. Handy M, Haase M, Timmermann D (2002) Low energy adaptive clustering hierarchy with deterministic cluster-head selection. In: IEEE international workshop on mobile and wireless communications network, pp 368–372. IEEE (2002)
20. Jafarizadeh V, Keshavarzi A, Derikvand T (2017) Efficient cluster head selection using naïve bayes classifier for wireless sensor networks. Spring Wirel Netw 23(3):779–785
21. Kaelbling LP, Littman ML, Moore AW (1996) Reinforcement learning: A survey. J Artif Intell Res 4:237–285
22. Karaboga D, Okdem S, Ozturk C (2012) Cluster based wireless sensor network routing using artificial bee colony algorithm. Spring Wirel Netw 18(7):847–860
23. Karl H, Willig A (2007) Protocols and architectures for wireless sensor networks. John Wiley & Sons
24. Khan F, Memon S, Jokhio SH (2016) Support vector machine based energy aware routing in wireless sensor networks. In: IEEE international conference on robotics and artificial intelligence (ICRAI), pp 1–4. IEEE (2016)
25. Kim JM, Park SH, Han YJ, Chung TM (2008) Chef: cluster head election mechanism using fuzzy logic in wireless sensor networks. In: IEEE international conference on advanced communication technology, vol 1, pp 654–659. IEEE (2008)
26. Kulkarni RV, Forster A, Venayagamoorthy GK (2010) Computational intelligence in wireless sensor networks: A survey. IEEE Commun Surv Tutorials 13(1):68–96
27. Kulkarni RV, Venayagamoorthy GK (2010) Particle swarm optimization in wireless-sensor networks: A brief survey. IEEE Trans Syst Man, Cybern, Part C (Appl Revi) 41(2):262–267
28. Li K, Huang H, Gao X, Wu F, Chen G (2019) Qlec: A machine-learning-based energy-efficient clustering algorithm to prolong network lifespan for IoT in high-dimensional space. In: ACM international conference on parallel processing, pp 105:1–105:10
29. Mukherjee A, Goswami P, Yang L, Tyagi SKS, Samal U, Mohapatra S (2020) Deep neural network-based clustering technique for secure IIoT. Springer neural computing and applications pp 1–9 (2020)
30. Mukherjee A, Jain DK, Goswami P, Xin Q, Yang L, Rodrigues JJ (2020) Back propagation neural network based cluster head identification in mimo sensor networks for intelligent transportation systems. IEEE Access 8:28524–28532
31. Nehra NK, Kumar M, Patel R (2009) Neural network based energy efficient clustering and routing in wireless sensor networks. In: IEEE international conference on networks & communications, pp 34–39. IEEE (2009)

32. Park GY, Kim H, Jeong HW, Youn HY (2013) A novel cluster head selection method based on k-means algorithm for energy efficient wireless sensor network. In: IEEE International conference on advanced information networking and applications workshops, pp 910–915. IEEE (2013)
33. Qing L, Zhu Q, Wang M (2006) Design of a distributed energy-efficient clustering algorithm for heterogeneous wireless sensor networks. Comput Commun 29(12):2230–2237
34. Rao PS, Jana PK, Banka H (2017) A particle swarm optimization based energy efficient cluster head selection algorithm for wireless sensor networks. Spring Wirel Netw 23(7):2005–2020
35. Ray A, De D (2016) Energy efficient clustering protocol based on k-means (eecpk-means)-midpoint algorithm for enhanced network lifetime in wireless sensor network. IET Wirel Sens Syst 6(6):181–191
36. Russell S, Norvig P (2002) Artificial intelligence: a modern approach
37. Shi Y et al (2001) Particle swarm optimization: developments, applications and resources. In: IEEE Congress on evolutionary computation, vol 1, pp 81–86. IEEE (2001)
38. Singh B, Lobiyal DK (2012) A novel energy-aware cluster head selection based on particle swarm optimization for wireless sensor networks. Human-Centric Comput Inform Sci 2(1):13
39. Singh MK, Amin SI, Imam SA, Sachan VK, Choudhary A (2018) A survey of wireless sensor network and its types. In: IEEE International conference on advances in computing, communication control and networking (ICACCCN), pp 326–330. IEEE (2018)
40. Su TJ, Huang MY, Sun YJ (2011) An adaptive particle swarm optimization for the coverage of wireless sensor network. In: Springer international conference on computer science, environment, ecoinformatics, and education, pp 386–391. Springer (2011)
41. Thakkar A, Kotecha K (2014) Cluster head election for energy and delay constraint applications of wireless sensor network. IEEE Sens J 14(8):2658–2664
42. Wahba S (2020) World Economic Forum. https://www.weforum.org/agenda/2020/02/smart-cities-inclusive-future-technology-growth. Accessed 23 April 2020
43. Wang, G., Wang, Y., Tao, X.: An ant colony clustering routing algorithm for wireless sensor networks. In: IEEE International Conference on Genetic and Evolutionary Computing, pp. 670–673. IEEE (2009)
44. Watkins CJ, Dayan P (1992) Q-learning. Mach Learn 8(3–4):279–292
45. Yi S, Heo J, Cho Y, Hong J (2007) Peach: Power-efficient and adaptive clustering hierarchy protocol for wireless sensor networks. Comput Commun 30(14–15):2842–2852
46. Younis M, Akkaya K, Kunjithapatham A (2003) Optimization of task allocation in a cluster-based sensor network. In: IEEE symposium on computers and communications, pp 329–334. IEEE (2003)
47. Yu M, Mokhtar H, Merabti M (2006) A survey of network management architecture in wireless sensor network. In: Sixth annual post-graduate symposium on the convergence of telecommunications, networking and broadcasting (2006)
48. Zeb A, Islam AM, Zareei M, Al Mamoon I, Mansoor N, Baharun S, Katayama Y, Komaki S (2016) Clustering analysis in wireless sensor networks: The ambit of performance metrics and schemes taxonomy. Int J Distribut Sens Netw 12(7)

A Multi-level Ontology to Manage Service Level Agreements in Smart Cities

Cathryn Peoples, Mamun Abu-Tair, Bin Wang, Kashif Rabbani, Adrian Moore, Philip Morrow, Joseph Rafferty, Sally McClean, M. Zoualfaghari, Z. Cui, and P. Kulkarni

Abstract Internet of Things (IoT) services, to date, have been managed through affordances made by service providers (SP) to data provider (DP) customers who supply data for hosting in a shared repository. Services provided to data consumers (DC), on the other hand, are not managed in a similar way, with DCs being able to access datasets without providing detail to track them. Typically, DCs are not

---

C. Peoples (✉) · M. Abu-Tair · B. Wang · K. Rabbani · A. Moore · P. Morrow · J. Rafferty · S. McClean
School of Computing, Ulster University, Newtownabbey, UK
e-mail: c.peoples@ulster.ac.uk

M. Abu-Tair
e-mail: m.abu-tair@ulster.ac.uk

B. Wang
e-mail: b.wang1@ulster.ac.uk

K. Rabbani
e-mail: rabbani-k@ulster.ac.uk

A. Moore
e-mail: aa.moore@ulster.ac.uk

P. Morrow
e-mail: pj.morrow@ulster.ac.uk

J. Rafferty
e-mail: j.rafferty@ulster.ac.uk

S. McClean
e-mail: si.mcclean@ulster.ac.uk

M. Zoualfaghari · Z. Cui
BT, Adastral Park, Martlesham Heath, Ipswich, UK
e-mail: mh.zoualfaghari@bt.com

Z. Cui
e-mail: zhan.cui@bt.com

P. Kulkarni
Computer & Network Engineering Department, College of IT, UAE University, Al Ain, UAE
e-mail: parag@uaeu.ac.ae

M. Khan et al. (eds.), *Smart Cities: A Data Analytics Perspective*,
Lecture Notes in Intelligent Transportation and Infrastructure,
https://doi.org/10.1007/978-3-030-60922-1_7

paying customers, and subsequently receive a best-effort Quality of Service (QoS)—thus they are vulnerable in the current system to change in data availability. To promote continued growth of the IoT, it is anticipated that changes are required to the business model. This may result in greater levels of protection for DC customers and more guaranteeable levels of service. In this chapter we present an ontology which responds to the challenge of managing customer information and providing a service autonomously in response. An application of the ontology is contextualized using the smart city waste management domain.

Keywords Internet of things (IoT) · Interoperability · Ontology · Service level agreement (SLA) · SLA life cycle · Smart city · Smart waste management

1 Introduction

To date, data collected from smart cities has been submitted to Internet of Things (IoT) repositories on the free will of clients, data providers (DP), engaging in this environment. There is no obligation to make uploaded datasets publicly available, and there are no consequences if the data, once public, becomes private, is removed, or is no longer updated. In general, IoT deployments supporting smart cities provide few guarantees of what citizens may expect. This approach, however, leaves data consumers (DC), as another type of smart city IoT customer, who may or may not also be DPs, in a vulnerable position. To contextualize the IoT players described from the perspective of a smart city, consider the open data made available by Transport for London (TFL) [1]. Detail on train arrivals at all stations, potentially provided by sensors onboard trains (acting as DPs) can allow passengers (DCs) to predict the service for a particular route. In the case of line failure and subsequent congestion on the line, sensor data could potentially be communicated directly between trains, one acting as a DP and one as a DC, to determine the duration of time it should wait at the preceding station. Where DCs are external to TFL, there is little guarantee that the data feeds will remain in a constant state, a fact which may be a contributor to the low levels of trust observed in IoT technologies today [2]. The concept of trust, and how to improve it, continues to be a key IoT research area today [3].

Vulnerability in smart city IoT service provision need not, however, be the case: A Service Level Agreement (SLA) defines the duties which each participant in the scenario must fulfil—this includes both clients, typically DPs, and service providers (SP). Obligations managed through the SLA generally define SP behaviors. These are considered from the perspective of guaranteeing a certain level of service provision according to platform uptime or service availability. The SLA defines a service to which a customer agrees, in addition to remuneration if the service is not fulfilled. DPs are not obligated in their activity according to the SLA, other than paying for their subscribed to service. Furthermore, their activity is not restricted and this may result in legal concerns such as providing copy written, sensitive, illegal, or restricted data. In a flexible IoT environment, current practices do not control the behavior of DP customers.

This model has largely served smart city IoT needs to date. However, changing SP priorities, as described in [4], may indicate that it is no longer adequate. The SP *prodea* [5] proposes to "*unleash the internet of services*" in response to problems experienced by SPs. Problems include non-interoperable IoT solutions and high service costs, with infrastructure needing to be replicated with each service launch. *prodea* additionally recognizes the complex challenge of hosting and managing solutions, and a need for personalized services. Together, these factors slow service provision and reduce transparency.

Interoperability in the IoT is limited and challenging as a result of variation in the data used between providers in device, application, data hub, and management solutions; it is also distinct in the semantic approaches adopted. Agreement must be achieved on the syntax and semantics to ensure that they are consistent in all service deployments. Syntax defines what information is used to support the operation, while semantics refer to how the operation is executed. With inconsistency prevalent, devices and protocols are limited in their capacity to process information received from other devices. Furthermore, it is not guaranteed that data communicated to a centralized hub are translatable. As transparency is absent in the ways in which services are provided, there is operational and service provisioning complexity, and a general lack of service accessibility.

SPs typically offer services where they control the devices providing data. For example, Etisalat, a telecommunications SP in the United Arab Emirates, provides IoT services via a reseller [6]. In the reseller model, the SP helps in application development and then sells that application to end users. There is no demarcation between DPs and SP in this scenario.

Looking into the capabilities of and demands placed on SPs, and considering customer roles in the smart city IoT, it may be reasonable to expect that more behaviors and activities can be controlled using SLAs. A SLA could, for example, have a SP reimburse a DC if they are unable to access a dataset for a pre-agreed period of time. A SLA could similarly penalize a DP who has breached their contractual obligation with the SP to provide data for the agreed duration. Currently, services and their supporting SLAs are not provided in such detail, and instead largely measure SP behavior according to platform uptime. Facilitating such operation may, however, provide a more consistent, predictable, and therefore trustworthy environment.

Transparency in the IoT service provisioning process is important. In this context, transparency refers to customers not needing to interact with SPs during service setup. Instead, they would have a provisioning process which autonomously responds to detail collected during the service request. A standardized approach is anticipated to improve the transparency of IoT operations, ensuring that all SPs operate consistently [10]. The agreement of these terms can be facilitated using an ontology, which defines a hierarchical mapping of the language and the relationships between attributes. With agreement of a standardized ontology definition, new technologies can provision capabilities in their systems to ensure that the common approach to operation is adopted, and thus opportunities for interoperability are facilitated.

Developing standards to support IoT operations, however, is challenging due to the diversity across the smart city IoT. A range of domains and device types are connected to the IoT, with different data collectable from each. There is therefore little guarantee that a consistent set of information can be collected from all devices so that each can be managed in a common way. This is challenging to support autonomously and, if not supported autonomously, leads to a lack of service provisioning transparency. It also limits the extent to which interoperability is possible. In an unstandardized environment, a more personalized approach to service setup and management becomes necessary, with less autonomous flexibility and more interaction with SPs. It is therefore to the challenge of smart city IoT standardization that this chapter contributes. Standardization in this context refers to the use of a single ontology to respond to the needs of all scenarios in the smart city IoT; we do not suggest that this should plug into outputs from the standardization bodies, but rather that it exists as a standalone approach.

To facilitate a standardized approach, we propose an ontology for use by IoT SPs. An ontology is a formal framework which supports knowledge representation. An ontology was selected due to the expressivity and ability to support automated processing using semantic reasoners. Currently, an ontological SLA format which is universally accepted does not exist. We therefore aim to respond to this through our work. The ontology will be applied when a customer's device is connected to a SP's platform, ensuring that a consistent set of information is collected regardless of who the SP is. This information can influence their SLA through automating service provision. It is also anticipated that a customer can re-negotiate their SLA once it becomes active, to ensure that customers continue to be satisfied with the service received in a dynamic environment. The extent to which this occurs will be influenced by information collected during device set up.

The remainder of the chapter is organized as follows: The state-of-the-art in SLA life cycles and IoT ontologies are presented, with a view to highlighting the gaps in relation to an evolving IoT business model. This is followed with our research proposal, which includes both a revised SLA life cycle and ontology which is applicable for the smart city IoT. Its application is contextualized using a SLA provisioning process. This is followed with a consideration of an application of the ontology and life cycle in relation to smart waste collection within a smart city. The chapter finally concludes and presents future work.

2 State-of-the-Art in IoT SLA Operations

An IoT service is managed using a SLA life cycle which refers to the creation of SLAs, possible re-negotiation to more suitably meet customer needs, and potential termination when the SLA period expires, or is violated. The SLA life cycle is important in relation to IoT ontology definition, dictating the phases which can be supported with the data collected.

2.1 SLA Life Cycles

Several SLA life cycles are available. The IBM SLA life cycle in [7], as one example, involves five stages:

Initial state, SLA identified: This state is entered when a consumer requests a dependency on a service offering the required service level definition (SLD).
Request SLA, SLA requested: The SLA provider must approve the request for it to become active. This may involve approving, rejecting, or asking for the request to be revised.
Approve SLA request, SLA inactive: The consumer does not yet have authorization to access the service requested.
Revise SLA request, SLA identified: The SLA is negotiated by the service provider, manipulating details of the SLA provided by the consumer.
Activate SLA, SLA active: The SLD is now operational between the end-points, and the consumer is able to access the provider's service.
Deactivate SLA, SLA inactive: SLA can be suspended if operational issues exist.
Terminate SLA, SLA terminated: SLA reaches the end of its lifetime.

This life cycle includes a SLA inactive stage, during which violations can be avoided by disabling the SLA when service levels cannot be sustained. It is questionable, however, how effective this is while violations have been prevented, the customer need for service may not have been removed.

A SLA life cycle for cloud environments is presented in [8] and involves three core phases:
Acquisition:

SLA assessment: A pre-assessment of the service needs.
SLA preparation: The first contact with cloud SPs.
SLA negotiation & contracting: Making a deal with a cloud SP.
Operation:
SLA execution & operation: Setting up and using the service.
SLA updates & amendments: Response by cloud SP to support amended needs.
SLA escalation: Deals with any breaches or incidents during the contractual period that have resulted in a dispute.

Termination:

SLA termination: Manages the end of the customer relationship, dealing with data export and deletion.

Updates are accommodated during the SLA life cycle, and include those made by the cloud SP, but not by the customer. In a dynamic cloud IoT environment, however, it may be assumed that a customer will wish to make changes to their SLA themselves.

A further SLA life cycle presented in [9] involves:

SLA management: Involves SLA negotiation, enforcement, and evolution.

SLA definition: The SP agrees SLA terms in terms of metrics, quality, price and penalties.
SLA modelling: The SLA is presented in an agreed format.
SLA negotiation: The SLA is negotiated with the customer.
SLA monitoring: Identifies if the service meets QoS requirements defined in the SLA.
SLA violation: This phase is reached if the SLA fails to achieve the agreed QoS in relation to reliability, availability, dependability, security, and performance.
SLA evolution: The SLA is adapted in response to change in the requirements of either producer and/or consumer.

This life cycle is specific to the IoT and includes an important stage which is distinct from other models, that of SLA evolution. This phase provides ability to cope with a changing operational environment, avoiding potential violations when customer service needs change. As customers change the terms of their SLA, however, and their SLA evolves, this may impact other customers. This requires separate management, and could potentially influence a further phase in the life cycle.

Due to the IoT's dynamism, we believe that SLA evolution, as presented in [9], is important. This will support flexibility and, consequently, customer trust. The notion of SLA evolution, however, increases complexity of the general SLA model, not only from the perspective of change to a customer's service, but due to the potential knock-on effect impacting other customers. Further phases to the SLA life cycle supporting the smart city IoT may therefore be necessary, going beyond the capabilities of models proposed to date. It is therefore to this gap that we additionally contribute in our ontology.

2.2 IoT Ontologies

Ontologies explicitly conceptualize a domain and define IoT data unambiguously. This is important in the IoT, where attributes may be defined by the customer who collected it, and may be indistinguishable to DCs with access to publicly available hosted datasets. A standardised ontology in this context helps to overcome such limitations. Defining a classification for the IoT, however, is complicated by the fact that a single definition of what the IoT is does not exist.

Nonetheless, ontologies are evolving to support different aspects of the IoT; they do not respond to all needs but focus instead on a subset. In [11], for example, the authors propose Lilliput, an ontology-based platform to represent relationships in social networks. It captures details on profiles, relationships between profiles, and additional context which includes the profile owner's location. The overall goal of the platform is to facilitate intelligence regarding a community of social network citizens. This platform explores the fact that an online social network can be used to create an IoT social network, and that smart objects can be accessible dependent on the social relationships between people, objects and places. Smart objects in this context for example refer to, "*air-conditioner, dimmer, temperature sensor*" [11]. Users can exploit their social relationships to execute operations on smart things. As this work responds to a specific domain, it has limited utility in that it does not meet wider IoT needs. In [12] the ontology models IoT services; it consists of a Thing context (e.g., a status), a Device context (e.g., device availability), an IoT Service context (e.g., functional properties of the service), and a User context (e.g., user status such as eating or walking). The ontology is contextualized using a smart building. Similar to [11], this approach is applicable to a subset of the IoT only. In [13], the ontology is organized around the Service, User, and Reference Components. These refer to the IoT services in a smart city, details about the user of the service, and references to specific IoT resources. The ontology interacts with ex-ternal IoT platforms and provides semantic interoperability between them. In [14], on the other hand, the service is more specifically organized around needs within a smart home and accommodates spatial and temporal concepts. Each of these ontologies responds to a very specific need, a fact particularly true in the case of the pre-IoT examples.

The authors of [15] propose a more thorough ontology for the IoT, with a goal of being lightweight. Classes include a domain parent class, a domain class, domain subclass, a set of IoT devices, a domain location, business data (the attribute value), and metadata (its frequency), among others. In the ontology, devices store device-specific information and an ontology server stores global information of domain relevance. Maintaining the lightweight quality of the ontology involves the ontology update, which is determined using the device category and its capacity. Data is filtered based on frequency of access, with an assumption that the less frequently accessed data is less important. While such a scheme will have utility in certain contexts, the loss of detail may be undesirable in all.

IoT-Lite is another IoT-specific ontology. It captures Metadata associated with a Sensor built into a Sensing Device on a Device within a System. The authors' objective in developing this model is to create a schema which can be adopted widely across the IoT due to its simplicity. The authors describe their schema as, "*a core part of a semantic model in which … different semantic modules can be added to provide additional domain and application specific concepts and relationships*" [16]. They inform, however, that "*IoT-Lite does not intend to be a full ontology for the IoT*" [16]. There is therefore scope for an ontology to respond to this gap.

The authors in [19] present an IoT ontology which merges aspects of different ontologies already in existence. This approach incorporates related attributes from a variety of sources thus increasing common expressivity. The rationale given for the approach is that, as features from different ontologies are incorporated, inter-operability can be achieved. Such an approach, however, remains to be used in practice.

A review of IoT ontologies highlights that a complete solution does not exist. Contributions can plug gaps in ontological needs for the IoT, however, a single approach does not provide a comprehensive solution to capture the needs of all. One of the more complete solutions, IoT-Lite, advises that additional modules can be added on a domain-specific basis. As a result, they are unable to support autonomous SLA provision. There is therefore a gap in the technologies available which we respond to with this work.

3 Research Proposal

The ontology and service management proposal presented in this paper is driven by a statement from Mubeen et al. (2017) that "*Current cloud technologies offer a limited support for dynamic renegotiation of SLAs between participants*" [9]. We agree that re-negotiation is attractive in the IoT: In this dynamic environment, peoples' needs change. Re-negotiation leads to a more complicated management process however, due to the potential impact on SLA achievement. Of significance is the fact that violations can occur both for users re-negotiating their SLA terms (most likely DPs) and, unknowingly, other users (most likely DCs). We therefore anticipate requirements in both the ontology design (to capture the information), and the SLA life cycle (to manage the process).

3.1 Proposed IoT SLA Life Cycle

In this chapter, we make an assumption in relation to the life cycle through which SLAs may pass, and consider a scenario where a SLA can transition between an *active* and *re-negotiation* state during the SLA contract. Figure 1 presents a graphical overview of the proposed lifecycle. This enables the SLA to be reconfigured in real-time. As an example of a scenario where such activity may be useful: Change to the public availability of a DP's dataset may mean that a DC is unable to access the dataset for as long as anticipated. The ontology supports customers changing the terms of their SLA once it moves into the active state. It is also possible that a change to the SLA terms by one customer will have a subsequent effect on other customers. It is therefore important that there are phases available in the life cycle to provide adequate support. Rapid recovery will be essential to ensure that the DC's SLA is not breached, which may involve the SP making recommendations on equivalent datasets to access instead. This will be facilitated by activities carried out during the SLA Recommendation phase of the life cycle proposed in this chapter.

3.2 Proposed IoT Ontology

The ontology presented in Fig. 3, is proposed from the perspective of scenarios where DPs contribute data for hosting on a SP's platform. From the context of a smart city, this could involve autonomous vehicles contributing data to a centralized repository for later consumption by citizens interested in traffic flows. There is an opportunity

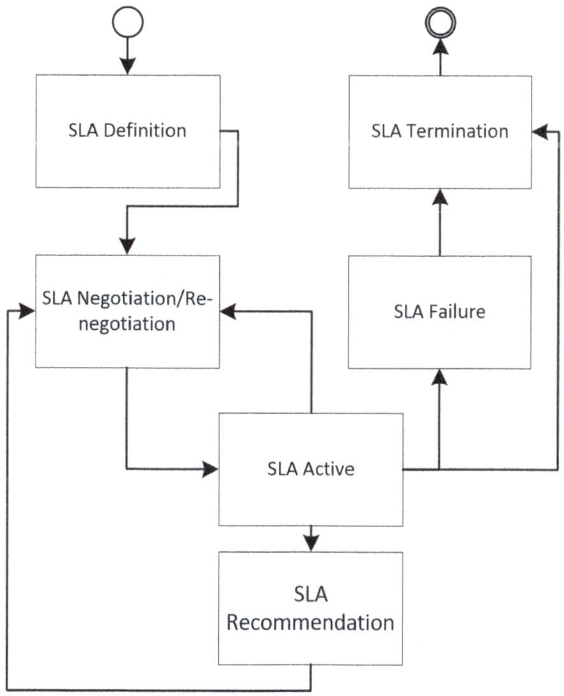

Fig. 1 Proposed SLA life cycle to be adopted by this work

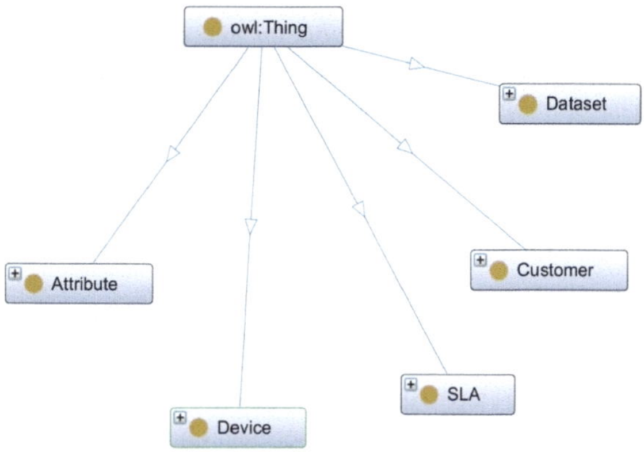

Fig. 2 Key Classes in IoT SLA Ontology (in Protégé [17])

for the data to be publicly available for consumption by DCs through agreement with the SP. The SLA's objective is to protect the interests of both DPs and DCs. It is the intention that this ontology supports the SLA provisioning process, used when a customer registers for a service with a platform.

Distinct from ontologies in the literature, we capture the characteristics of customers using the ontology. These aspects will indicate the overall role which customers play in the environment (i.e., data provide or data consume), together with

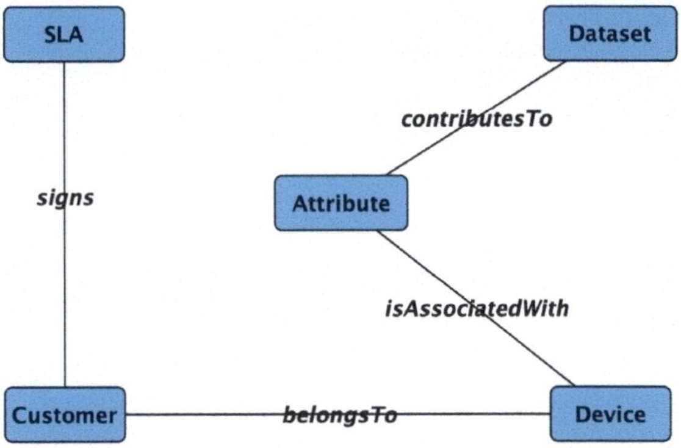

Fig. 3 IoT Ontology Proposal—Top Level

detail on the ways which their service may change in a constrained environment, in a manner which is acceptable to them.

It is our objective to present a scheme in which:

1. *The ontology is independent of external factors.*

The ontology is not restricted by the domain (e.g., an office block or shopping mall in a smart city) in which it is used through the ability to specify domain-specific attributes. It is also independent of device-specific aspects, and therefore the data it is communicating. Furthermore, the ontology is independent of the type of customer supported, accommodating both DPs and DCs.

2. *The ontology is extensible for new applications.*

The ontology is adaptable for new scenarios through the integration of domain-specific attributes, the number of which is variable on a domain-specific basis.

Metadata specific to DCs and DPs is associated with the fact that DCs may be buying into a service dependent on data made available by DPs. This ensures that the needs of both are met with the service provided. DC metadata is used by the SP in service provisioning decisions: The `Renegotiable` attribute, for example, allows SLA terms to be adapted while it is *active*.

Using the proposed ontology, DCs and DPs will subscribe to SLAs which offer a different service to each `Customer` type. The DC service will support data access from the platform; the DP's service, on the other hand, will provide dataset hosting on the platforms, and ability to potentially consume their own data.

The relationships between these primary classes are presented in Fig. 2. This framework describes that: A `Device` belongs To a `Customer`, a `Customer` signs a `SLA`, a `Device` is Associated With an `Attribute`, and an `Attribute` contributes To a `Dataset` (Fig. 2).

In terms of the design adopted, it could be reasonable to expect that `Device` and `SLA` might be subclasses of `Customer`, and that `Attribute` and `Dataset` might be subclasses of `Device`. A decision has been made, however, to keep these as separate classes: `Attribute` settings, for example, may be configured dependent on properties of the `Customer` and `SLA` classes. A DP customer could decide that delivery of their datastream to the hub should not be interrupted, in which case an attribute within the `Attribute` class will be set. This assignment must be made specific to the attribute being generated from the device. If the SLA enters a state of *failure*, attributes within the `Attribute` class should be adapted to indicate that the data is no longer being collected. While the `SLA` will be primarily dependent on the `Customer` class, its relationship with `Attribute` means that it is inappropriate to capture it as a subclass of `Customer`. Similar reasons prevent the other classes being assigned as sub-classes.

Each class is explained in more detail below. The ontology begins with a `Customer`.

3.2.1 Customer

A `Customer` is required to specify their type, either a DP or DC, and their desire to potentially re-negotiate the SLA (`Renegotiable`) after moving into an *active* state. `Anonymization` is captured by the ontology from two perspectives, for both DCs and DPs simultaneously:

1. `Customer:Anonymization:True/False`
2. `Dataset:Anonymization:True/False`

This detail helps to indicate:

1. If the DP wants data to be anonymized
2. If the DC can tolerate anonymized data

From the first perspective, a `Customer` will indicate their ability to tolerate anonymized datasets (DC), or desire for their dataset to be anonymized (DP). From the second perspective, the ontology will report if the dataset has been anonymized. The ability to tolerate anonymization will influence the datasets made available and recommended to DCs.

DP `Location` supports identification of any data protection aspects that data should be exposed to, in addition to data transferability between countries. It can also be used to influence the datasets which are recommended to DCs, dependent on their preferences.

Customers will indicate desire to potentially renegotiate SLA details once moving into an *active* state (`Customer:Renegotiable:True/False`). This potential desire will give DCs and DPs the option of flexible service provision. (In addition to the DC and DP, a SP may also want to renegotiate the SLA; this aspect is described in more detail later in the chapter.)

Only certain aspects of the SLA may be renegotiated, including `Dataset: PubliclyAvailable`, `Dataset:Anonymization`, `Attribute:Frequency`, `Attribute:Interruptible`, and `Attribute:CollectionDuration`. The objective of renegotiation is to support flexible service provision, and to allow customers to remain satisfied in a dynamic environment. As an example of how renegotiation is used, a DC could initially specify before their SLA becomes *active* that they wish to access datasets collected in a certain `Location`. With time, they may wish to change the `Location`, a change which can be documented during the SLA *re-negotiation* phase.

`Customer:Domain` will indicate the domain that the `Customer` is working in (DP) or is interested in (DC). This information can influence the datasets made available to a DC. A `Customer` is required to indicate their ability to tolerate risk—`Customer:Risk:True/False`. From a DC's perspective, this indicates tolerance of a dataset becoming unavailable or out-of-date, which could occur if the DP re-negotiates SLA terms. (Based on the ruleset supporting service provision, a customer would only be offered datasets with potential to change in the event that they indicate `Customer:Risk:True`.)

A `Domain` is presented in this model from the perspective of a `Customer`, as opposed to being a standalone class of `Thing`. The `Domain`, when considered from this angle, will communicate if it is one in which the `Customer` is working (DP) or one in which the `Customer` is interested in working in (DC) (Fig. 2). This knowledge can be used to fuel later decision-making, by potentially recommending datasets for DC customers when the dataset/s to which they originally subscribed become unexpectedly unavailable (when the SLA enters the *re-negotiation*, and potentially the *recommendation*, states).

In addition to `Customer` metadata, information is collected on the `Device` being registered.

3.2.2 Device

Customers will capture characteristics of the device used to provide (DP) and consume (DC) data. The DC is required to indicate the `Device:Category` of data they wish to consume, while the DP will provide the category of the device used to generate the dataset. This attribute helps to ensure that, where a recommendation is made by the SP (which may occur due to a DP renegotiating SLA terms once the SLA is *active*), it references a device which belongs to the relevant `Device:Category`. Customers are similarly asked to indicate the `Device:Location` of datasets which they may consume; this is important in the event that they are interested in datasets belonging to certain regions. In relation to the device on which they wish to consume data, they are asked to indicate the `Device:Manufacturer` to ensure that compatible datasets are provided. Additionally, they should indicate their device capacity (`Device:Capacity`) so that estimations of performance achievable may be determined. `Device:Mobility` is collected due to the impact it can have on the reliability and accuracy of data transferred across the network. `Attribute` values can be device-specific in that the attributes collected on one device may not be collectable on another. Furthermore, the attribute values collected across different devices are generally not interoperable with or usable by other systems. We therefore specify that `Device:Manufacturer` is part of the context set to ensure that attributes from specific manufacturers may be provided to DCs. Situations where alternative datasets are accepted (i.e., not collected using devices of the same manufacturer as the device on which they will be used) will be at the DC's risk.

The overall goal of the ontology is to determine the domain and domain-specific attributes in an autonomous way, such that a SLA may be recommended, without the need for customers to have this technical knowledge. Some customers are discouraged from participating in the IoT due to complexity of the way which services are offered [18]. The goal of this mechanism is therefore to improve the accessibility of IoT services across society. Determining the `Domain` autonomously requires interoperation between the `Device:Category` and `Device:Sensor` attributes. Domain-specific attributes will be subsequently definable.

The `Device` is responsible for generating `Attribute` data.

3.2.3 Attribute

In relation to `Attribute:DataCategory`, DCs can identify the category of datasets they wish to consume, and DPs will indicate the category of datasets being contributed. The DC has an opportunity to indicate the preferred `Attribute:Frequency` at which attribute values are collected. The DP can specify if data delivery to the centralized hub is interruptible (`Attribute:Interruptible`).

Domain-specific information is considered separately in the detail collected from customers. This allows the ontology to be independent of external factors, such as the domain. A variable range of `Attribute:DomainSpecific` elements can be recorded, dependent on the scenario. In a waste collection use case, domain-specific attributes might include the bin's weight and a threshold fill level of the bin. To be usable, it is important that the `Attribute:DomainSpecific` fields are supported with effective descriptors. From the domain knowledge, additional detail may be inferred; for example, a combination of domain knowledge in relation to the `Customer:Domain` with `Attribute:DataCategory` will reveal information which includes the priority and urgency of the data transfer.

`Attribute` values will be organized into `Datasets`.

3.2.4 Dataset

A DP can indicate the minimum duration for which the `Dataset` should be available using `Dataset:Availability`. In relation to DP desire, the information collected captures whether the data will be available to the public (`Dataset:PubliclyAvailable`), or if it is private to the DP.

Several generic attributes characterize a customer's `SLA`.

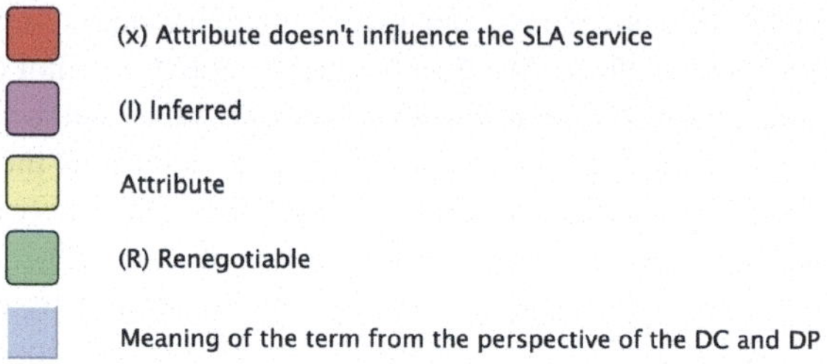

Fig. 4 IoT Ontology Proposal—Colour Key

3.2.5 Sla

Generic attributes will be assigned to the `SLA`, including `SLA:Name`, `SLA:Description`, `Customer:Identifier`, and `SLA:Period`. A selection of the generic `SLA` attributes influences the `SLA` terms and conditions. For example, the consequence of a breach of the agreed terms (`SLA:Breach`), together with the maximum acceptable disruption period (`SLA:Disruption`). The number of rules which a customer may apply (`SLA:Rules`) and number of messages which a customer may send (`SLA:Messages`) are also captured. `SLA:State` will be updated throughout the `SLA` lifetime.

The maximum disruption period is an important attribute in the event that a `SLA` *recommendation* is being made. A `SLA` *recommendation* will be made in the event that a customer has changed their `SLA` terms, and the change has a knock-on effect on other customers, which may lead to a violation of their `SLA`. The `SLA` *recommendation* is made in an attempt to prevent `SLA` breaching. However, this may cause a disruption to the client's service. The maximum disruption period therefore defines the longest period of time for which the customer can tolerate their service being interrupted.

The ontology supports service flexibility by allowing a SP to enforce `SLA` *renegotiation* (`Customer:SLA:SP:True/False`). This could occur if an SP detects that a consumer has changed location to a different jurisdiction through device mobility. There may be implications from a data protection regulations perspective given the new jurisdiction that the DC and/or DP now belong to. It is therefore reasonable to expect that a SP should be able to shut both off from potentially contributing and accessing data to comply with regulations (Figs. 4, 5, 6, 7, 8, 9).

4 Building the Ontology in Protégé

The ontology has been implemented in Protégé. From the perspective of the superclass and subclass relationship, exemplar values are given for instances within each subclass. This defines the way in which the smart city IoT has been contextualized for support within the ontology, taking into account the domains, devices and sensor types accommodated.

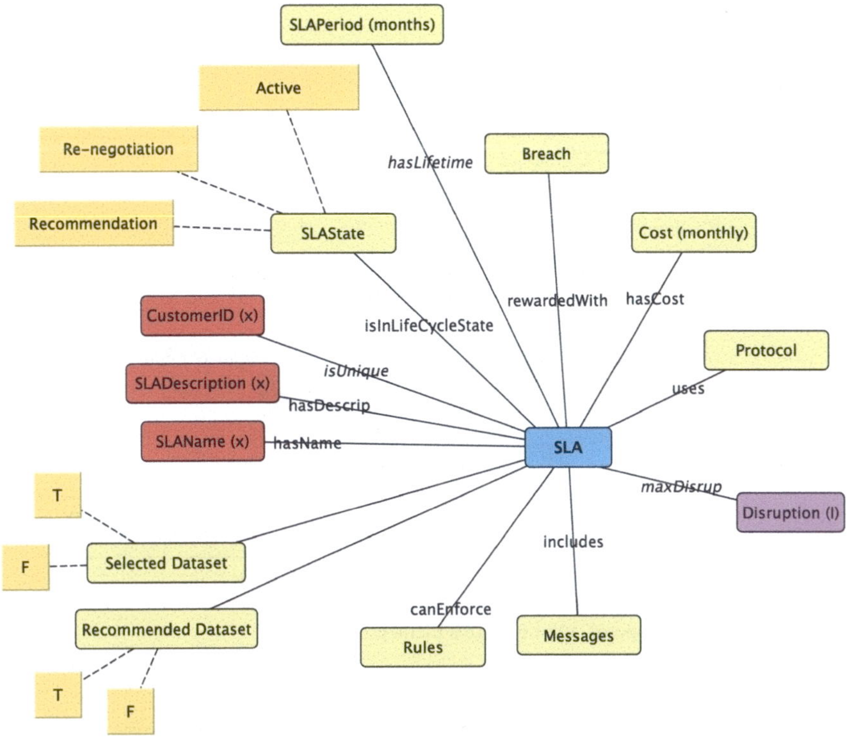

Fig. 5 IoT Ontology Proposal—SLA Element

4.1 Ontology Domains

The ontology implementation is based on a domain set identified as being somewhat consistent across several SP platforms and standardization bodies. Domains specific to smart cities explicitly considered include:

Home: e.g., bed, bin, fridge, smoke detector, door, lighting
Building: e.g., smoke detector, door, lighting
City: e.g., waste management, lighting
Personal: e.g., smart watch, smart phone
Industrial Manufacturing, including packaging (e.g., product authentication, product tampering, package tracking, package temperature, digital labelling), asset management (e.g., where the asset is, who is using it, condition, maintenance schedule, user manual), and operations (e.g., machine sensor, environment sensor)
Medical, including wearable sensors (e.g., heart activity, glucose levels, blood pressure, oxygen level)
Vehicle: e.g., monitoring surroundings, obstacle detection, navigation

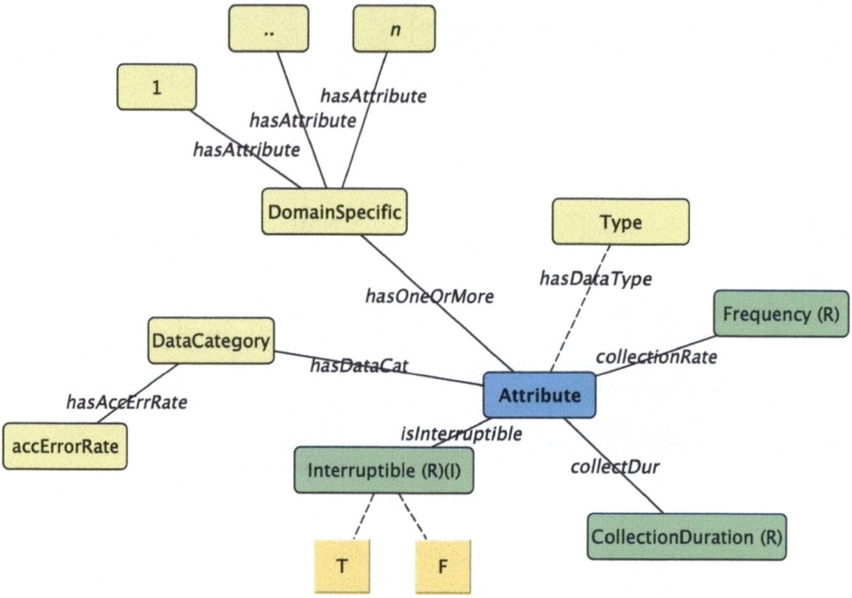

Fig. 6 IoT Ontology Proposal—`Attribute` Element

Fig. 7 IoT Ontology Proposal—`Dataset` Element

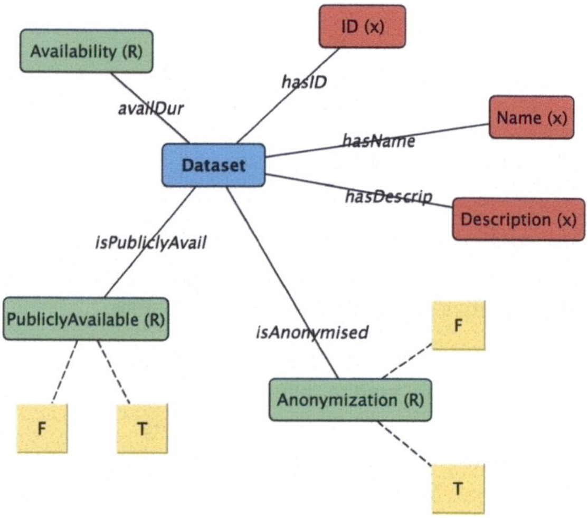

Transportation, including car parking: e.g., fleet management, public transit management, smart inventory management, optimal asset utilization
Retail: e.g., foot traffic counter, inventory tracking
Leisure: e.g., sports analytics, player safety, fan engagement
Utilities: e.g., smart metering of water, gas and electric
Energy: e.g., equipment monitoring, smart thermostats

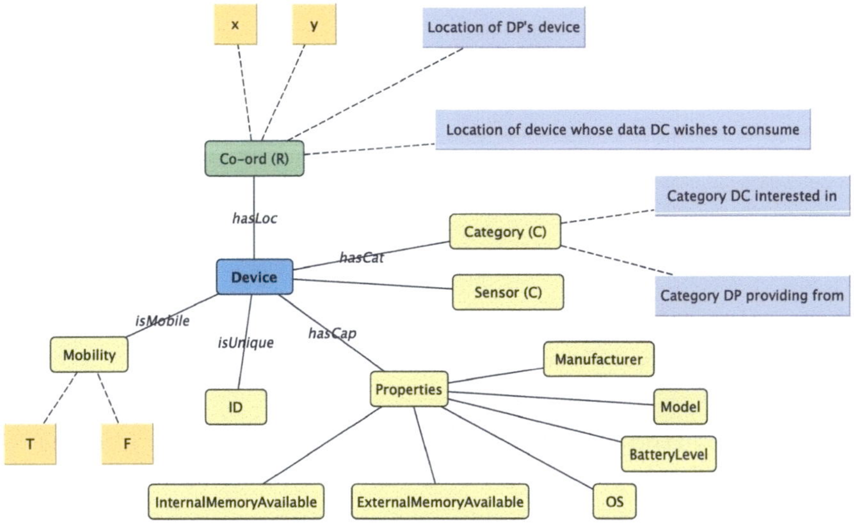

Fig. 8 IoT Ontology Proposal—`Device` Element

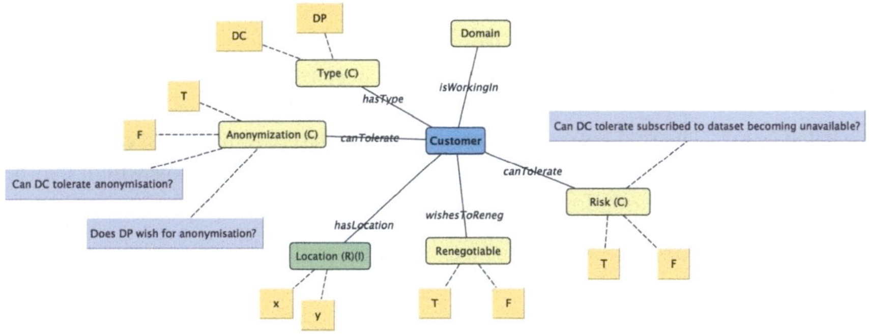

Fig. 9 IoT Ontology Proposal—`Customer` Element

Supply chain: e.g., goods location, speed of goods movement, storage conditions monitoring

While a range of specific domains are presented for contextualization purposes, this model is defined with an understanding of its applicability for all domains. Extensibility is supported through ability to specify domain-specific attributes.

`Devices` specific to `Domain:Home`, as an exemplar, are presented in Fig. 10. A `Domain` is defined from the perspective of a `Customer` (the `Domain` which a DP is generating data from within or the `Domain` which a DC is interested in consuming data about).

`Device:Category` is insufficient in all cases to identify the `Domain`, as highlighted in Fig. 10 where the overlap between `Domain:Home` and

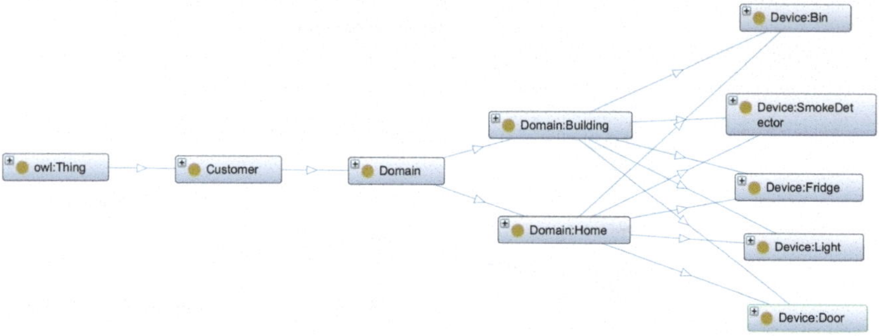

Fig. 10 Properties Associated with `Domain:Home`

`Domain:Building` can be observed. The `Domain` is therefore identified by supplementing detail on `Device:Category` with detail on `Device:Sensor`.

4.2 *Ontology Sensors*

The sensor communicating data from a device can be represented as either a standalone sensor (`Device:Sensor`), or a sensor integrated into a device (`Device:Category`). A selection of sensors included within this implementation are listed below. The first few sensors are contextualized in terms of the combination of domain and sensor type information to represent the information which can be communicated via the ontology.

Air quality: detects the presence of air pollution (e.g., `Customer:Domain:Environment`, `Device:Category:Sensor:AirQuality`)
Altitude: of an object above a fixed level (e.g., `Customer:Domain:Environment`, `Device:Category:Sensor:Altitude`)
Accelerometer: measures acceleration forces (e.g., `Customer: Domain:Personal`, `Device:Category:Smartphone:Accelerometer` or `Customer:Domain:Vehicle`, `Device:Category:Car:Accelerometer`)
Chemical: transforms chemical information into a useful signal
Flow: measures the amount of liquid, gas or vapor passing through the device
Gas: monitors changes in air quality
Gyroscope: measures angular velocity
Humidity: measures air moisture (Fig. 11)
IR: sense characteristics in the surroundings by either emitting or detecting infrared radiation
Level: measures the height of liquids and solids
Image: converts an optical image into an electronic signal
Motion: detects movement in an area

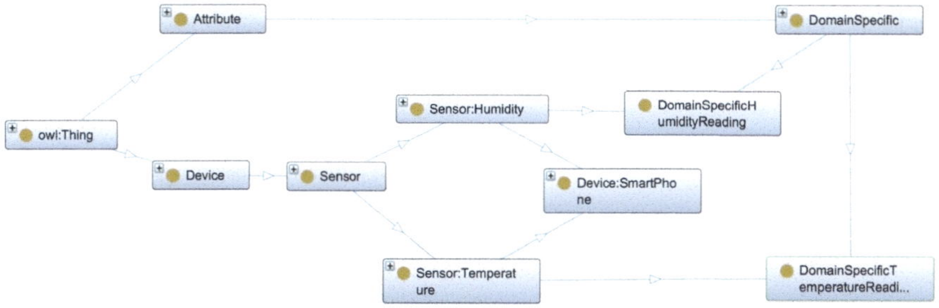

Fig. 11 `Device:Sensor`

Optical: measures the quantity of light
Photoelectric: the presence or absence of an object
Position: supports position measurement
Pressure: detects a force exerted on a surface
Proximity: detects the presence of nearby objects
Smoke: senses smoke and its level
Temperature: measures the amount of heat energy generated by an object (Fig. 11)
Water quality: ion monitoring in water distribution

The sensor will dictate several values within the ontology. For example, dependent on the sensor, autonomous decisions can be made on the communication protocol (`SLA:Protocol`), tolerance of the data stream to interruption (`Attribute:Interruptible`), and an advisable number of sensor readings per second (`Attribute:CollectionFrequency`).

5 Contextualizing the Service Provisioning Process in a Smart City

SLAs are typically offered on the basis of platform availability, in addition to service provisions which consider storage space, number of messages, and number of rules. The service offering is selected by the customer. This process can instead be automated, as we propose. This SLA provisioning approach is not specific to smart cities, but it is applicable to smart cities, given its diversity to support domains contributing to the IoT. We recognize DP and DC customers in smart cities. The goal of the SLA provisioning process is to encourage certain behaviors and manage network load. In the scenario presented (Fig. 12), customers are considered in the smart city in terms of being either a DP or DC customer. DP customers are weighted lower than DCs, with a view to encouraging data providing behaviors. The domain within which the customer is operating is considered next, with a view to encouraging the contribution of data from popular domains from the DP perspective, and regulating load on the network from the DC perspective. The service which the cus-tomer will participate

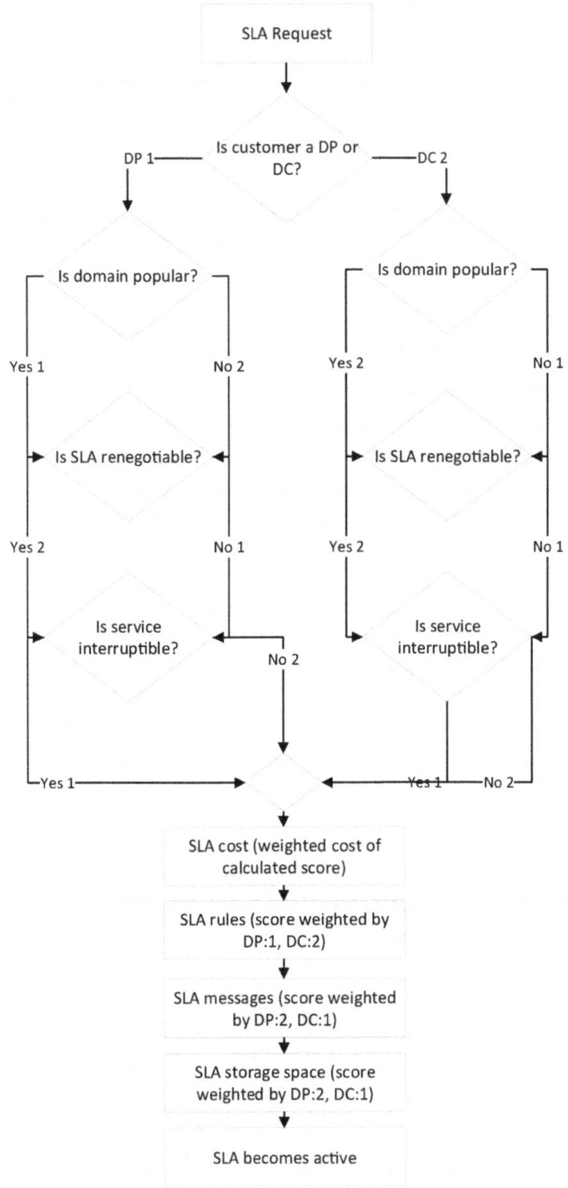

Fig. 12 SLA Provisioning Process

in is considered next: If a customer wishes for a potential opportunity to renegotiate their SLA terms, they will be charged more highly for the privilege of doing so. If, on the other hand, they can accept a service which is interruptible, they will be rewarded. Based on this detail, a score can then be used to calculate a SLA cost, the number of rules a customer can apply on hosted data, the number of messages they can send, and the total storage space they can use. At this stage, the SLA becomes *active*, and the service becomes operational.

6 Ontology Use Case in a Smart City

Application of this ontology is contextualized in relation to waste collection as part of a smart city IoT scenario:

A DP returns bin fill levels (DomainSpecific Attribute (Fill), DomainSpecific Attribute (Weight)) and Device Location on a 12-hourly rate (Frequency). The Waste Collection Domain DP has relatively relaxed communication requirements: Data delivery may be interrupted (Interruptible), and 100% data reliability is not required (accErrorRate). The DP will also communicate domain-agnostic information with the SP which includes: Device:Location, Device:Manufacturer, Device:Category, Device:Capacity, Device:Mobility, Attribute:DataCategory, Attribute:Frequency, Dataset:Availability, Dataset:Anonymization, Customer:Renegotiable, and Dataset:PubliclyAvailable. Each time a fill level (DomainSpecific (Fill)) and fill volume (DomainSpecific (Weight)) is returned to the SP it will be communicated with the Device:ID. When Dataset:Anonymization is required, detail not made available to DCs outside the City Council includes Device:Location. The Customer:Renegotiable attribute captures the DP's potential desire to renegotiate SLA terms with the SP; this may involve deciding that their dataset, once anonymized, no longer needs to be, or that during busier periods, data collection rates could increase. Metrics communicated from the DP to SP with domain-specific detail include Customer:Domain (e.g., Waste Collection) and Attribute:DomainSpecific (e.g., Threshold Bin Fill Level and Threshold Bin Fill Weight).

As a DC, the Waste Collection Depot subscribes to the SP platform to access bin fill level datasets stored on the Hub. The DC will execute rules on the dataset to identify clusters where fill levels are above the threshold. Summary results are accessed through a dashboard, and collection points and routes are disseminated to waste collectors. Bin locations within a Cluster ID can be determined offline by the DP.

As an example of a rule which may be executed on bin fill level datasets: If > 50% of Device:IDs out of all Device:IDs in Cluster:ID have a Sensor Reading (Fill) > Threshold Bin Fill Level and a Sensor Reading (Weight) > Threshold Bin Weight Level, then empty all bins in Cluster:ID. The rate of rule execution will depend on the DC's agreement with the SP, as defined in their SLA.

In relation to data sharing, other councils, as DCs, may wish to access datasets to observe trends over time. This will require that datasets have been uploaded as publicly available by the DP. It will also require that Anonymization of the dataset has been agreed, and that only those which meet the DC's Anonymization needs will be made available. DCs will execute rules on datasets to identify aspects of interest, which might involve the regions with the most waste collections.

A renegotiable aspect of the SLA by the DP is how long the dataset will be available (Dataset:Availability). In the event that a DC finds value in a dataset, yet the DP removes it after a period of time, the SP can recommend other datasets which could be relevant to the DC. Recommendations will be made based on Device:Location, Device:Manufacturer, Attribute:DataCategory, and Device:Category.

To protect the DC's QoS when a DP has re-negotiated SLA terms, which will potentially breach the DC's SLA, recommendation will be made of a dataset provided from a device of the same Device:Manufacturer, Device:Category, and with the same Attribute:DataCategory and Attribute:Frequency.

7 Conclusion

This paper has presented the design of an ontology to support a standardized approach to registering devices and managing services within the IoT. It helps to overcome the limitations of current technologies by providing ability to be domain-agnostic. We have contextualized application of ontology using the waste management domain, but it could as easily be applied to the e-health or transport domains. This is made possible through the use of domain-specific attributes, in addition to those which are generic across domains. We recognize the need for flexibility in the IoT infrastructure, and the fact that all elements of a schema may not be known apriori. Therefore, it is paramount that a schema designed for the IoT should have such flexibility; the proposed approach is catering to this requirement in that sense.

Further work involves development of the SLA knowledge base and rule set, which will enable automatic SLA provision.

References

1. Transport for London: Our Open Data. [Online] Available: https://tfl.gov.uk/info-for/open-data-users/our-open-data?intcmp=3671 (n.d.)
2. Sato H, Kanai A, Tanimoto S, Kobayashi T (2016) Establishing trust in the emerging era of IoT. In: IEEE symposium on service-oriented system engineering. https://doi.org/10.1109/sose.2016.50 (2016)
3. Teixeira FA, Pereira FMQ, Wong H-C et al (2019) SIoT: securing internet of things through distributed systems analysis. Systems 92:1172–1186. https://doi.org/10.1016/j.future.2017.08.010
4. Android Developers Blog: An Update on Android Things. [Online] Available. https://android-developers.googleblog.com/2019/02/an-update-on-android-things.html (2019)
5. prodea Homepage [Online] Available: prodea.com/iot-service-x-change (n.d.)

6. Ericsson: Etisalat Digital and Ericsson Demonstrate Unified IoT Platform for Smart Traffic Management at GITEX 2018. [Online] Available. https://www.ericsson.com (2018)
7. IBM Knowledge Center: Service Level Agreement Lifecycle. [Online] Available: https://www.ibm.com (n.d.)
8. SLA Ready: SLA-Ready. [Online] Available. http://www.sla-ready.eu/cloud-sla-lifecycle (n.d.)
9. Mubeen S, Abbaspour Asadollah S, Papadopoulos AV et al (2017) Management of service level agreements for cloud services in IoT: a systematic mapping study. In: IEEE access, pp 30184–30207. https://doi.org/10.1109/access.2017.2744677 (2017)
10. Diao Y, Lam L, Shwartz L, Northcutt D (2014) Modelling the impact of service level agreements during service engagement. IEEE Trans Netw Serv Manage 11(4):431–440. https://doi.org/10.1109/TNSM.2014.2378779
11. Byun J, Kim SH, Kim D (2014) Lilliput: ontology-based platform for IoT social networks: In: IEEE international conference on services computing, pp 139–146. https://doi.org/10.1109/scc.2014.27 (2014)
12. Aziez M, Benharzallah S, Bennoui H (2017) An ontology based context model for the discovery of IoT services in the internet of things. In: international conference on mathematics and information technology, pp 209–213. https://doi.org/10.1109/mathit.2017.8259719 (2017)
13. Ryu M, Kim J, Yun J (2015) Integrated semantics service platform for the internet of things: a case study of a smart office. In: Sensors, pp 2137–2160 (2015)
14. Ning H, Shi F, Zhu T, Li Q, Chen L (2019) A novel ontology consistent with acknowledged standards in smart homes. Comput Netw 148:101–107
15. Kovács L, Czimás E (2018) Lightweight ontology in IoT architecture. In: IEEE international conference on future IoT technologies (Future IoT). https://doi.org/10.1109/fiot.2018.8325591 (2018)
16. Bermudez-Edo M, Elsaleh T, Barnaghi P et al (2016) IoT-Lite: a lightweight semantic model for the internet of things. In: IEEE international conference on ubiquitous intelligence & computing. https://doi.org/10.1109/uic-atc-scalcom-cbdcom-iop-smartworld.2016.0035 (2016)
17. Protégé Homepage [Online] Available. https://protege.stanford.edu/ (n.d.)
18. Ofcom. Access and Inclusion in 2018, Consumers' Experiences in Communications Markets. [Online] Available. https://www.ofcom.org.uk/__data/assets/pdf_file/0018/132912/Access-and-Inclusion-report-2018.pdf (2019)
19. Agarwal R, Gomez Fernandez D et al (2016) Unified IoT ontology to enable interoperability and federation of testbeds. In: IEEE 3rd world forum on internet of things, pp 70–75. https://doi.org/10.1109/wf-iot.2016.7845470 (2016)

Smart Transportation and Mobility

Smart Transportation: A Reference Architecture for Big Data Analytics

Camilo Castellanos, Boris Perez, and Dario Correal

Abstract Smart transportation systems (STS) play a key role in smart cities as an underlying infrastructure that supports cities' core services, such as public transport, mobility, and logistics. Information sources and analytics in STS include datasets with high volume, speed, and heterogeneity characteristics. Next-generation architectures must deal with the quality attributes of big data analytics (BDA) applications to store and process this kind of data. The combination of the extensive catalog of new technologies and multiple big data sources implies a large number of possible software solutions. Hence, organizations managing STS are facing many challenges when requiring fast adoption of these technologies, due to the lack of guidance on finding the right solution architecture. Previous works have proposed reference architectures (RA) for BDA or STS, but not both. The current STS RAs do not pay special attention to analytics capabilities over large volumes of data. This chapter presents a RA to address the challenges of big data analysis in STS, including architectural patterns and tactics. We use three case studies in accident analysis, real-time delay prediction, and mobility to illustrate and validate our proposal

Keywords Big data · Analytics · Reference architecture · Tactics · Patterns · Smart transportation · Smart cities · Intelligent transport systems

---

C. Castellanos (✉) · B. Perez · D. Correal
Department of Computing and System Engineering, Universidad de Los Andes, Bogotá, Colombia
e-mail: cc.castellanos87@uniandes.edu.co

B. Perez
e-mail: br.perez41@uniandes.edu.co; borisperezg@ufps.edu.co

D. Correal
e-mail: dcorreal@uniandes.edu.co

B. Perez
Department of Systems and Informatics, Universidad Francisco de Paula Stder, Cúcuta, Colombia

M. Khan et al. (eds.), *Smart Cities: A Data Analytics Perspective*,
Lecture Notes in Intelligent Transportation and Infrastructure,
https://doi.org/10.1007/978-3-030-60922-1_8

1 Introduction

Smart City have been shown a winning urban strategy using technology to improve the citizen's life in terms of quality of environment and services. One of the crucial fields in smart city is transportation because it enables the functioning and communication within the urban area. Smart transportation systems (STS) are essential in smart cities as underlying infrastructure that supports other services, such as public transport, mobility, and logistics [1–3]. Achieving a high level of automation in STS requires the exchange of data with increasingly intelligent systems, which improve the efficiency of infrastructure and other transport resources. According to Benevolo et al. [2], STS impacts several smart city's dimensions and the quality of life of citizens by reducing pollution, reducing traffic congestion, increasing people's safety, improving transfer speed, and reducing transfer costs. The collection, transmission, storage, and processing of bulks of STS data are crucial for uninterrupted and seamless operation of smart city [4]. Hence, new technological trends such as big data analytics (BDA) and the Internet of Things (IoT) are redefining business processes and generating new challenges in STS.

Big data is traditionally described as a set of high-volume, speed, and variety of information characteristics that imply new processing models. Analytics techniques are based on statistics and machine learning to predict trends of phenomena within a particular domain and extract valuable insights from these data. An extensive catalog of new technologies, along with multiple big data sources and diverse analytical techniques, provides a large number of options to implement BDA applications [5]. Previous studies and industry surveys [6–8] have shown challenges on BDA applications deployment which include delayed time-to-market, interoperability, immature deployment procedures, and technology heterogeneity. The complexity to design and develop BDA projects in the STS domain requires a set of software architecture guidelines and catalogs to facilitate and accelerate the adoption of these cutting edge technologies.

Reference architectures (RA) aims at guiding the implementation of software architectures within a particular domain such as STS and BDA. An RA allows the definition of good practices, provides a common vocabulary on which to discuss different implementation proposals, and it is a technology-neutral. Hence, an RA helps guide software design processes and make them efficient, effective, and interoperable [9, 10]. Although previous works have proposed RAs for BDA [5, 11–15] or STS [16–23], all of these works address each domain separately, so an integrated RA is required to have an applied guidance focused on this context. Therefore, smart cities yearn to discover new realms and appealing strategies to deal with its Big Data generation and analysis. As a result, organizations which manage STS demand smooth adoption of scalable, intelligent, and low-latency solutions to extract value from big data; however, there is a lack of guidance to design and develop them.

This chapter aims to bridge this gap by proposing a RA called ArcBDA-STS to address the architecture challenges of BDA in STS. Our proposal includes functional, integration, and technology selection views, along with a catalog of architectural patterns and tactics. The contributions of this chapter are summarized as follows:

- A review of software and reference architectures for BDA and STS.
- A proposal of a reference architecture for BDA in STS.
- A catalog of BDA patterns and tactics in the context of STS.
- The validation of this proposal feasibility and usability with three case studies.

The rest of the chapter is organized as follows: Sect. 2 reviews reference architectures for STS and BDA. Section 3 proposes the RA for BDA in STS. Section 4 summarize patterns and tactics used in this domain. Section 5 describes the case studies. Finally, Sects. 6 and 7 present discussion and conclusions.

2 Related Work

Bass et al. [24] define an RA as the set of functionalities of a known domain mapped to software elements, along with their information flows. An RA allows the definition of good practices, provides a common vocabulary on which to discuss different implementation proposals, and it is a technology-neutral guide for the implementation of software architectures within a particular domain. Angelov et al. [9] classify RAs based on the following dimensions: (i) Goal, *standardization* of software architectures, or *facilitation* of implementation by providing inspiration and design guidance. (ii) Context, organizations that design or use it, i.e. *single* or *multi-organizations*. (iii) Timing, *preliminary* when the technology demanded its application does not yet exist in practice, or *classical* when the artifacts already exist, and they are tested in practice. (iv) Level of detail, the RA can be *detailed*, *semi-detailed* or *aggregated*. (v) Level of formality, regarding the language or notation used the RA can be defined *informally*, *semi-formally*, or *formally*.

In the literature review conducted, RA on "Smart Transportation" were not found, but on intelligent transport systems (ITS). Besides, previous RA proposals independently describe smart cities' capabilities, ITS, internet of things (IoT), or big data analytics, but a specific and integrated RA for big data analytics in STS has not been proposed. In the rest of this section, the previous proposals are summarized.

2.1 Reference Architectures for Intelligent Transport Systems and Smart Cities

RAs for ITS have been proposed for different goals, and at different levels of abstraction. The ISO standard 14813-1:2015 [16] defines the ITS reference model, service domains, groups, and services as a framework to develop ITS architectures and deployments. Osorio et al. [17] present an RA focused on intelligent IT-based infrastructure in a collaborative network offering ITS services. The US Department of

Transportation defines the RA for Cooperative and Intelligent Transportation (Arc-IT) [18], which states standard foundations for practitioners to design, implement, and deploy ITS. Arc-IT comprises enterprise, functional, and communication views to separated these specific concerns. Datex II [19] is the European standard for ITS information exchange. The Transportation Association of Canada's ITS Architecture [20] is a framework for planning, defining, and integrating ITS, defining component interactions, and data exchange requirements. Passchier and Van Sambeek [21] present the Dutch Cooperative ITS RA (C-ITS) with a focus on how standards can be applied in existing legacy traffic management systems, and on emerging ways of cooperation between stakeholders. The European ITS Framework Architecture (FRAME) [22] contains high-level requirements and use cases for ITS applications to be used as a reference by ITS architects, but a technology catalog is not included. Clement et al. [23] proposes a Service-Oriented RA for smart cities, but in their proposal, smart transportation is superficially addressed within the infrastructure domain.

These proposals focus on the capabilities offered by ITS, smart cities, and on the interoperability between their subsystems. According to the classification by Angelov et al. [9], most of these RAs are classical and aim at ITS standardization for multiple organizations with varying levels of detail and formalization to use.

2.2 *Reference Architectures for Big Data Analytics*

There are several RA proposals for technology-neutral big data analysis and others as part of the technology providers' product portfolio. The United States Institute of Standards and Technology [11], to reach a consensus on RA of big data, establishes the Big Data Public Working Group, which brings together academia, industry, and government for the definition of an RA of big data interoperability. Sawant and Shah [12] seek to answer the question of how to select the best architecture in the context of big data. Chan's [13] architecture integrates traditional analytical solutions for data warehousing with new analysis strategies for big data, including details of some technologies to generate decision-making capabilities. Geerdink [5] develops in his work, an RA, as an innovation guide using big data and predictive analytic techniques to improve organizations' performance. Ramesh [14] tackles an RA for big data from the value that analytics should generate to the business, and not from the costs associated with the underlying infrastructure. Weyrich and Elbert [15] review RAs evolution of Internet of Things and how they impact industry.

The different proposals agree on identifying phases of data ingestion, debugging, analytics, and visualization to exploit the raw data and add value to the business. Since most technologies and concepts of big data are still discussed, there is no consensus on the definition of RA on big data. Therefore, and according to Angelov et al. [9], these RAs can be classified as classic, being the NIST the most advanced in the way of seeking a unified vision among suppliers, government and academia. Due to this situation, the reviewed RAs aim to facilitate the implementation of solutions in multiple organizations with varying levels of detail and formalization.

After reviewing the literature, we found current RAs for STS do not address the management of scalable infrastructure to store and process large volumes of data, nor do they describe the analytical capabilities required for the extraction of valuable information for decision making in smart transportation. On the other hand, RAs for BDA are designed in terms of capabilities to support the 3Vs, but they are too general to cover all industries, without considering particular conditions of STS and their interoperability requirements. These shortcomings are the gap this proposal aims to bridge.

3 ArcBDA-STS: A Reference Architecture for Big Data Analytics in Smart Transportation

This Section presents ArcBDA-STS, an RA for BDA in STS, to guide the planning, development, and deployment of applications in this domain. According to Angelov et al. [9], ArcBDA-STS could be classified as classic, and its purpose is to facilitate the implementation of big data analysis solutions for multiple transport organizations. ArcBDA-STS comprises detailed views and catalogs, including the definition of candidate technologies and integration protocols. ArcBDA-STS includes the stages of data acquisition, storage, extraction, processing, and exploitation aligned to STS domain services, offering an integrated view and guidance for a transit authority to use in building big data analytics solutions.

RAs are described by views, each one representing specific systems' concerns. These views are of particular interest to a stakeholder and facilitate the communication and discussion [24]. In particular, the views included in ArcBDA-STS are functional, technology, and integration.

The functional view, see Fig. 1, presents services (set of common-features), connectors (communication between services), and service groups. The *Presentation Zone* (A) defines how the services provide information through channels. The zone (B) details the *Backend Services*, which gather the reviewed RAs of STS, along with integration and geographic information services (GIS). GIS services are explicitly included due to their relevance to the STS sector.

The *Analytics Zone* (C) indicates the types of analysis that can be performed, which are supported by the services exposed in the *Data Landing Zone* (D) and *Analytical Persistence Zone* (E). *Analytical Persistence Zone* offers services such as distributed file systems, SQL, and Non-SQL storage. Finally, the *Ingestion Zone* (F) has the responsibility of collecting external data sources and redirecting them to the corresponding consumers.

The integration view is based on the Trivadis Integration Architecture Blueprint [25] to describe each type of connector (numbered from 1 to 11 in Fig. 1) of the transport, distribution, mediation and application layers. Figure 2 details connector

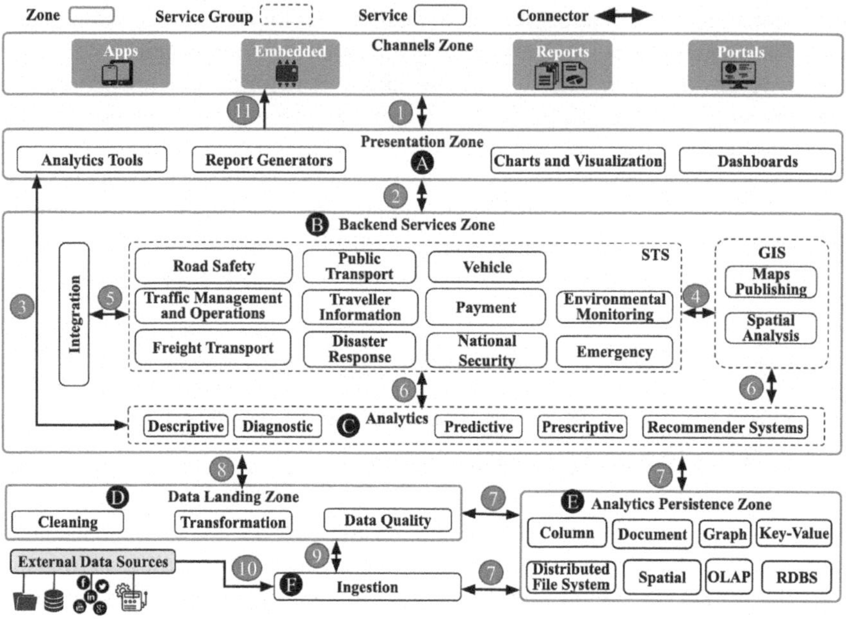

Fig. 1 ArcBDA-STS—functional view

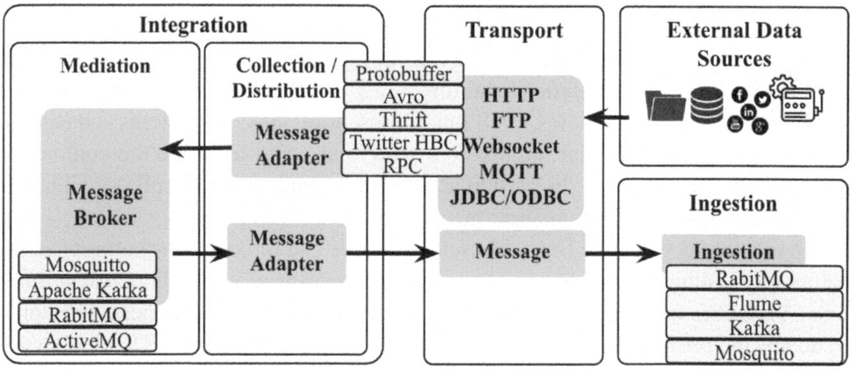

Fig. 2 Integration view of connector type 10

type 10, describing the integration between external data and the ingest service. This connector details the protocols recommended to integrate external sources in the integration and transport layer such as HTTP, FTP, WebSocket, and MQTT. These protocols have to be selected regarding communication schema (connection-oriented or Pub/Sub) and their endpoints (web application, database, IoT device, FTP server, etc.).

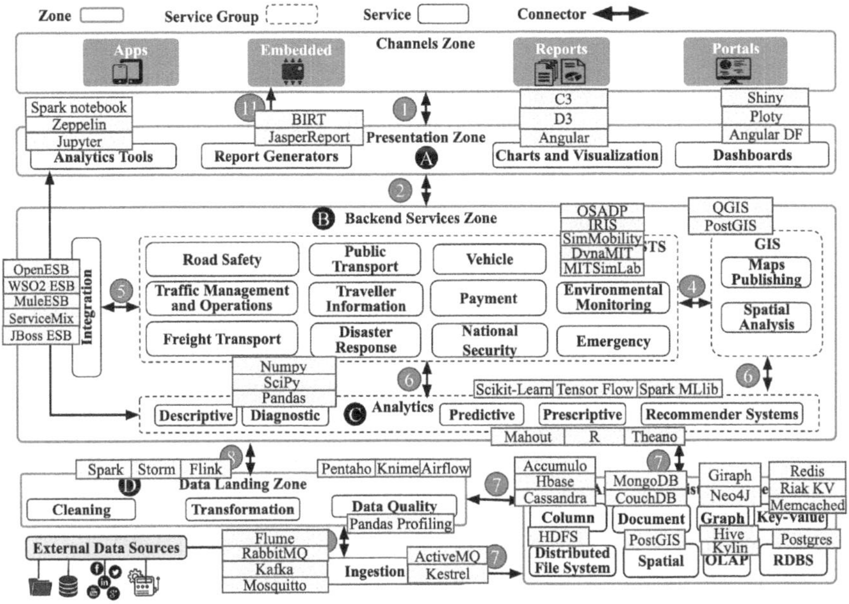

Fig. 3 ArcBDA-STS—technology selection view for open source

The technology selection view provides a list of recommendations on specific products that offer the capabilities required by a service or group of services to be instantiated. This view works as a guide in the implementation of specific solution architectures. Figure 3 details the technology view for open source technologies; other versions of this view can focus on specific vendor products.

4 Architectural Tactics and Patterns Catalog for BDA

An RA guides the software architecture design to fulfill the system quality attributes, and this design is built on building blocks called patterns and tactics. Architectural patterns are well-established packaged strategies for solving some recurrent problems. Tactics are design decisions that influence the achievement of a quality attribute response such as performance, scalability, availability, modifiability, and security [24]. In this section, catalogs of patterns and tactics are compiled to support a systematic and guided architecture design. These catalogs are not intended to be comprehensive, but a representative collection of key and common architectural building blocks found in the literature, which is especially applicable in BDA software on STS.

4.1 A Catalog of BDA Tactics

Previous works have reviewed and proposed a set of tactics that are highly applied to BDA systems [24, 26], or designed specifically for big data cybersecurity [27] and big data databases [28]. Hence, in this section, we identify and classify these tactics in a unified catalog. Table 1 summarize tactics to achieve the quality attribute(s), and their source references. This catalog is a distilled and synthesized collection of proposed tactics for the BDA specific-domain. We have not included tactics for security and modifiability deliberately because we consider them cross-cutting decisions for overall systems. Regarding these considerations, the collected tactics are described below, but more information can be found in the respective reference.

Table 1 Tactics for BDA

Tactic	Quality attributes	References
Increase resources	Performance	Bass et al. [24, 26]
Introduce concurrency	Performance	Bass et al. [24, 26]
Maintain multiple copies of data	Performance	Bass et al. [24, 27]
Maintain multiple copies of computations	Performance	Bass et al. [24, 26]
ML algorithm optimization	Performance	Ullah and Babar [27]
Unnecessary data removal	Performance	Ullah and Babar [27]
Feature selection and extraction	Performance	Ullah and Babar [27]
Parallel processing	Performance	Rozanski and Woods [26, 27]
Reuse previous results	Performance	Rozanski and Woods [26]
Result polling and optimized notification	Performance	Ullah and Babar [27]
Data cutoff	Performance	Ullah and Babar [27]
Shard data set across multiple servers	Performance	Klein and Gorton [28]
Asynchronous processing	Performance, Scalability	Rozanski and Woods [26]
Load balance across replicas	Performance, Scalability	Klein and Gorton [28]
Relax transactional consistency	Performance, Scalability, Availability	Rozanski and Woods [26]
Automatically rebalance data across nodes	Performance, Availability	Klein and Gorton [28]
Monitor data ingestion	Availability	Ullah and Babar [27]
Automatically maintain cluster membership list	Scalability, Availability	Klein and Gorton [28]
Elastic resources provision	Scalability	Klein and Gorton [28]
Combine multiple analytics models	Accuracy	Ullah and Babar [27]

- *Increase resources*: The scaling up or out can reduce latency by including faster processors, additional processors, additional memory, and faster networks [24].
- *Introduce concurrency*: Concurrency can be introduced to reduce blocked time by processing on different threads, increasing the overall throughput [24].
- *Maintain multiple copies of data*: It involves keeping copies of data with different access speeds. Data replication implies keeping separate copies to reduce conflicts over concurrent accesses [24, 27].
- *Maintain multiple copies of computations*: Multiple processors to reduce the contention of computation compared to single server [24].
- *ML algorithm optimization*: Selecting the machine learning algorithm that is most efficient or tunes the algorithm to achieve a lower computational complexity [27].
- *Unnecessary data removal*: Remove data, records, or features that not contribute to the analytics goal, e.g.., duplicated data or data with low variance [27].
- *Feature selection and extraction*: Select and extract data characteristics that contribute more to explain the target variable [27].
- *Parallel processing*: Split large and lengthy processes into subprocesses, and consolidate them into a single result when the last subprocess is complete [26, 27].
- *Reuse previous results*: Reuse the results of the costly processing, by caching and recycling them whenever needed in the following computations [26].
- *Result polling and optimized notification*: The mapper node notifies the reducer as soon as values change to a defined threshold, and accordingly, the mapper forwards the updated results [27].
- *Data cutoff*: Use data samples to reduce the size of the dataset, improving the overall system performance [27].
- *Shard data set across multiple servers*: Sharding improves the system's capacity by splitting the data set over a cluster, reducing the amount of data stored on each node, and improving performance by reducing I/O contention [27, 28].
- *Asynchronous processing*: Improve response times using asynchronous processing to reduce resource contention [26, 28].
- *Load balance across replicas*: Create multiple replicas of each object improve read performance by allowing read requests to be balanced across multiple servers, and readers can access a closer copy, reducing read latency.
- *Relax transactional consistency*: The need for immediate system-wide consistency can be relaxed by allowing parts of the system to receive updates at slightly different times to be consistent eventually [26].
- *Automatically rebalance data across nodes*: Automatically reallocate the data across the nodes according to the number of available nodes, reassigning or resizing data shards [28].
- *Monitor Data ingestion*: Monitor the flow of streaming data to detect and block traffic peaks to prevent system crashes [27].
- *Automatically maintain cluster membership list*: A gossip protocol to manage the cluster membership list automatically that provides an eventual consistency of cluster across all nodes [28].

- *Elastic resources provision*: Allocate and deallocate nodes based on the computing demand enables infrastructure to fit dynamically with fluctuating workloads in a cost-effective way [28].
- *Combine multiple analytics models*: Integrate the results of multiple analytic models to reduces false positives and increase the accuracy [27].

4.2 A Catalog of BDA Architectural Patterns

The difference between architectural style and patterns is a still open discussion. Aligned to previous works [23, 29], this paper also uses the term "Architectural Pattern" as a particular recurring design problem that arises in specific design contexts and presents a well-proven architectural generic scheme for its solution. Previous works [29–33] have proposed BDA patterns, and we have compiled them in this catalog summarized in Table 2, specifying the pattern and the problem to be addressed. The architectural patterns are described below, and information can be found in the respective reference.

Table 2 Architectural patterns for BDA

Pattern	Problem	References
Lambda	Combine stream and batch processing	Marz and Warren [30]
Kappa	Combine stream and historical data processing	Kreps [31]
Poly storage	Persist high-volume, high-velocity and high-variety data	Thomas et al. [32]
Data refinery	Integrate relational and no-relational data sources	Cervantes and Kazman [29]
Data lake	Store multi-structured raw data centrally	Miloslavskaya and Tolstoy [33]
Interactive query engine	Ad hoc analytics queries over big data	Cervantes and Kazman [29]
Distributed message broker	Decouple producer and consumers with high availability and reliability	Cervantes and Kazman [29]
Data collector	Collect high volume, speed, variety datasets from different formats	Cervantes and Kazman [29, 32]
Big data pipeline	Break out complex processing operations	Thomas et al. [32]

- *Lambda Architecture*: Combine batch processing and stream processing applied to unbound datasets to obtain real-time views and aggregated views with higher accuracy in the long term. Hence, this pattern includes speed, batch, and serving layers [30].
- *Kappa Architecture*: This is a simplification of Lambda without the batch layer, reusing the speed layer to reprocess historical data with a longer retention policy [31].
- *Poly Storage*: This pattern stores data with high volume, velocity, variety enabling availability for streaming, and random access [32].
- *Data Refinery*: This pattern cleans and structures non-relational data to load it in a relational data warehouse for further integration and analysis [32].
- *Data Lake*: A massively scalable repository that stores raw data centrally in their native format to be used by different processing/querying clients [33].
- *Interactive Query Engine*: This pattern executes distributed batch and interactive analytic queries over high-volume data [29].
- *Distributed Message Broker*: Intermediary connector which offers high scalability by distributing messages along the cluster using a publish/subscribe approach [29].
- *Data Collector*: This pattern ingests, transforms, and delivers data for later use, supporting a wide variety of formats, velocities, and sources [29, 32].
- *Big Data Pipeline*: This is a compound pattern that comprises ingress, processing, and egress of big data integrating multiple sources and stages [32].

5 Case Studies

This section illustrates and validates the feasibility and usability of ArcBDA-STS with three BDA case studies and their concrete architectures framed in smart transportation: CS1, CS2, and CS3. Table 3 details each case study along with tactics and patterns applied from the catalogs described in Sect. 4. CS1 is currently used by the Transit Authority of Bogotá Colombia to analyze road accidents and their relationship with mobility, and traffic tickets. The datasets used in case studies 2 and 3 are open data, so they are publicly available. CS2 is a near real-time analytical application to predict bus delays in the public transportation system of Vancouver, Canada. In the last case, CS3 addresses smart mobility analysis regarding pollution metrics collected by sensor networks in Aarhus, Denmark. The following sections will detail each case study, including concrete software architectures, and their components mapped to ArcBDA-STS service zones (letters in black circles) to validate their compliance. In addition, each concrete architecture component specifies the open-source technologies used to implement it according to the technology selection view.

Table 3 Case studies with applied tactics and patterns

Code	Description	Tactics	Patterns
CS1	Accidents analysis	Unnecessary data removal, feature selection and extraction	Data refinery, poly storage
CS2	Bus delay prediction	parallel processing, shard data set across multiple servers, reuse previous results,	Lambda architecture, interactive query engine, distributed message broker
CS3	Shortest paths based on pollution	Unnecessary data removal, feature selection and extraction, parallel processing, shard data set across multiple servers	Data collector, data refinery

5.1 CS1: Accidents Analysis in Bogotá, Colombia

A web application was implemented to offer a descriptive and predictive analysis of accidents and the monitoring of public transportation operation in the transit authority of Bogotá. Accidents module analyzes historical accident data along with their relationship with traffic tickets to support decision-making on road safety. The operation monitoring module offers the historical tracking of buses in the public transport system to understand mobility patterns. These modules are integrated within a dashboard that allows users to interact with accident history, bus trips, and their dimensions such as geolocation, type, vehicle, date, route. The predictive analysis included the calculation of accident risk indexes by road segment and correlations of accidents and traffic tickets associated with the ticket's cause. As a result, the transit authority of Bogotá can now analysis historical data to discover insights, define safety policies supported in this data, and measure the results of these decisions.

Figure 4 describes the concrete software architecture, which was designed following ArcBDA-STS, and Table 4 details the mapping between components and ArcBDA-STS services. Accidents, traffic tickets, and road networks are the external data sources loaded by the AccidentsETL component, which is part of service zones *F* and *D* of ArcBDA-STS, and is implemented using Python, Pandas, and PostGIS. In terms of volume and variety, data were composed of structured data such as traffic tickets (1,598,199 records), accidents (2,385,000 records), and bus GPS traces (263,937,822 records), and also road network (137,962 segments) in a graph network. These data sources are cleaned, filtered, integrated, and stored in batch in a MongoDB database once a new dataset arrives, these steps correspond to service zone *E*. Besides, historic bus operation data are loaded and integrated by the component OperETL implemented in Spark in zones *F* and *D*. The resulting aggregated operational data is also stored in MongoDB. In the zone *B*, the components AccidentBackend and OperBackend access and aggregate pre-processed

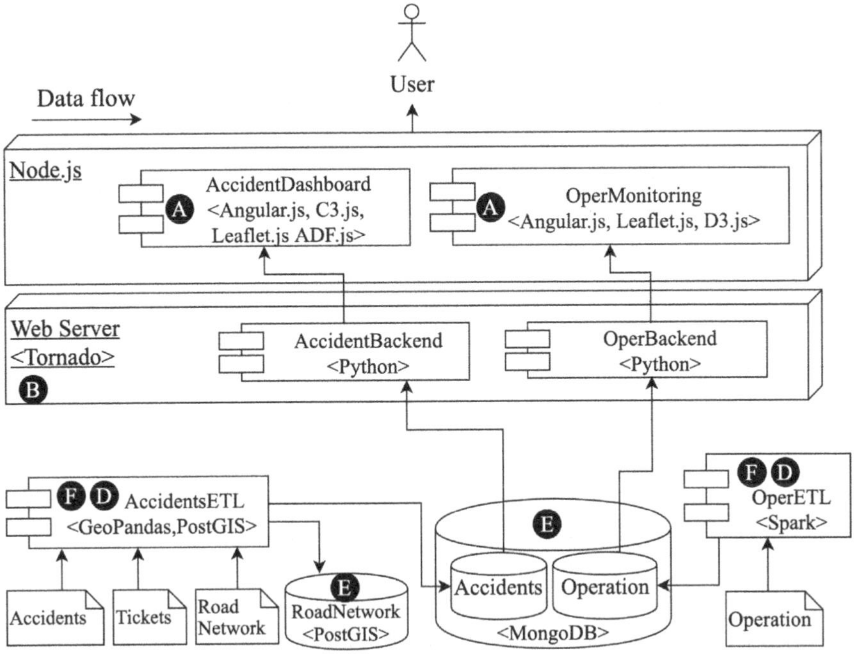

Fig. 4 CS1: accident analysis in Bogotá

Table 4 CS1 components mapped to ArcBDA-STS functional view

Components	Zone	Service
AccidentsETL	F, D	Ingestion, cleaning, transformation
OperETL	F, D	Ingestion, cleaning, transformation
Accidents, Operations	E	Document store
RoadNetwork	E	Spatial DB
AccidentBackend	B, C	Descriptive analysis in road safety
OperBackend	B, C	Descriptive analysis in traffic management and Ops.
Accident and operation UI	A	Dashboards visualization

data of accident and operations using Python, and then expose analysis services via REST consumed by frontend components. Finally, frontend components Accident-Dashboard and OperMonitoring are built within a dashboard using technologies such as Angular Dashboard Framework (ADF), AngularJS, C3, D3, and Leaflet within service zone *A*.

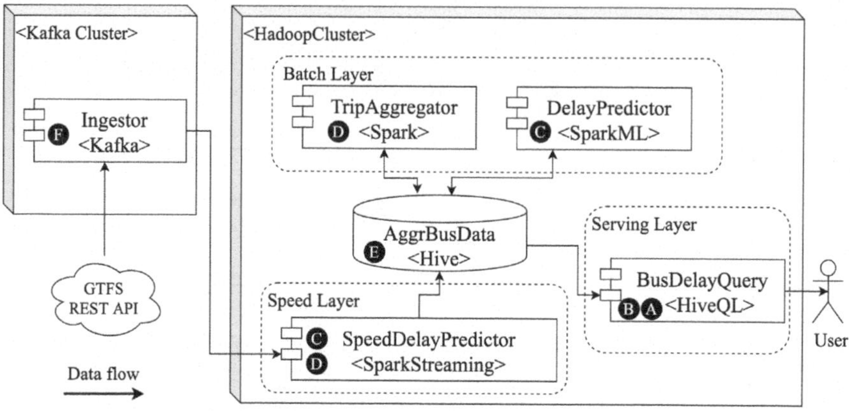

Fig. 5 CS2: delay prediction at Vancouver

Table 5 CS2 components mapped to ArcBDA-STS functional view

Components	Zone	Service
Ingestor	F	Ingestion
SpeedDelayPredictor	D, C	Transformation, predictive analysis in traveler information
TripAggregator	D	Transformation
DelayPredictor	C	Predictive analysis in traveler information
AggrBusData	E	OLAP
BusDelayQuery	B, A	Traveler information, interactive analytics tools

5.2 *CS2: Bus Delay Prediction in Vancouver, Canada*

The case study 2 implements a Lambda architecture to analyze near real-time data of bus trips in the public transport system of Vancouver.[1] A detailed description of this case study implemented in a Software as a Service model can be found in [34]. Trip updates are collected and aggregated by bus route to calculate the delay average per time window, five minutes. Data processing has low-latency analytical constraints and limited accuracy due to the restriction to micro-batches of data. On the other hand, batch processing, although it allows building more robust analytical models, does not offer adequate response times. This application allows potential users to have a more accurate prediction of the bus delays to better plan their trips, and the transport authority better operation management.

A Lambda architecture was designed to be compliant with ArcBDA-STS as detailed in Fig. 5, and Table 5 details the mapping between components and ArcBDA-

[1] https://developer.translink.ca.

STS services. The Vancouver real-time travel update data were delivered in GTFS format (General Transit Feed Specification). Each GTFS trip update is generated every 60 s and processed in the pipeline downstream. Each trip update is a binary protobuffer file (56KB on average), which has to be deserialized to JSON format. Trip update feeds are consumed by an ingestion component using Apache Kafka (service zone *F*) to provide low latency and high availability. In the speed layer (zone *C*), the delay predictor component consumes data from the Kafka topic, to aggregate and predict the delay expected per time window using Spark Streaming and storing the speed views in HDFS. Besides, each raw trip update is stored in HDFS to be further processed by the batch layer. The batch layer was implemented using Spark and SparkML Lib, where TripAggregator component (zone *D*) processes historical data to be stored in HDFS. DelayPredictor (zone *C*) estimates the expected delay per time window and bus route, storing the outputs as batch views in HDFS. In the serving layer (zones *B* and *A*), both speed and batch views are integrated using interactive queries by the users through a query component using HiveQL.

5.3 CS3: Smart Mobility in Aarhus, Denmark

Smart mobility is a core concept in smart city strategies, and it supports the management of traffic jams in cities with increasingly high urbanization rates. To illustrate our proposal's generalization to other smart mobility case studies, we align ArcBDA-STS to the architecture presented in [35], which analyzes road traffic and pollution using sensors data. The datasets range from pollution, road traffic, weather, to parking. Road traffic and pollution were collected every five minutes from semi-structured datasets in CSV format Open Data Aarhus. A graph is built with the sensors network dataset to calculate the shortest paths between locations. Then, the least polluted paths are calculated based on the pre-calculated shortest paths and pollution data collected by sensors network using a Hadoop cluster. This application enables final users to select healthier paths, and transport authorities to monitor the environmental variables to improve informed decision-making.

Figure 6 details the software architecture CS3, as an adaptation of the one presented in [35], and Table 6 details the mapping between components and ArcBDA-STS services. Data sets of parking, weather, road traffic, and pollution are ingested and pre-processed by a file loader implemented in R (zone *F*). These pre-processed data are taken by two components (zone *D*) to build a graph network and calculate the shortest paths in HDFS, zone *E*. These shortest paths are processed by Hadoop to calculate the healthiest path between two points, zone *C*. In Shiny, backend components AnalysisServer and SmartMobilityServer consume the pre-processed and healthiest paths data to be delivered to UI components. These backend components are located in the services zone *B*) of ArcBDA-STS, specifically, they are mapped to *Traveller Information*. Regarding presentation zone *A*, AnalysisUI, and MobilityUI components provide a dashboard within a Shiny server.

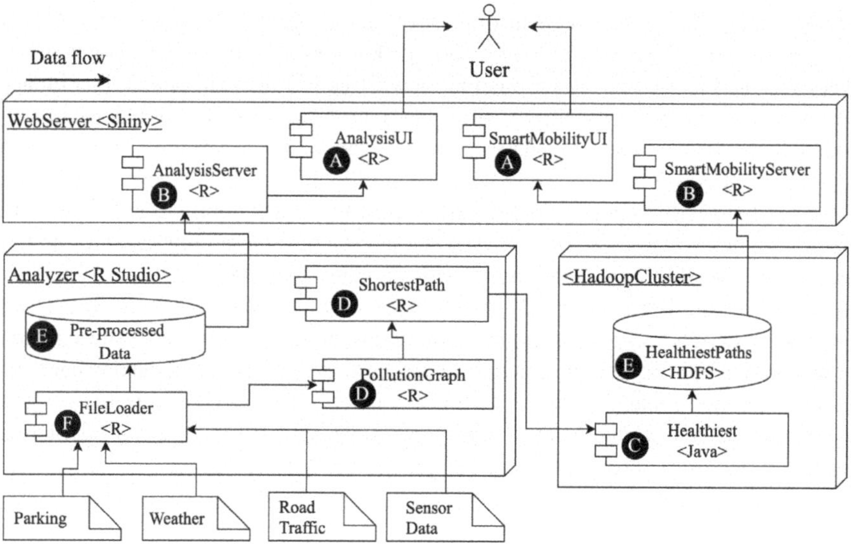

Fig. 6 CS3: smart mobility in aarhus

Table 6 CS3 components mapped to ArcBDA-STS functional view

Components	Zone	Service
FileLoader	F	Ingestion
PollutionGraph	D	Cleaning, transformation
ShortestPath	D	Transformation
Healthiest	C	Descriptive analysis
HealthiestPaths, Pre-processed data	E	Distributed file system
AnalysisServer, SmartMobility	B	Traveler information, environmental monitoring
AnalysisUI, SmartMobilityUI	A	Dashboard visualization

6 Discussion

Big data analytics projects present particular challenges, such as communicating with business users to identify objectives, selecting from a wide variety of novel tools, and using analytical techniques that require specific knowledge. These challenges imply steep learning curves that delay the construction of these types of solutions. In this work, RA proposals were reviewed, which allowed summarizing the concerns and expected capacities in STS and BDA common to different authors. Although previous works have proposed reference architectures for ITS and BDA separately, we argue that those proposals lack of detailed integration blueprint, technology, patterns, and tactics catalogs to guide the development of these projects in practice. Our proposal aims to tackle these challenges by offering a reference architecture, technology blueprints, catalogs to guide and facilitate these projects, and their application using three case studies.

The BDA tactics found in the reviewed literature are mainly focused on performance, scalability, and availability to tackle big data characteristics. These catalogs have been used in the reviewed case studies to tackle quality attributes by rebalancing data and computation across clusters and introducing concurrency and parallelism. Patterns catalog facilitated the reuse of known solutions to recurrent problems in combining stream and batch processing, collecting and storing multi-structured data, and decoupling producer and consumers. Smart cities projects with a strong requirement of IoT applications could require specific tactics and patterns in this domain. Therefore, extensions of this catalogs will be needed, hence the iterative nature that we assign to our proposal.

Our case studies have shown that ArcBDA-STS can cover a wide range of BDA applications in the STS domain by instantiating concrete architectures, patterns, and tactics. Although the technology selection view is not exhaustive, it has been sufficient to support all technologies used in these case studies, but other projects can also require other open-source technologies. Hence both technologies and catalogs are incremental artifacts. In addition, these case studies have demonstrated how smart transportation can positively impact the lives of smart cities' citizens by providing healthy, safe, and efficient routes for travelers, and offering data tools to support the transport authority's decision-making. Open data and geographic information are key elements to foster BDA applications in STS, as shown in presented case studies, so public policies in these directions can accelerate the adoption of smart city strategies.

7 Conclusions

In this chapter, we have presented ArcBDA-STS, a reference architecture to guide the design and deployment of BDA applications in STS. ArcBDA-STS guided the definition and structuring of services in this domain and their integration. This proposal also facilitated the design and development of smart transportation architectures by offering patterns, tactics, and technology catalogs. The use of this proposal has been illustrated in three case studies to validate its feasibility and usability: accident analysis, bus delays prediction in public transportation, and the healthiest path estimation. Despite the application in the presented case studies, further experimentation is required to validate its generalization. The impact of STS in transport mobility, safety, and operation management in smart cities has been discussed, and their improvement using data analytics in the case studies has been demonstrated.

Finally, additional research can address other STS domains such as electronic payment or shipping using ArcBDA-STS to extend and refine it. On the other hand, this proposal does not address concerns such as security and governance, which are cross-cutting issues in any solution, but future projects could consider specific implications in BDA applications.

Acknowledgements This research is supported by Fulbright Colombia and the Center of Excellence and Appropriation in Big Data and Data Analytics (CAOBA), supported by the Ministry of Information Technologies and Telecommunications of the Republic of Colombia (MinTIC) through the Colombian Administrative Department of Science, Technology, and Innovation (COLCIENCIAS) within contract No. FP44842-anexo46-2015.

References

1. Schlingensiepen J, Nemtanu F, Mehmood R, McCluskey L (2015) Autonomic transport management systems: enabler for smart cities, personalized medicine, participation and industry grid/industry 4.0. Studies in Systems, Decision and Control. Springer International Publishing AG, Switzerland, pp 3–35
2. Benevolo C, Dameri RP, D'auria B (2016) Smart mobility in smart city. In: Empowering organizations, vol 11, pp 13—-28
3. Lom M, Pribyl O, Svitek M (2016) Industry 4.0 as a part of smart cities. In: 2016 smart cities symposium prague (SCSP), pp 1–6
4. Silva BN, Khan M, Han K (2018) Towards sustainable smart cities: a review of trends, architectures, components, and open challenges in smart cities. Sustain Cities Soc 38:697–713
5. Geerdink B (2013) A reference architecture for big data solutions introducing a model to perform predictive analytics using big data technology. In: 8th international conference for internet technology and secured transactions (ICITST-2013), pp 71–76
6. Chen H-M, Kazman R, Matthes F (2015) Demystifying big data adoption: beyond IT fashion and relative advantage. In: Twentieth DIGIT workshop, Texas, US, pp 1–14
7. Castellanos C, Pérez B, Varela CA, Villamil MDP, Correal D 2019 A survey on big data analytics solutions deployment. In: Bures T, Duchien L, Inverardi P (eds) Software architecture. Springer International Publishing, Cham, pp 195–210
8. Rexer K, Gearan P, Allen H (2016) 2015 data science survey. Technical report, Rexer Analytics
9. Angelov S, Grefen P, Greefhorst D (2012) A framework for analysis and design of software reference architectures. Inf Softw Technol 54(4):417–431
10. Martínez-Fernández S, Medeiros Dos Santos PS, Ayala CP, Franch X, Travassos GH (2015) Aggregating empirical evidence about the benefits and drawbacks of software reference architectures. In: 2015 ACM/IEEE international symposium on empirical software engineering and measurement (ESEM), pp 1–10
11. N. I. of Standards and Technology, Nist big data interoperability framework: reference architecture. https://bigdatawg.nist.gov
12. Sawant N, Shah H (2013) Big data application architecture. In: Big data application architecture Q & A. Springer, pp 9–28
13. Chan JO (2013) An architecture for big data analytics. Commun IIMA 13(2):1
14. Ramesh B (2015) Big data architecture. In: Big data. Springer, pp 29–59
15. Weyrich M, Ebert C (2016) Reference architectures for the internet of things. IEEE Softw 33:112–116
16. ISO/TC 204 (2015) ISO 14813-1:2015 - Intelligent transport systems – Reference model architecture(s) for the ITS sector – Part 1: ITS service domains, service groups and services. Technical report, International Standard Organization
17. Osório AL, Afsarmanesh H, Camarinha-Matos LM (2010) Towards a reference architecture for a collaborative intelligent transport system infrastructure. In: Camarinha-Matos LM, Boucher

X, Afsarmanesh H (eds) Collaborative networks for a sustainable world. Springer, Berlin, pp 469–477
18. US Department of Transportation, Architecture Reference for Cooperative and Intelligent Transportation
19. DATEX II, The standard for its on european roads. http://www.datex2.eu/sites/www.datex2.eu
20. Transportation association of Canada (2017) Its architecture for Canada. https://www.itscanada.ca/about/architecture/index.html. Accessed 1 May 2017
21. Passchier I, Van Sambeek M (2016) Architecture for C-ITS applications in the Netherlands. Technical report, DUTCH ROUND TABLES FOR SMART MOBILITY
22. Jesty PH, Bossom RA (2011) Using the FRAME Architecture for planning integrated intelligent transport systems. In: 2011 IEEE forum on integrated and sustainable transportation systems. FISTS 2011, pp 370–375
23. Clement SJ, McKee DW, Xu J (2017) Service-oriented reference architecture for smart cities. In: Symposium on service-oriented system engineering, pp 81–85
24. Bass L, Clements P, Kazman R (2012) Software architecture in practice. Addison-Wesley Professional
25. Schmutz G, Liebhart D, Welkenbach P (2010) Service-oriented architecture: an integration blueprint: a real-world SOA strategy for the integration of heterogeneous enterprise systems: successfully implement your own enterprise integration architecture using the trivadis integration architecture blueprint. Packt Publishing Ltd.
26. Rozanski N, Woods E (2012) Software systems architecture: working with stakeholders using viewpoints and perspectives. Addison-Wesley
27. Ullah F, Babar MA (2019) Architectural tactics for big data cybersecurity analytic systems: a review. J Syst Softw 151:81–118
28. Klein J, Gorton I (2015) Design assistant for nosql technology selection. In: Proceedings of the 1st international workshop on future of software architecture design assistants, FoSADA '15. ACM, New York, NY, USA, pp 7–12
29. Cervantes H, Kazman R (2016) Designing software architectures: a practical approach. Addison-Wesley Professional
30. Marz N, Warren J (2015) Big data: principles and best practices of scalable realtime data systems. Manning Publications Co.
31. Kreps J (2014) Questioning the lambda architecture
32. Thomas E, Wajid K, Paul B (2016) Big data fundamentals: concepts, drivers & techniques. Prentice Hall Press
33. Miloslavskaya N, Tolstoy A (2016) Big data, fast data and data lake concepts. Proc Comput Sci 88(300–305):63
34. Perez Arteaga PF, Castellanos C, Castro H, Correal D, Guzman LA, Denneulin Y (2018) Cost comparison of lambda architecture implementations for transportation analytics using public cloud software as a service. In: 13th international conference on software technologies - ICSOFT, pp 889–896
35. Zenkert J, Dornhöfer M, Weber C, Ngoukam C, Fathi M (2018) Big data analytics in smart mobility: modeling and analysis of the aarhus smart city dataset. In: 2018 IEEE industrial cyber-physical systems (ICPS)

Designing for Urban Mobility: The Role of Digital Media Applications in Increasing Efficiency of Intelligent Transportation Management System

Stephen T. F. Poon

Abstract How to improve traffic efficiency has become a challenge for the planning, design, construction of smart cities. The large-scale adaptation to digital networks and mobile platforms provide smart solutions to urban management issues such as transportation services, traffic congestion and road navigation. The aim of research is to demonstrate how digital media is effectively integrated into the implementation of efficient *intelligent transport management systems* (ITMS). Case studies analyse DiDi, a mobile app which provides transportation services such as ride-hailing and ridesharing in China's urban cities, and Amap, a location-based digital mapping solutions provider for smartphone users. Analysis shows that digital networks play a crucial economic role by producing robust, agile, interactive data that improves users' awareness of traffic conditions, from drivers to passengers and travellers. Findings demonstrate that innovative network-based features and capabilities such as interactivity and real-time response enable users to plan for smoother journeys to destinations, suggesting that urban mobility is increasingly dependent on technological solutions to provide services accessible to all. Other emerging smart cities should emulate China's urban mobility design model utilising digital platforms to reduce resource use and time wastage through better informed, better connected and more efficient interactive systems.

Keywords Digital networks · Intelligent traffic management system · Mobile platform · Smart city · Urbanisation

---

S. T. F. Poon (✉)
Asia Pacific University of Technology & Innovation, Kuala Lumpur, Malaysia
e-mail: stephentfpoon@aol.com

M. Khan et al. (eds.), *Smart Cities: A Data Analytics Perspective*,
Lecture Notes in Intelligent Transportation and Infrastructure,
https://doi.org/10.1007/978-3-030-60922-1_9

1 Introduction

China's massive urban transportation system affects hundreds of millions of people daily and has resulted in inevitable problems such as stressful living and working conditions, traffic congestions, inefficient transport, commuting safety and fuel consumption wastage.

For urban traffic management efficiency, Eduardo Quiñones from Barcelona Supercomputing Center [1] state that the adaptation of digital media technologies is a necessary factor that improves urban mobility needs based on *real-time* and *reactive* responses and systematic *analytics* of complex data (also known as Big Data), as it carries benefits such as computational analysis, storage and processing of information, for example, what drivers and passengers need to know before navigating traffic conditions in busy city centres, accessing parking or detecting obstacles and accidents in advance to improve safety, or to prevent and reduce driving risks.

The *objectives* of this paper is to explore the role of new media platforms such as mobile apps and location-based navigational networks in the implementation of ITMS; to understand how the platforms contribute to smart city design by reducing traffic congestions during peak (rush) hours and enabling smarter navigation; and to demonstrate how two-way communication flow plays a role in reducing urban mobility challenges.

This paper has two-fold significance: *first*, it frames the way urban planners redefine the relationship between two-way communication flows and daily life; *second*, it explores the potential of interactive communication platforms such as *applications* (apps) to promote and foster more efficient and sustainable urban economic systems by gratifying the transportation needs of densely populated metropolises. To achieve these research objectives, key questions to be addressed are: *How does digital and new media platforms such as mobile applications improve transportation system efficiency? What role does digital media utilisation play in the reduction of urban traffic congestion to enhance intelligent transport management systems (ITMS) for smart cities?*

2 Literature Review

Katz et al. [2] developed the classical theory that the use of media is driven by *uses* and *gratification* based on *personal needs*. Traditional communication modes of "one-to-one" or "one-to-many", are represented by unidirectional flow of information, and the passive reception or acceptance of messages. This form of communication is viewed as a hindrance to connectivity and communication.

Ruggiero [3] and Neuman [4] theorise on the core attributes of *interactivity*, *demassification*, *hypertextuality* and *asynchronicity* in digital media, regarding interactivity as a prerequisite feature of *two-way information systems*, the assumption being that immediate, multi-channel and constant information flows improves users' awareness and increases participation in social communication. Interactivity of system networks enhances social connections and use gratification. Neuman [4: 62–65] further stated that as media evolves, the volume of data transmissions increase in the form of "information explosion", as diversification of technological systems, digitisation processes and electronic integration become the new "norm", describing users' ability to access, adapt, control, interconnect, extend, filter and gain self-knowledge, essentially, the opposite of passive, unidirectional communication flows. Author-researcher Logan [5] concurs with these pioneering researchers regarding the *two-way flow of information* as an important feature of new media, represented by a basic model of "*many-to-many*", where information is generated and produced for mass audiences - coordinated, managed, and responded to, in real-time, resulting in high interactivity and connectivity in communication orientation.

Among urban researchers, debates have arisen on how urban populations use, manage and share data as *cultural resources*, similar with sharing living spaces and economic resources, and regarding information as valuable assets. Salvia et al. [6] for instance, reviewed fieldwork on the "urban fabric" of connectivity, spatial and environmental sustainability, and social consumption patterns of 30 cities. The authors found the phenomenon of "shapring" a key factor in urban social dynamics, such as in timeshare properties, bike- or ride-sharing services, co-working spaces, cloud platforms, mobile transport, and online community services. Dr Currid-Halkett [7] presents empirical evidence to demonstrate that society's dependency on *cultural capital* such as media, shared networks and online resources, similar to real estate and investments, have become important extensions of the cultural economy. At the same time, she argues that it may lead to stratification in the distribution of resources among the economically less-privileged social classes. Another theoretical position on urbanisation of global communication systems comes from cultural geography. Based on studies on urban economic growth by University of Chicago sociologist Sassen [8], designations of "world city" status among metropolises of North America, Asia Pacific and Western Europe are notably moving away from population concentration trends, to aspirational cultural advancement indexes ranked as *Alpha*, *Beta*, *Gamma* and *Emerging*, in accordance to their cultural advancements through enterprise markets, financial services and service innovations. While urban centres aspire to be ranked among global economic hierarchies, these rankings are not stagnant but updated frequently since policymaking on transportation development based on population sizes are shifting criteria in the processes of urbanisation at different junctures of social, economic, spatial and technological capacities.

Engineering professor Ahmed Abdel-Rahim [9] from The University of Idaho, explores Beaverstock et al. [10] theories in describing the cosmopolitan characteristics of world cities. Abdel-Rahim examines the role of *intelligent transportation systems* (ITS) in the rise of "full service" (*Alpha*), "major" (*Beta*), "minor" (*Gamma*) and "emerging" cities, giving them first-, second and third-tier rankings based on the extensiveness of wireless and wired communication networks working cohesively and sustainably with road grid-planning, digitisation of facilities (e.g. metro-trains, light rail transits), computerised interactions and online e-Commerce services, as well as social adaptiveness of ITS technologies and applications. Example of ranked world cities include London, New York and Paris (*Alpha*); Brussels and Mexico City (*Beta*); Kuala Lumpur and Shanghai *(Gamma)*; Mumbai and Dubai (*Emerging*).

Virginia Tech architectural and urban scholar Paul L. Knox in *Cities and Design* [11: 236] stated that *true urbanism* is a balance achieved between the complex networks (nodes) for consumption, commuting, working, living, recreation, etc. The design of urban environments are classed as "liveable" when economic functions and social experiences in urban living flourish as part of the cultural economy. In this regard, the effects of well-thought out traffic management system and efficient communication of urban designs are considered a crucial factor of liveability. Criticisms are noted too. Maynooth University's Kitchin [12] believes that increasingly pervasive, panoptic environments in developing smart city infrastructure would result in higher manipulation through technocratic governance by central authorities, echoing concerns of policymakers, innovators and developers who wish the human dimensions to be better amplified in social adaptation to technologies of the future [13]. As urban planner Sevtsuk [14] argues, for efficiency to shape metropolitan growth, reducing the number of cars may not be the easiest solution. Instead, social adaptability enables more choices in gratifying road user needs or to solve temporal conditions, such as traffic jams. Digital or mobile platforms (e.g. WeChat) resolve these issues by providing interactive methods to gain real-time information on e-Commerce services (food delivery, ticket booking), trends, local or community events. The following section reviews the development of smart systems for transport management in China's urban metropolises.

2.1 Background of Intelligent Traffic Management System (ITMS)

Urban expansion has triggered housing prices, expanding choices of suburban residential development yet adding to the burden of commuting time between workplaces and residential areas; the hour-long commute time seems an inevitable daily feature (Fig. 1).

There are three main reasons for long commuting time: increased distance, inefficient use of transport facilities and traffic congestion. The marked increase in commuting distances and time is an urban planning problem (Fig. 1), which will be discussed in the following section in addressing urban policymaking and urban migration trends. The second problem is mainly the lack of public information and awareness of traffic facilities and road conditions which result in time and resource inefficiency.

Beijing, for instance, suffered economic losses of 105.593 billion yuan caused by traffic congestion in 2010, accounting for 7.5% of Beijing's annual GDP. As traffic jams choke China's metropolitan areas during rush hour (Fig. 2), these issues' increasing impacts including environmental pollution, higher fuel and energy use, commuting time delay costs, affecting performance and productivity. In previous

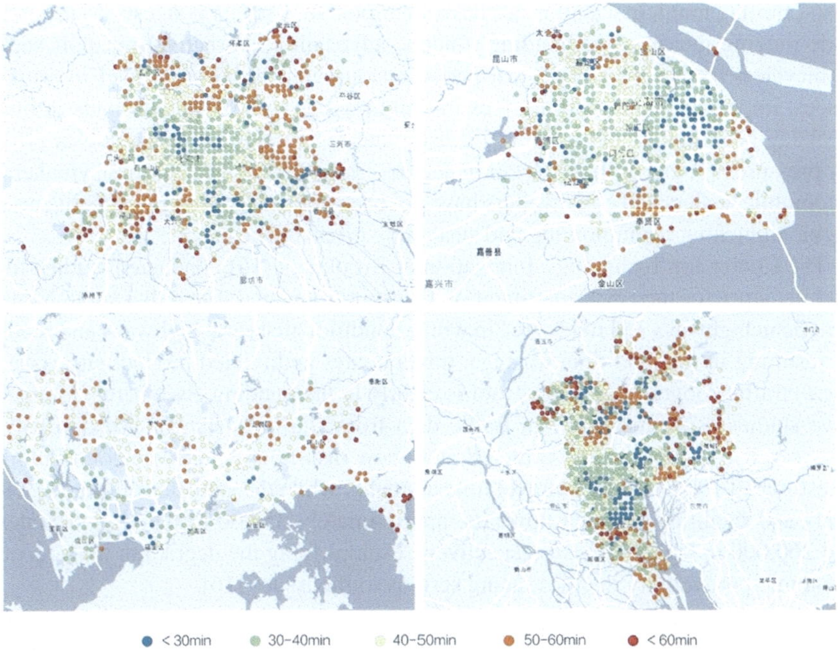

Fig. 1 Average commuting time in Beijing, Shanghai, Guangzhou and Shenzhen Nov 2017. *Intelligent Research Group* [Online]. https://wenku.baidu.com/view/3ca4e67d590216fc700abb68a98271fe910eaf3c.html [08 January 2020]

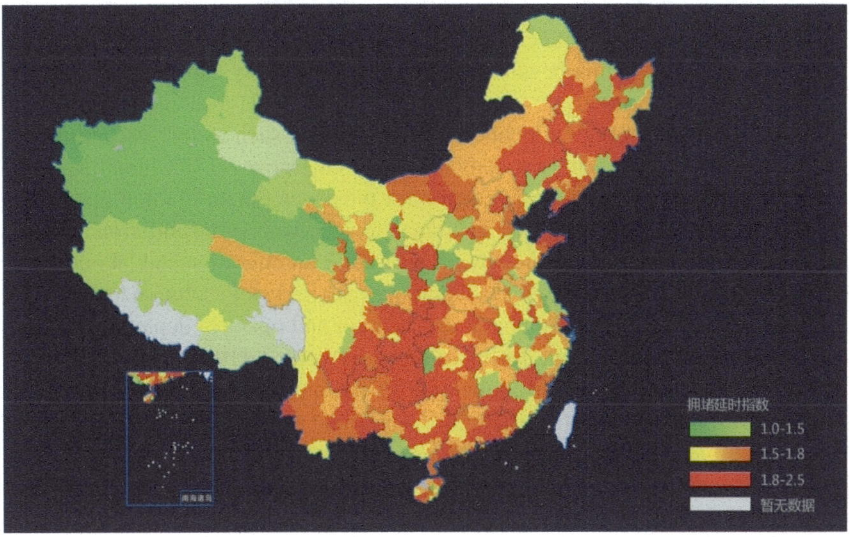

Fig. 2 Rush hour congestion in China's urban areas. DiDi (2017) *Annual City Travel Report*

years, Beijing municipal authorities have attempted to ease city congestions through wide-ranging measures as building tunnels, advocating alternative transport such as bicycles, increasing subway car-parks, encouraging proximity travel to workplaces, imposing higher parking fees, prohibiting cars on downtown roads during weekday peak hours, and limiting license plate registration of new vehicles [15]. As pressures mount on transportation sectors, stakeholders such as policymakers, automobile dealers and city dwellers have not been able to agree on the effectiveness of rule tightening for improving and sustaining urban traffic use.

The Center for Technology Innovation at Brookings [16] conducted studies for a white paper for the US Department of Transportation, and found that some North American highways and roads are "in worse condition that rural highways and roads … resulting in fuel and time wastage, travel delays and related productivity issues due to traffic congestions". The proof is clearly in the pudding elsewhere: Comparative studies using metropolitan traffic data from Beijing, Istanbul, Mexico City, Moscow, Rio de Janeiro and United States show rush-hour congestion contributing an estimated 100 h of wasted travel time annually, while the statistics of traffic accidents and fatalities spiralled upwards: approximately 35,000 in the United States and 260,000 in China in 2016 respectively, exemplifying the degrading impacts on economic productivity in cities facing serious traffic issues [16].

2.2 *Urban Transport Management Policies and Growth of Enabling Technologies in China*

As early as 2006, the Chinese government identified ITMS as the future direction of transportation sector. As the effects of the global financial crisis devastated economies across Asia in 2008, China began exploring urban mobility capabilities to transition from middle-income to developed status. This include awareness and educational campaigns encouraging technological utilisation such as live streaming, use of mobile apps for cashless transactions and online enterprises [17]. China's Ministry of Transport [18] provides policy direction for the development of intelligent transportation via frameworks establishing connections between transportation and data utilisation. This is further informed by social studies on migration and labour distribution patterns, where technological adaptation was a necessary solution to overcome serious unemployment, environmental degradation and resource depletion that saw once-bustling townships shrink, losing millions of workers [19].

Lang [20: 37] discusses the application of moratoria in impacting large-scale transportation and urban infrastructural design, and reviews the roles of council developers, project builders, planners, legal agencies and enforcement units involved. His key argument is that if roads leading to and from suburbs to downtown areas are "beyond [the city's] capacity to cope, it could implicate [the size of investments into] economic development zones".

Dr Rojas Sandford [21] observed that the large-scale flow of locals from outlying towns, as well as the influx of foreign working sectors and social migrants into cities like Beijing, Guangzhou, Shenzhen and Shanghai has made the integration of state policies governing the construction of transportation infrastructure, housing, industrial and commercial zones the intense focus of local economic policies in attaining maximum benefits of sustainable growth, while tussling with the centralised management of China's macroeconomics model.

Little wonder that the urbanisation of China has become a favourite case study for urban designers trying to find ways to move forward in rising above congestion and environmental pollution issues through advancements in data management technologies. Some domestic experts reserve judgement on such potential. Professor Long Ying at Tsinghua University [22], for instance, used inkstone indexing of night light intensity to map economic activity in cities. He cautions that the "breakneck" urbanisation scenario in 3,300 Chinese metropolises may not viably support ITMS if city planners continue to "detach local council ambitions" from real demographic trends showing shrinking town populations dependent on few resources, low-birth rates, aging populations and less-agile technological adaptiveness. Regardless of these issues, the design of *intelligent transport management systems* (ITMS) is interlinked with the development of smart cities.

3 Research Method and Materials

The methodology of research for this paper will consider two cases in a qualitative review, applying the two-way flow communication model described in literature. Case study research in urban design studies is a challenging yet opportunistic undertaking.

As Kim et al. [23] state, the qualitative review of each case presented adds value to researchers working in urban design planning, particularly for ITMS policymaking. Qualitative methods enable a range of urban mobility patterns and issues to be understood, and how positive and negative impacts of social, economic and environmental issues affect ITMS implementation, while the connections between big data capabilities, social adaptation and real-time response to two-way information flow systems in populated urban areas can be critically examined.

Theories of urban mobility provide urban researchers a descriptive, qualitative dimension to demonstrate the advantages of digital media technologies such as real-time data and how its relationship with city traffic management facilitates more efficient transport services. It also focuses the analysis on specific regional situations to understand the broader issues involved in successful implementation of ITMS and its challenges based on the economic development and communication needs of populations facing daily urban mobility challenges.

Advanced machine-to-machine technologies behind *Artificial Intelligence* (AI), cloud-based computing and the *Internet of Things* (IoT) has changed human conceptual understanding and perceptions about distances in space through deep learning modes of intelligent systems [13]. The increase in the volume, density, frequency, and flow of communication enables increased speeds of real-time interaction and connections in urban areas. Data such as satellite images have long been in use to monitor meteorological, vehicular, traffic conditions and public transportation network, but machine-generated data must be integrated effectively with human-generated data to solve problems of urban transportation.

Clearly, the evolutionary and cumulative impacts of ITMS need to be studied via relevant case examples from the field and critical solutions found, as they reflect on the ability of nations to achieve balanced socioeconomic growth by providing optimal utilisation of facilities such as transportation, infrastructure, traffic management, utilities and information to enable the efficient movements of goods and people in the shortest period of time [24].

For urban mobility systems to work effectively, the efficiency of ITMS must support population growth. Empirical evidence and analysis play its role to demonstrate the potential of digital media networks and online platforms such as mobile platforms. Research which demonstrate its advantages in regards efficiency factors such as timeliness, proximity and seamless connectivity should also be quantified and presented [25].

4 Discussion of Case Studies

In this inductive thematic analysis, the development of ITMS potential in China's urban cities will factor in *functionality*, *interactivity* and *growth trends* as key frameworks. The use of two-way information flow systems is applied to understand the effectiveness of mobile transportation platform DiDi, and the online mapping solutions provider Amap. This is followed by a discussion of how these digital platforms, utilising big data capacities provided by municipal governments, enhance China's urban mobility through reducing traffic congestions and improving users' road navigation needs.

4.1 DiDi: "More Than a Journey"

DiDi exemplifies digital media integration into urban traffic operations and driver management through mobile transportation platforms based on big data and interactive media [26]. Founded in 2012, DiDi Chuxing specialises in ride-hailing and ridesharing solutions globally. Its online services are delivered through a mobile *application* (app) which provides urban transportation services ranging from taxis, buses, coaches, e-bicycle, food delivery services and a range of enterprise logistics and solutions.

As the world largest operator of mobility networks, the vision of DiDi, known as China's version of Uber, is to lead in smart transportation technologies using mobile apps linked to auto technologies [27]. It was estimated that 1.1 billion rides were completed via DiDi in 2017, through partnerships with more than 500 Chinese taxi operators [28].

The volume of traffic data is performed through cloud computing and Artificial Intelligence that monitors traffic flows, signals, maintenance, alternative route suggestions, and *estimated time of arrival* (ETA). The smart transportation strategy via mobile app platform has enabled an estimated savings of more than 11 million commuting hours, reducing ETA delays by up to 20%, offering convenience to passengers in over 20 metropolises in China [28].

For service and fare payment, both cash and e-wallet system are available, making it a "huge hit" among locals and visitors who download the free app in English and Chinese [29]. Drivers going in the same direction may be booked at a lower price, resulting in tangible savings for public transport costs, decreasing passengers' commuting and transfer durations, while improving urban transportation efficiency.

To achieve these outcomes, information must *flow in both directions*. DiDi applies these characteristics by providing a user-friendly and interactive platform for drivers and passengers to get information about each other [23]. Passengers input specific journey needs as data, nearby drivers are prompted, and may choose to accept the booking according to destination scheduling and specific or preferred routes.

In the process, the advantages of digital networks' management of big data are well-utilised [30]: online platforms remove time and space barriers, seamlessly connecting passengers and drivers quickly [24, 25]. The two-way flow of information and interaction enables users to choose optimal routes. From "one person one car" to "two (or more) to one car", transportation efficiency results in better traffic congestion management by having an interactive system which *establishes smart connections between users and vehicles.*

4.2 Amap: Mapping China's Urban City Traffic

Amap, developed by Alibaba-owned software company AutoNavi, applies two features of intelligent data management to set up its "multi-channel traffic condition release system": specifically, satellite monitoring to measure congestion at intersections, and real-time situational reporting and updates via 3D simulation of actual traffic conditions via the proprietary technology platform owned Alibaba Cloud computing that collects transport data from over 700 million users.

The database is operable nationwide throughout China, covering approximately 3.6 million km of roads and giving locations for more than 20 million points of interests and landmarks [31]. With more than 100 million users of the free mobile map app, users of all major smartphone operation systems have access key location-based features that help simplify driving needs for real-time traffic, navigating using online three-dimensional views of road conditions, personalised choice of routes, ETA, etc. [32]. The digital map database technology platform is not uncommon, being the main competitor of the popular Baidu Map application, with the difference being the variety of modes of message communication and delivery through calls, apps or social media channels to map out the smoothest and fastest way to navigate around city congestions [33, 34].

5 Analysis of Findings

In recent years, the maturing market for web-based mobile applications solutions, big data and new media, has resulted in more efficient integration between traffic management systems, urban mobility and in solving traffic congestion problems in metropolitans [35]. The development of ITS took off from exploratory to practical stages, with factors such as urban market size relative to public transportation usage and technological penetration key in determining the design, planning, implementation and uptake of ITS. China demonstrated a healthy ITS development trend in a span of six years, with the market value of approximately RMB116.71 billion in 2017, a growth of 15 percent from RMB101.48 billion in 2016 (Fig. 3).

China's ITMS networks mainly cover urban areas, with the rate of urbanisation at 53.73%, a figure expected to rise to 70% in 2030 [36]. Rapid urbanisation and urban migration patterns undoubtedly serve to promote the relative efficiencies from well-managed ITS, but this also creates unresolved issues in the transitional period. China's ITS aims to achieve three main objectives, namely: to improve transportation efficiency; reduce traffic jams; and boost residents' participation, technological engagement, and service satisfaction.

Fig. 3 Development trends of China ITMS market between 2011 and 2017. DiDi (2017) *Annual City Travel Report*

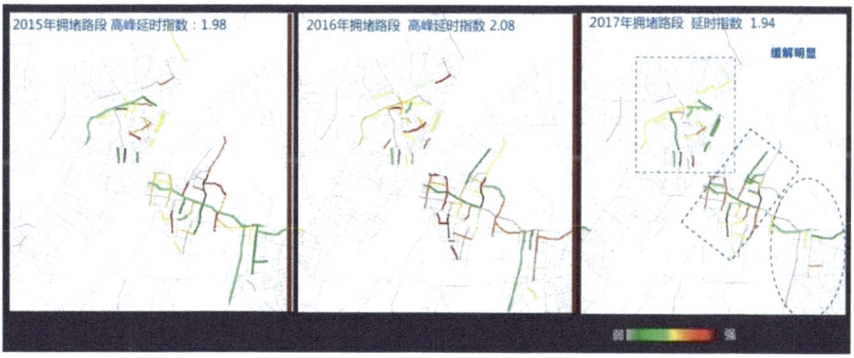

Fig. 4 Traffic congestion in Wuhan has improved significantly since Amap. Amap (2017) *Annual Traffic Analysis Report of Major Chinese Cities*

The first two objectives have contributed to a corresponding rise in traffic operations and management efficiency in China's megacities such as Wuhan (Fig. 4), the capital of Hubei in Central China, with a population of 11 million [35]. A study on efficiency evaluation of Beijing's ITMS was conducted based on the DEA model, analysing data from 2000–2010. It showed a correlative link between traffic management, energy efficiency and socio-economic development in sustaining Beijing's decade-long robust economic growth [37].

6 Proposed Recommendation

Urban mobility tools such as ride-sharing platforms, location mapping and vehicle navigation systems such as GPS have emerged as the smart models of information-based systems in large cities to solve domestic transportation problems. They must continue to focus on long term goals of providing mobility options through innovating services by investing in user-friendly apps development, safe networks and designing better predictability through large-scale data analytics of passenger behaviour such as preferred pickup and drop-off locations.

These emerging mobility businesses must also aim to further contribute to long term lowering of carbon footprint in urban cities through environmental mapping of alternative vehicular transportation modes, e.g. cycling paths, pedestrian walkways and public transportation [38]. Growth trends that draw on urban mobility solution include sharing data insights through apps and social networks for economic sustainability, which tracks business communities' information such as offices. Outlets or showroom locations, allowing customers to find the fastest routes to reach their intended destinations.

It is recommended that smart cities globally should seek to emulate the advantages seen in the China's ITMS digitisation model, where digital media networks integrate with environmental sustainability policies through incentivisation of online modes of transportation management. In emerging cities around the world, where deteriorating effects of traffic congestions and pollution on health, living quality and the ecology are observed due to continual assaults on human productivity, the experiences of technologically-advanced nations such as China could provide significant knowledge and integrated into local public administration's mobility information policymaking. Albeit varying cultural contexts and conditions such as social adaptation and competitiveness of the mobile app enterprise sectors do not make it easy to study the parameters of liveability, the common denominator which makes cities attractive should be carefully mapped by urban planners and designers at open forum sessions and public dialogues, in order that stakeholders may provide opinions and insights on how to build robust, agile and efficient urban transportation systems while resolving mobility issues.

As findings in case examples indicate, automated statistical analysis and network visualisation for data flows through interactive apps and digital navigation systems are designed to provide individualised choices of smart solutions and integrated to enable local authorities and communities to implement agendas advocating productivity, efficiency and sustainability.

7 Conclusion

The role of new media as an enabler in ITMS implementation may be simply summarised as the development and implementation of robust, agile and intelligent characteristics of web-based digital systems to analyse transportation information and generate optimal solutions quickly for passengers and travellers. Its advantages in interactivity, connectivity and sustainable city planning enable more efficient communication between users and vehicles to be established. By producing better designed, better informed and better-connected systems, network efficiency becomes an economic driver of ITMS growth. Traditional one-way media platforms like radio and television are not completely abandoned in the present scenarios; however, it is obvious that a key characteristic of urban communication scenarios is *choice*.

Conditional to this is the use of computational data and data analytics technologies enabling populations to adapt to living in smart environments, surrounded by wireless access, digital platforms, online networks, public transportation and vehicular technology, in interconnected socioeconomic climates [13, 39]. In this regard, various technology enterprises are leading the way by leveraging the power of connectivity through digital tools such as apps designed to address urban mobility challenges. From setting up data centres, security networking, and developing *IoT* (Internet of Things) solutions, network enterprises are now increasingly depending on the use of Big Data to position their services and solutions for the benefit of businesses, enterprise and communities [30].

Nevertheless, challenges abound in the development of urban mobility solutions for emerging smart cities. Urban mobility guidelines which frame the practices of data interactivity, sharing and functionality must be carefully designed and regularly reviewed by municipal planners, environmental experts, business and residential stakeholders to safeguard sustainability, wellbeing and safe living and travel conditions for cities strained by rural–urban migration, inefficient goods and services distribution, waste management, industrial pollution, gentrification, and inner-city decay. Although the '*new media + transport*' smart city model faces substantial implementation challenges in the present decade, including low penetration levels in non-urban areas, where shrinking populations, and insufficient utilisation of transportation services and facilities render investment losses, there is no doubt the two way flow of information communication using new media and digital technologies represents the future direction of intelligent transport management solutions in the era of urbanisation.

References

1. Intelligent Transport (2019) Efficient distribution of big data analytics for urban mobility applications. https://www.intelligenttransport.com/transport-articles/87903/efficient-distribution-of-big-data-analytics-for-urban-mobility-applications/. Accessed 28 March 2020
2. Katz E, Blumler JG, Gurevitch M (1974) Uses and gratifications research. Pub Opin Quart 37(4):509–523
3. Ruggiero TE (2000) Uses and gratifications theory in the 21st-century. Mass Commun Soc 3(1):3–37
4. Neuman WR (1991) The future of the mass audience. Cambridge University Press, Cambridge, Massachusetts
5. Logan RK (2010) Understanding new media: extending Marshall McLuhan. Peter Lang Publications, New York
6. Salvia G, Morello E, Arcidiacono A (eds) (2019) Sharing cities, shaping cities. Urban Sci 3(1):23. [Special Issue] MDPI Books, Basel, Switzerland
7. Currid-Halkett E (2006) A competitive analysis of four theories on world cities. Econ Dev Quart 20(4):330–350
8. Sassen S (1999) Embedding the global in the national: implications for the role of the state. Macalester Int 7(1):Article 8. https://digitalcommons.macalester.edu/macintl/vol7/iss1/8. Accessed 09 Feb 2020
9. Abdel-Rahim A (ed) (2012) Intelligent transportation systems. In-Tech Publishing, Rijeka, Croatia
10. Beaverstock JV, Smith RG, Taylor PJ (1999) A Roster of world cities. Cities 16(6):445–458
11. Knox PL (2011) Cities and design: critical introductions to urbanism and the city. Routledge, Abingdon, Oxon
12. Kitchin R (2014) The real-time city? big data and smart urbanism. GeoJournal 79:1–14
13. Pew Research Centre (2018) Artificial intelligence and the future of human. In: J Anderson, L Rainie (eds) [Online]. https://www.pewresearch.org/internet/2018/12/10/artificial-intelligence-and-the-future-of-humans/. Accessed 10 Jan 2020
14. Sevtsuk A (2012) How we shape our cities, and then they shape us. MAJA: Estonian Architect Rev 2(72):10–15. https://www.sutd.edu.sg/cmsresource/idc/papers/2012_How_we_shape_our_cities_and_then_they_shape_us.pdf. Accesed 21 Dec 2019
15. Chen X (2010) Beijing's plan to steer clear of traffic jams. China Daily. 14 December. https://www.chinadaily.com.cn/china/2010-12/14/content_11696079.htm. Accessed 02 Nov 2019
16. Ge YM, Liu XM, Tang LB, West DM (2017) Smart transportation in China and the United States [Online]. Washington, DC: Center for Technology Innovation at Brookings. https://www.brookings.edu/wp-content/uploads/2017/12/smart-transportation_final.pdf. Accessed 13 Dec 2019

17. Hancock T (2019) China's shrinking cities: 'most of my classmates have left'. Financial Times. 26 April. https://www.ft.com/content/bd1e1bf2-661f-11e9-a79d-04f350474d62. Accessed 24 Jan 2020
18. Ministry of Transport of China (2016) The development plan for transportation informatization in the 13th five-year period. https://www.mot.gov.cn/zhuanti/shisanwujtysfzgh/guihuawenjian/201702/t20170213_2163792.html. Accessed 05 Feb 2020
19. World Economic Forum (2018) In an urbanizing world, shrinking cities are a forgotten problem. In: Biswas A, Tortajada C, Stavenhagen M, 8 March. www.weforum.org/agenda/2018/03/managing-shrinking-cities-in-an-expanding-world. Accessed 16 March 2020
20. Lang J (2005) Urban design: a typology of procedures and products. Architectural Press, Oxford, UK
21. Sandford RR (2011) Origin and development of modern urban planning and management in China. https://www.rrojasdatabank.info/citieschina1011-6.pdf. Accessed 27 Feb 2020
22. Leng S, Chen Q (2019) Inkstone index: China's shrinking cities. Inkstone. 19 March. https://www.inkstonenews.com/society/inkstone-index-almost-one-third-chinese-cities-are-shrinking/article/3002365. Accessed 08 Jan 2020
23. Kim M, Yoon MY, Furuya K (2019) A Comparison of a smart city's trends in urban planning before and after 2016 through keyword network analysis. Sustainability 11:3155
24. Wootton JR, García-Ortiz A, Amin SM (1995) Intelligent transportation systems: a global perspective. J Math Comput Modell 22(4–7):259–268
25. Djahel S, Doolan R, Muntean GM, Murphy J (2015) A communications-oriented perspective on traffic management systems for smart cities: challenges and innovative approaches [Conference Paper]. IEEE Commun Surv Tutor 17(1)
26. Ran YL (2017) Innovative travel under intelligent transportation: a conversation with the leader of DiDi's intelligent transportation project. Big Data Time 6:33–35
27. DiDi Global. About Us. n.d. https://www.didiglobal.com/about-didi/about-us. Accessed 20 Nov 2019
28. Rogerson S (2018) DiDi traffic management brain adopted by 20 Chinese cities. IoT M2M, 20 February. https://www.iotm2mcouncil.org/didibrai. Accessed 31 Dec 2019
29. Wilson J (2018) How to get a taxi in China in English with the DiDi app. Career China, 22 February. https://www.careerchina.com/blog/how-to-get-taxi-in-china-in-english-didi-app. Accessed 02 Jan 2020
30. Forsyth Communications (2012) For big data analytics there's no such thing as too big. In: White Paper C. Forsyth (ed). March. https://www.cisco.com/c/dam/en/us/solutions/data-center-virtualization/big_data_wp.pdf. Accessed 19 Dec 2019
31. AutoNavi (2011) AutoNavi launches new enhanced mobile map application, "Amap," on Android Operating System. Globe Newswire. 17 May (News Release). www.globenewswire.com/news-release/2011/05/17/447441/222198/en/AutoNavi-Launches-New-Enhanced-Mobile-Map-Application-Amap-on-Android-Operating-System.html. Accessed 13 Feb 2020
32. AutoNavi (2013) AutoNavi's free mobile map app tops 100 million users in China. Globe Newswire. 14 January (News Release). www.globenewswire.com/news-release/2013/01/14/516170/10017993/en/AutoNavi-s-Free-Mobile-Map-App-Tops-100-Million-Users-in-China.html. Accessed 24 March 2020
33. Du J (2019) Beijing, Amap sign agreement to ease traffic jam, promote green transportation. China Daily. November 6. https://www.chinadaily.com.cn/a/201911/06/WS5dc27aeca310cf3e35575c8d.html. Accessed 05 Feb 2020
34. Xinhua (2018) Amap, Alibaba roll out smart system to reduce traffic congestion. China Daily. June 25. https://global.chinadaily.com.cn/a/201806/25/WS5b30cbc1a3103349141dea94.html. Accessed 16 Jan 2020
35. Li ZX (2015) Industrial development trend of urban intelligent transportation system in China. Intell Transport 7(1):71–74
36. Guo Y, Liu H, Chai Y (2014) The embedding convergence of smart cities and tourism internet of things in China. Adv Hosp Tour Res 2(1):54–69. https://www.ahtrjournal.org/admin/dosyalar/6/54_69.pdf. Accessed 30 Apr 2020

37. Wei ZL, Zhao P, Ai SL (2012) Efficiency evaluation of Beijing intelligent traffic management system based on super-DEA. J Transport Syst Eng Inf Technol 12(3), June: 19–23
38. Zhang YF (2018) Map app can track users' low-carbon activities. China Daily. June 5. https://www.chinadaily.com.cn/a/201806/05/WS5b167cb7a31001b82571e4c9.html. Accessed 27 Dec 2019
39. León-Coca JM, Reina DG, Toral SL, Barrero F, Bessis N (2014) 'Intelligent transportation systems and wireless access in vehicular environment technology for developing smart cities. In Bessis N, Dobre C (eds) Big data and internet of things: a roadmap for smart environments (Series 546), pp 285–313. Springer International, New York

Validation Frameworks for Self-Driving Vehicles: A Survey

Francesco Concas, Jukka K. Nurminen, Tommi Mikkonen, and Sasu Tarkoma

Abstract As a part of the digital transformation, we interact with more and more intelligent gadgets. Today, these gadgets are often mobile devices, but in the advent of smart cities, more and more infrastructure—such as traffic and buildings—in our surroundings becomes intelligent. The intelligence, however, does not emerge by itself. Instead, we need both design techniques to create intelligent systems, as well as approaches to validate their correct behavior. An example of intelligent systems that could benefit smart cities are self-driving vehicles. Self-driving vehicles are continuously becoming both commercially available and common on roads. Accidents involving self-driving vehicles, however, have raised concerns about their reliability. Due to these concerns, the safety of self-driving vehicles should be thoroughly tested before they can be released into traffic. To ensure that self-driving vehicles encounter all possible scenarios, several millions of hours of testing must be carried out; therefore, testing self-driving vehicles in the real world is impractical. There is also the issue that testing self-driving vehicles directly in the traffic poses a potential safety hazard to human drivers. To tackle this challenge, validation frameworks for testing self-driving vehicles in simulated scenarios are being developed by academia and industry. In this chapter, we briefly introduce self-driving vehicles and give an overview of validation frameworks for testing them in a simulated environment. We conclude by discussing what an ideal validation framework at the state of the art should be and what could benefit validation frameworks for self-driving vehicles in the future.

F. Concas (✉)
University of Helsinki, PL 64, Helsinki, Finland
e-mail: francesco.concas@helsinki.fi

J. K. Nurminen · T. Mikkonen · S. Tarkoma
University of Helsinki, PL 68, Helsinki, Finland
e-mail: jukka.k.nurminen@helsinki.fi

T. Mikkonen
e-mail: tommi.mikkonen@helsinki.fi

S. Tarkoma
e-mail: sasu.tarkoma@helsinki.fi

M. Khan et al. (eds.), *Smart Cities: A Data Analytics Perspective*,
Lecture Notes in Intelligent Transportation and Infrastructure,
https://doi.org/10.1007/978-3-030-60922-1_10

Keywords Self-driving vehicles · Validation frameworks · Simulation · Intelligent infrastructure · Smart cities

1 Introduction to Self-Driving Vehicles

An intelligent infrastructure that helps humans in their daily activities is a key feature of smart cities. Traffic is one of the biggest challenges of cities; smart cities could greatly benefit from a traffic infrastructure that can intelligently and autonomously manage it. Self-driving vehicles, or autonomous vehicles, are vehicles capable of driving autonomously with little or no human input. A network of self-driving vehicles able to communicate with each other have the potential to tackle this challenge and greatly improve the mobility in smart cities [32]. In recent years, self-driving vehicles have become more and more commercially easily available. A large number of self-driving vehicles now populate American and European roads, and projections show that their number will likely keep increasing [15]. As of 2019, twenty-nine states in the United States of America have enacted legislation related to autonomous vehicles [31].

Autonomous driving is achieved by using a combination of sensing units that sense the surrounding environment, and an advanced control system that interprets the output from the sensing units and drives the vehicle. The Society of Automotive Engineers (SAE) defined a standard for the levels of automation of self-driving vehicles, shown in Table 1. In this chapter, we focus on self-driving vehicles belonging to levels of automation 4 and 5, where the system takes care of the driving commands, the monitoring of the environment, and has also full responsibility during emergencies. In other words, we focus on vehicles for which the driver is left completely out of the loop.

The software that governs self-driving vehicles is, by the nature of the problem, very complex. The software has to elaborate extremely complex input coming from numerous sensors present in the vehicle, including cameras, LIDAR, and RADAR. Furthermore, the software has to drive the vehicle through lanes, traffic, intersections, and different weather conditions. For these reasons, validating self-driving vehicles is also a very complex task. Testing self-driving vehicles on public roads is unpractical. Firstly, it would take millions of hours of driving for the vehicle to encounter every possible scenario that can happen on a public road. Secondly, if an unforeseen fault occurs while the vehicle is driving on a public road, the vehicle poses a hazard to human drivers in the traffic.

Validation frameworks are being developed to tackle the challenge of testing self-driving vehicles. They work by simulating self-driving vehicles, including all of their sensors, on simulated scenarios that a self-driving vehicle can encounter in a public road. This way, potentially hazardous situations can be tested in a simulated environment so that the behavior of a self-driving vehicle can be tested without introducing a hazard on public roads.

Table 1 Levels of driving automation, as defined by the SAE EJ3016 standard [36]

SAE level	Name	Narrative definition	Execution of steering and acceleration/deceleration	Monitoring of driving environment	Fallback performance of dynamic driving task	System capability (driving modes)
Human driver monitors the driving environment						
0	No automation	The full-time performance by the human driver of all aspects of the dynamic driving task, even when enhanced by warning or intervention systems	Human driver	Human driver	Human driver	n/a
1	Driver assistance	The driving mode-specific execution by a driver assistance system of either steering or acceleration/deceleration using information about the driving environment and with the expectation that the human driver perform all remaining aspects of the dynamic driving task	Human driver and system	Human driver	Human driver	Some driving modes
2	Partial automation	The driving mode-specific execution by one or more driver assistance systems of both steering and acceleration/deceleration using information about the driving environment and with the expectation that the human driver perform all remaining aspects of the dynamic driving task	System	Human driver	Human driver	Some driving modes
Automated driving system monitors the driving environment						
3	Conditional automation	The driving mode-specific performance by an automated driving system of all aspects of the dynamic driving task with the expectation that the human driver will respond appropriately to a request to intervene	System	System	Human driver	Some driving modes
4	High automation	The driving mode-specific performance by an automated driving system of all aspects of the dynamic driving task, even if a human driver does not respond appropriately to a request to intervene	System	System	System	Some driving modes
5	Full automation	The full-time performance by an automated driving system of all aspects of the dynamic driving task under all roadway and environmental conditions that can be managed by a human driver	System	System	System	All driving modes

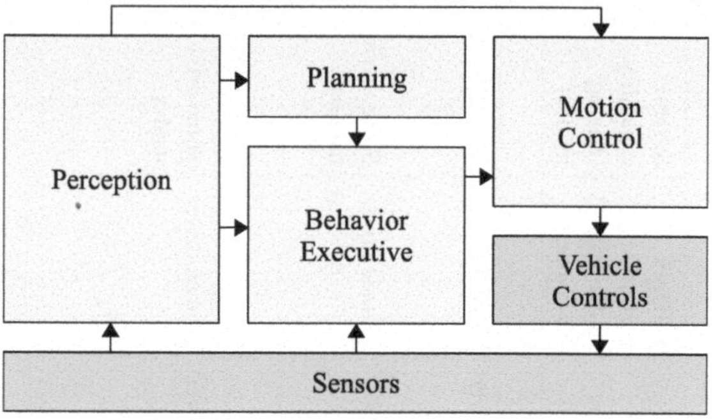

Fig. 1 Generic architecture of self driving vehicles

1.1 Components of Self-Driving Vehicles

A self-driving vehicle, in addition to the mechanical components present in all vehicles, includes sensors, processing units, and the self-driving unit. A self-driving vehicle perceives its surrounding environment through multiple sensors, such as cameras, LIDAR, RADAR, and GPS components [22]. The outputs of the sensors are processed by on-board processing units, such as CPUs, GPUs, and FPGAs [22]. To communicate with the processing units, the sensors are connected to them via multiple buses, which are in turn connected through gateways [43]. The self-driving unit is the software that, using the information given by the sensors, controls the mechanical components to drive the vehicle. Usually, the self-driving unit of a self-driving vehicle consists of four main components: (a) A perception unit; (b) A planning unit; (c) A behavioral executive unit; (d) and a motion control unit [43]. A generic architecture of self-driving vehicles is shown in Fig. 1.

Perception. This unit receives the outputs from the sensors and produces information regarding the road shape, the position of the vehicle, road signals, and obstacles such as other vehicles, pedestrians, and animals. The output of the perception unit is then sent to every other component of the self-driving unit. This is a very complex computer vision problem, which in turn requires a very complex machine learning algorithm. Deep learning has shown promising results in computer vision; a related survey can be found in Voulodimos et al. [40].

Planning. This unit receives information from the perception unit and plans the route, trying to minimize the time considering variables such as route length, speed limits, and traffic. An example of an algorithm that takes into account all these variables to plan an optimal route is the Google Maps route planner [9]. The output of this unit is sent to the behavioral executive unit.

Behavioral executive. This unit receives information from the perception unit, the planning unit, and sensors, and is in charge of making driving decisions such as in which lane to drive, how to behave in intersections, how to move through traffic, and where to park and how to behave in parking lots. The output of this unit is sent to the motion control unit. To make these decisions, the behavioral executive unit needs to take into account also traffic laws. In the robotics field, this is known as the roadmap planning problem, in which statistical methods have shown promising results [10].

Motion control. This unit receives information from the perception and behavioral executive units and physically executes the commands by controlling the mechanical parts of the vehicle. This unit needs to take into account the comfort of the passengers, to reduce the chance of motion sickness and injuries caused by sudden accelerations. Internal sensors located in the mechanical parts are used to send feedback to the units that receive inputs from the sensors, namely the perception and the behavioral executive units.

1.2 Software-Related Issues

The software needed to govern a self-driving vehicle is very complex, due to the nature of the autonomous driving problem. Let us discuss some common issues of this kind of software.

Black-box logic. Self-driving cars rely on complex machine learning models, as discussed in Sect. 1. These models, usually belonging to the class of artificial neural networks (ANNs), rely on very complex connections in the hidden layers, which cannot be manually set but are instead "trained" using the backpropagation algorithm. Due to the high complexity of the architecture of these models, it is very hard, if not impossible, to understand what every single layer of the model is exactly doing. This is why it is very difficult to debug these models, and why we can refer to complex ANNs as black-boxes.

Vulnerability to attacks. ANNs and deep models consisting of many ANNs can be vulnerable to attacks. These attacks consist in providing some input in the input space so that the network produces a wrong output. For example, in the self-driving vehicle problem, a pattern or some black rectangles arranged in some way on a street sign, apparently harmless to the human eye, can disrupt the detection of the signal by the perception unit. An attacker can put stickers on a road sign so that self-driving vehicles misread the signal, or ignore it altogether. This can lead to catastrophic consequences in a road traffic scenario, and it is the reason why the vulnerability to this kind of attack should be considered when validating a self-driving vehicle. a survey on currently known vulnerabilities to attacks of computer vision models based on deep learning can be found in Akhtar and Mian [1].

2 Safety Requirements of Self-Driving Vehicles

Self-driving vehicles are supposed to navigate through traffic, where a large number of risks are ever-present. Therefore, there is a need for a set of safety requirements that they should meet. To the best of our knowledge, a standard that defines safety requirements for self-driving vehicles does not yet exist. The ISO 26262 [17], which

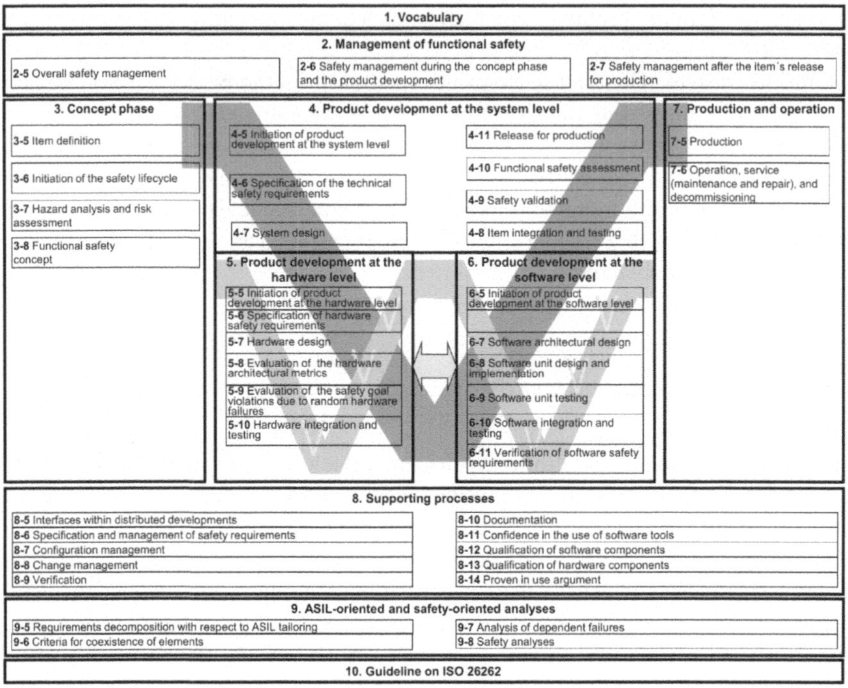

Fig. 2 V-model of the ISO 26262 [17]. This model is a good starting point for defining a validation model for self-driving vehicles

is the standard that defines the safety requirements of electric and electronic systems of vehicles, is a good starting point for defining an ad-hoc standard. ISO 26262 covers every phase of the development and production process using the V-model (Fig. 2) [5] as a reference validation model [17].

To measure the risk of failure of a specific component, ISO 26262 uses a risk classification scheme called Automotive Safety Integrity Level (ASIL). There are four ASIL levels from A to D, A indicating the lowest risk and D indicating the highest. The risk is determined by three factors: severity, which indicates the possible damage to passengers and property from the failure; exposure, which indicates the probability of a failure to occur; and controllability, which indicates how much the situation can be controlled when the failure occurs. Once these factors are determined, the ASIL level can be determined from a table that can be found in the ISO 26262 [17].

The ISO has recently been developing a standard specifically for self-driving vehicles, named ISO/PAS 21448 [37]. It covers the cases in which a safety hazard can arise without it being caused by a system failure. The difference with the ISO 26262 is that, instead of focusing on system failures, it focuses on malfunctions that can be caused by bad system design. Both ISO 26262 and ISO/PAS 21448 should be taken into account when designing a self-driving vehicle, as they complement each other. The limitation of ISO 26262 and ISO/PAS 21448, however, is that they are designed for self-driving vehicles supervised by a human.

Koopman and Wagner [24] also proposed a new standard for self-driving vehicles, named UL 4600. This standard, instead of focusing on hardware and software specifications, remains technologically neutral, focusing instead on the safety principles. The difference with the ISO 26262 and ISO/PAS 21448, is that the UL 4600 is specifically designed for vehicles with levels 4 or 5 of driving automation, where the driver is completely left out of the loop. The limitation of this standard is that it does not provide benchmarks for testing on a road, or criteria for safety or performance.

2.1 Driver Out of the Loop

To validate vehicles with levels 4 or 5 of driving automation (Table 1), the driver should be left out of the loop during the testing phase. This is because the vehicle should be able to recognize and deal with potentially catastrophic situations by itself, without the intervention of a human driver. During the validation phase, it is useful to observe how a self-driving vehicle reacts to dangerous situations by itself, to discover in which kind of scenarios the vehicle needs more training. This is the main reason why it is dangerous to validate a self-driving vehicle on a public road—it is unacceptable to release a vehicle which is a potential hazard.

2.2 Accepted Failure Rate

Failure rate is the frequency at which a catastrophic failure in a fleet of self-driving vehicles occurs [25]. In a hypothetical fleet of one million vehicles operated one hour per day, if the accepted failure rate is one catastrophic failure every 1000 days, Koopman et al. [25] argue that the average time between catastrophic failures should be one billion hours. This means that to validate such a failure rate, several billion hours of testing should be carried to be statistically significant, which is clearly unfeasible in the real world.

2.3 Hazards

Driving presents numerous hazards arising from nature and the behavior of other road users, such as bad weather, drivers that break rules either by distraction or intentionally, or animals suddenly crossing the street. A fully autonomous vehicle should be able to identify these hazards, should drive carefully to avoid a potentially catastrophic scenario, and should know how to intervene on the controls if it finds itself in such a scenario. Since many of these hazards occur randomly and cannot be predicted, the only practical way to validate the vehicle whenever they occur is to simulate them using a validation framework.

2.4 *Traffic Laws*

A fully autonomous vehicle should observe traffic laws when driving on a public road. Since traffic laws cannot be simply inferred from training alone, they should be manually hard-coded. It is worth mentioning that traffic laws change over time and from country to country, which means that a self-driving vehicle should be regularly kept up to date, and should also be able to switch from one set of laws to another when it crosses a border between countries.

A self-driving vehicle should also be able to handle exceptions and conflicting road signs. For example, if the vehicle sees a road marking that signals that overtaking is allowed (dashed line), but it also sees a vertical sign that forbids overtaking, the vehicle should know that the vertical sign has a priority on the road marking. Hence, the vehicle should decide that it is not allowed to overtake in that particular road section. This situation is very common during roadworks.

Since breaking traffic laws can cause potentially catastrophic scenarios, e.g. crossing a busy intersection with a red light, validating the lawful behavior of a self-driving vehicle should be done with a validation framework. This also raises the question: who has the penal responsibility for a fatal crash caused by a self-driving vehicle because it broke a traffic law? New laws and regulations written specifically for self-driving vehicles are needed.

3 Validation Approaches and Techniques

As discussed in the previous section, testing is impractical to do in the real world. We would need billions of hours of testing, in the traffic and with the driver out of the loop, where so many things can go wrong. Therefore, simulations on validation frameworks seem to be the best solution at the state of the art for testing self-driving vehicles. Validation frameworks are usually based on simulating frameworks.

Many frameworks exist for simulating the behavior of robots in a three-dimensional environment, e.g.. Gazebo in conjunction with the ROS interface [23], or more specific simulators for self-driving vehicles. Specific simulators include SUMO [26], an urban mobility simulator; Pro-SiVIC [14], a simulation framework that can easily integrate sensors of self-driving vehicles; CarSim,[1] a vehicle and traffic simulator that accurately simulates physics; and SCANeR by AVSimulation,[2] a simulation framework especially focused on making realistic scenarios.

[1] https://www.carsim.com/.

[2] https://www.avsimulation.fr/solutions/.

3.1 *Simulating Scenarios*

A self-driving vehicle, once deployed, needs to drive through complex urban environments, where it can encounter signals, intersections, and traffic. Simulating realistic scenarios is, therefore, an important part of testing self-driving vehicles. These scenarios should represent real-world streets as accurately as possible, and should also simulate the traffic as realistically as possible.

3.1.1 Closed World Mapping

Closed-world mapping is a useful method for simulating a realistic urban scenario, including all its physical features and its traffic laws [46]. In the context of validating self-driving vehicles, a closed world model should include a 3D model of the closed world to be mapped, its physical laws, and its traffic laws. The 3D model should include everything that is found in the real world, such as horizontal signs (e.g. lanes, parking signs), vertical signs (e.g. traffic lights, stop signs), temporary barriers in case of roadworks, random obstacles, etc.

Zofka et al. [46], for the 2016 AUDI Autonomous Driving Cup, demonstrated how to simulate a closed world for testing self-driving vehicles. Their framework is capable of simulating all the physical features of a realistic urban scenario, and to verify that the self-driving vehicles also follow the traffic laws. Lattarulo et al. [27] also created a framework that contains a simulator of a closed world, in an urban setting consisting of intersections and roundabouts.

3.1.2 Simulating the Traffic

The scenarios that a vehicle can encounter in the traffic are virtually infinite. There is a variety of roads, intersections, roundabouts with many different shapes, and vehicles of different shapes and sizes that navigate through them every day. Plus, it is not to exclude that other drivers or self-driving vehicles can violate traffic laws, unintentionally or even on purpose. A good framework should test self-driving vehicles with a comprehensive set of traffic scenarios so that the vehicle can be tested in as many traffic situations as possible. Simulating the traffic scenarios, therefore, is very time consuming and costly in terms of computational resources.

To cover every possible scenario, simulations of parts of scenarios can be optimally combined so that as many scenarios as possible are covered. This can be done by splitting scenarios into multiple sub-tasks, where every sub-task is fulfilled by its own simulation. Zofka et al. [45] have shown that, with this approach, a framework can achieve a larger coverage of possible scenarios than simulating whole scenarios. This way, testing scenarios can be done in a shorter time and with lower requirements of computational resources.

Traffic scenarios can also be represented by game-theoretic modeling. Li et al. [29] proposed a model based on the idea that different agents follow different levels of reasoning [7, 38]. A level-0 agent behaves without considering the actions of other agents in the system, a level-1 agent assumes that other agents are level-0 and behaves accordingly, a level-2 agent assumes that other agents are level-1 and so on. As reported by Li et al. [29], this model is good because this multi-level reasoning has been observed in humans through multiple experiments [6, 13].

3.2 Image Generation

Adding artificial rain, or fog, or distortions to the images perceived by self-driving vehicles, can have a dramatic effect on their behavior, as shown by Pei et al. [33]. This was confirmed later by Tian et al. [39], which have also shown that the same thing happens when changing the lightning conditions of such images. This implies that self-driving vehicles should be trained to be able to deal with all kinds of meteorological scenarios and possible lighting conditions. This is practically unfeasible, and even thousands of hours of recording of driving in the real world cannot cover every possible scenario [28]. To increase the coverage of scenarios, both for training and testing purposes, real images can be artificially transformed into different scenarios. For example, a recording of a bright sunny day can be transformed into an image of a foggy or a rainy day. To transform images and still keep their realistic look, complex machine learning models based on deep learning can be used [33, 39, 42].

Generative adversarial networks (GANs), a framework originally developed by Goodfellow et al. [11], has successfully been applied to the generation of artificial photo-realistic images [8, 41]. A state-of-the-art approach by Zhang et al. [42] successfully applied GANs for generating high-quality photo-realistic images for training self-driving vehicles. This includes, for example, transforming the image of a road taken in summer into a realistic-looking winter snowy road. The generation of such images allows the generation of comprehensive training and test datasets in every possible meteorological and lighting condition, which is important for validating self-driving vehicles.

3.3 V2V and V2I Networks

Vehicle-to-vehicle (V2V) and vehicle-to-infrastructure (V2I) communication architectures, generalized as V2X, can be used to tackle some problems related to traffic. As we have discussed in the introduction, this is especially important in smart cities, where self-driving vehicles have the potential to improve the traffic flow. V2X has been proposed for numerous applications, such as intelligent traffic management at intersections [4], cooperative driving systems [21], and vehicular density estimation systems [3]. This, however, presents numerous challenges: the timing of the messages is affected by the surrounding environment; the wireless connection between moving vehicles is by its nature unstable; the connection might be vulnerable to attacks.

To increase the safety and efficiency of self-driving vehicles, different V2X architectures have been proposed [12]. To provide a simulation framework for these challenges, Zheng et al. [44] proposed a cross-layer modeling, exploration and validation framework called CONVINCE. It can be used to simulate how external situations can impact the performance of V2X communications. Specifically, in their study, they show a case study on how a flooding attack can cause packet loss and cause problems in V2X communications. Another framework based on Dynacar, developed by Lattarulo et al. [27], contains a module that enables the testing of V2X communications.

3.4 Fault Injection

Fault injection is a technique that consists in stressing a system in an unusual way, to check how it reacts. For example, if we wish to test the hardware of a system we can subject it to different temperatures or different voltages. If we wish to test the software, we can feed it unusual inputs. Usually, fault injection on software is carried in a way that maximizes the coverage of code paths.

Self-driving vehicles can have many kinds of faults, both of hardware or software nature. For example, a sensor or one of the cameras can fail, or the car can react in the wrong way to a scenario never encountered before. Another possibility is that the perception unit is disrupted by a pattern placed on a traffic sign by an attacker. Validating self-driving vehicles can be done, for example, by injecting its sensors with distorted inputs [19, 33, 35]. Faults and errors could also be injected into the self-driving system itself [19, 20].

Jha et al. [18] proposed a clever approach to perform fault injection on self-driving vehicles, named "Bayesian fault injection". It consists in using machine learning to automatically mine situations and faults that can critically affect a self-driving vehicle.

4 Roadmap

Let us now conclude the chapter, discussing what the ideal framework for the validation of self-driving vehicles should be, and a possible roadmap for solving the open problems.

4.1 State-of-the-Art

The main purpose of validation frameworks for self-driving vehicles should be testing them in every possible urban and extra-urban scenario. Each presents its own challenges: the car should be able to navigate through the traffic in urban settings, and should also be able to drive fast on the highway while maintaining a high level of safety by monitoring the surrounding environment.

The best approaches at the state of the art for the mapping of a closed world can be found in Zofka et al. [46] and Lattarulo et al. [27]. Specifically to the generation of traffic scenarios, the approaches found in Li et al. [29] and Zofka et al. [45] should be considered, namely modeling the traffic using a game-theoretic approach, but also optimally considering different scenarios to optimize the coverage of all possible scenarios. It could be useful to equip every self-driving vehicle with a black box and use the data recorded during crashes to infer which are the real world situations most likely to lead to an accident.

Since the sensors of a self-driving vehicle include also cameras, they also need to be tested with images as photo-realistic as possible. The best approach at the current state of the art seems to be the generation of images through GANs [42].

V2X should be an important component of self-driving vehicles. As we discussed, though, it can be subject to problems and attacks. At the current state of the art, the best validation approaches can be found in Zheng et al. [44] and Lattarulo et al. [27].

The software of self-driving vehicles should be thoroughly tested for faults and errors, and its behavior when receiving distorted inputs should also be thoroughly tested. As we discussed, there is a framework at the state of the art that combines these two important issues, that can be found in Jha et al. [18].

Once issues in a self-driving vehicle are spotted by the validation framework, they need to be addressed by making changes to the hardware or software of the vehicle. Whenever a change is made to the hardware or software that governs a self-driving vehicle, the vehicle should be validated again. This raises the question, should the whole validation process be repeated, or should only a subset of the whole validation process be carried? If the second case is possible, an ideal validation framework should be able to select smartly the subset of tests that should be carried out after a change in one of the components.

4.2 Future Work

The validation of self-driving vehicles is still an open research problem. Let us discuss some issues that future research could focus on.

4.2.1 Safety Requirements

Currently, no special requirements exist for periodical inspections of self-driving, they have to pass the same criteria that normal vehicles have to. A well-defined set of safety requirements for self-driving vehicles can also benefit the development of validation frameworks that will be needed for this specific technical inspection.

4.2.2 V2X

Communication between self-driving vehicles increases their safety and efficiency. To the best of our knowledge, no standard exists on V2V and V2I communications between self-driving vehicles. A standard should be developed so that different manufacturers can develop vehicles able to communicate with each other. In turn, this will also allow easier development of validation frameworks for self-driving vehicles, since frameworks will not need to have different communication modules for different manufacturers.

4.2.3 Vulnerabilities

Only a few of the validation frameworks at the state of the art focus on possible vulnerabilities of self-driving vehicles [33, 44]. Future validation frameworks can benefit from more thorough testing of vulnerabilities, based on recent works on sensor and model vulnerability [1, 2, 16, 30, 34].

References

1. Akhtar N, Mian A (2018) Threat of adversarial attacks on deep learning in computer vision: a survey. IEEE Access 6:14410–14430. https://doi.org/10.1109/ACCESS.2018.2807385
2. Amoozadeh M, Raghuramu A, Chuah C, Ghosal D, Zhang HM, Rowe J, Levitt K (2015) Security vulnerabilities of connected vehicle streams and their impact on cooperative driving. IEEE Commun Mag 53(6):126–132. https://doi.org/10.1109/MCOM.2015.7120028
3. Barrachina J, Sanguesa JA, Fogue M, Garrido P, Martinez FJ, Cano J, Calafate CT, Manzoni P (2013) V2x-d: a vehicular density estimation system that combines v2v and v2i communications. In: 2013 IFIP wireless days (WD), pp 1–6
4. Bento LC, Parafita R, Nunes U (2012) Intelligent traffic management at intersections supported by v2v and v2i communications. In: 2012 15th international IEEE conference on intelligent transportation systems, pp 1495–1502
5. Clark JO (2009) System of systems engineering and family of systems engineering from a standards, v-model, and dual-v model perspective. In: 2009 3rd annual IEEE systems conference, pp 381–387. https://doi.org/10.1109/SYSTEMS.2009.4815831
6. Costa-Gomes MA, Crawford VP (2006) Cognition and behavior in two-person guessing games: an experimental study. Am Econ Rev 96(5):1737–1768. https://doi.org/10.1257/aer.96.5.1737

7. Costa-Gomes MA, Crawford VP, Iriberri N (2009) Comparing models of strategic thinking in Van Huyck, Battalio, and Beil's coordination games. J Eur Econ Assoc 7(2–3):365–376. https://doi.org/10.1162/JEEA.2009.7.2-3.365
8. Denton EL, Chintala S, Szlam A, Fergus R (2015) Deep generative image models using a laplacian pyramid of adversarial networks. In: Cortes C, Lawrence ND, Lee DD, Sugiyama M, Garnett R (eds) Advances in neural information processing systems, vol 28. Curran Associates, Inc., pp 1486–1494
9. Geisberger R (2014) Route planning. US Patent 14/157,913. https://patents.google.com/patent/US20140200807
10. Geraerts R, Overmars MH (2004) A comparative study of probabilistic roadmap planners. Springer, Berlin, pp 43–57. https://doi.org/10.1007/978-3-540-45058-0_4
11. Goodfellow I, Pouget-Abadie J, Mirza M, Xu B, Warde-Farley D, Ozair S, Courville A, Bengio Y (2014) Generative adversarial nets. In: Ghahramani Z, Welling M, Cortes C, Lawrence ND, Weinberger KQ (eds) Advances in neural information processing systems, vol 27. Curran Associates, Inc., pp 2672–2680
12. Harding J, Powell G, Yoon R, Fikentscher J, Doyle C, Sade D, Lukuc M, Simons J, Wang J (2014) Vehicle-to-vehicle communications: readiness of v2v technology for application. Technical report, United States National Highway Traffic Safety Administration
13. Hedden T, Zhang J (2002) What do you think i think you think?: strategic reasoning in matrix games. Cognition 85(1):1–36. https://doi.org/10.1016/S0010-0277(02)00054-9
14. Hiblot N, Gruyer D, Barreiro JS, Monnier B (2010) Pro-SiVIC and roads, a software suite for sensors simulation and virtual prototyping of ADAS. In: Proceedings of the driving simulation conference, pp 277–288
15. IHS Markit Ltd. (2018) Autonomous vehicle sales to surpass 33 million annually in 2040, enabling new autonomous mobility in more than 26 percent of new car sales, ihs markit says. https://technology.ihs.com/599099/autonomous-vehicle-sales-to-surpass-33-million-annually-in-2040-enabling-new-autonomous-mobility-in-more-than-26-percent-of-new-car-sales-ihs-markit-says. Accessed 17 Feb 2020
16. Jafarnejad S, Codeca L, Bronzi W, Frank R, Engel T (2015) A car hacking experiment: When connectivity meets vulnerability. In: 2015 IEEE globecom workshops (GC Wkshps), pp 1–6. https://doi.org/10.1109/GLOCOMW.2015.7413993
17. Jeon S, Cho J, Jung Y, Park S, Han T (2011) Automotive hardware development according to iso 26262. In: 13th international conference on advanced communication technology (ICACT2011), pp 588–592
18. Jha S, Banerjee S, Tsai T, Hari SKS, Sullivan MB, Kalbarczyk ZT, Keckler SW, Iyer RK (2019) Ml-based fault injection for autonomous vehicles: A case for bayesian fault injection. In: 2019 49th annual IEEE/IFIP international conference on dependable systems and networks (DSN), pp 112–124. https://doi.org/10.1109/DSN.2019.00025
19. Jha S, Banerjee SS, Cyriac J, Kalbarczyk ZT, Iyer RK (2018) AVFI: fault injection for autonomous vehicles. In: 2018 48th annual IEEE/IFIP international conference on dependable systems and networks workshops (DSN-W), pp. 55–56. https://doi.org/10.1109/DSN-W.2018.00027
20. Jha S, Tsai T, Hari S, Sullivan M, Kalbarczyk Z, Keckler SW, Iyer RK (2019) Kayotee: a fault injection-based system to assess the safety and reliability of autonomous vehicles to faults and errors. arXiv:1907.01024. Accessed 17 Feb 2020
21. Jia D, Ngoduy D (2016) Enhanced cooperative car-following traffic model with the combination of v2v and v2i communication. Transp Res Part B: Methodol. 90:172–191. https://doi.org/10.1016/j.trb.2016.03.008
22. Jo K, Kim J, Kim D, Jang C, Sunwoo M (2015) Development of autonomous car-part ii: a case study on the implementation of an autonomous driving system based on distributed architecture. IEEE Trans Ind Electron 62(8):5119–5132. https://doi.org/10.1109/TIE.2015.2410258
23. Koenig N, Howard A (2004) Design and use paradigms for gazebo, an open-source multi-robot simulator. In: 2004 IEEE/RSJ international conference on intelligent robots and systems (IROS) (IEEE Cat. No.04CH37566), vol 3, pp 2149–2154. https://doi.org/10.1109/IROS.2004.1389727

24. Koopman P, Ferrell U, Fratrik F, Wagner M (2019) A safety standard approach for fully autonomous vehicles. In: Romanovsky A, Troubitsyna E, Gashi I, Schoitsch E, Bitsch F (eds) Computer safety, reliability, and security. Springer International Publishing, Cham, pp 326–332
25. Koopman P, Wagner M (2016) Challenges in autonomous vehicle testing and validation. SAE Int J Transp Saf 4(1):15–24
26. Krajzewicz D, Erdmann J, Behrisch M, Bieker L (2012) Recent development and applications of SUMO - simulation of urban mobility. Int J Adv Syst Meas 5(3&4):128–138
27. Lattarulo R, Pérez J, Dendaluce M (2017) A complete framework for developing and testing automated driving controllers. IFAC-PapersOnLine 50(1):258–263. 20th IFAC World Congress. https://doi.org/10.1016/j.ifacol.2017.08.043
28. Leudet J, Christophe F, Mikkonen T, Männistö T (2019) Ailivesim: an extensible virtual environment for training autonomous vehicles. In: 2019 IEEE 43rd annual computer software and applications conference (COMPSAC), vol 1. IEEE, pp 479–488
29. Li N, Oyler DW, Zhang M, Yildiz Y, Kolmanovsky I, Girard AR (2018) Game theoretic modeling of driver and vehicle interactions for verification and validation of autonomous vehicle control systems. IEEE Trans Control Syst Technol 26(5):1782–1797. https://doi.org/10.1109/TCST.2017.2723574
30. Lim BS, Keoh SL, Thing VLL (2018) Autonomous vehicle ultrasonic sensor vulnerability and impact assessment. In: 2018 IEEE 4th World Forum on Internet of Things (WF-IoT), pp 231–236. https://doi.org/10.1109/WF-IoT.2018.8355132
31. National Conference of State Legislatures (2019) Autonomous vehicles|self-driving vehicles enacted legislation. http://www.ncsl.org/research/transportation/autonomous-vehicles-self-driving-vehicles-enacted-legislation.aspx. Accessed 17 Feb 2020
32. Nikitas A, Kougias I, Alyavina E, Njoya Tchouamou E (2017) How can autonomous and connected vehicles, electromobility, brt, hyperloop, shared use mobility and mobility-as-a-service shape transport futures for the context of smart cities? Urban Sci **1**(4). https://doi.org/10.3390/urbansci1040036
33. Pei K, Cao Y, Yang J, Jana S (2017) Deepxplore: automated whitebox testing of deep learning systems. In: Proceedings of the 26th symposium on operating systems principles, SOSP '17. ACM, New York, NY, USA, pp 1–18. https://doi.org/10.1145/3132747.3132785
34. Petit J, Shladover SE (2015) Potential cyberattacks on automated vehicles. IEEE Trans Intell Transp Syst 16(2):546–556. https://doi.org/10.1109/TITS.2014.2342271
35. Rubaiyat AHM, Qin Y, Alemzadeh H (2018) Experimental resilience assessment of an open-source driving agent. In: 2018 IEEE 23rd Pacific rim international symposium on dependable computing (PRDC), pp 54–63. https://doi.org/10.1109/PRDC.2018.00016
36. SAE international (2008) Taxonomy and definitions for terms related to driving automation systems for on-road motor vehicles. Technical report, SAE International
37. Schnellbach A, Griessnig G (2019) Development of the iso 21448. In: Walker A, O'Connor RV, Messnarz R (eds) Systems, software and services process improvement. Springer International Publishing, Cham, pp 585–593
38. Stahl DO, Wilson PW (1995) On players' models of other players: theory and experimental evidence. Games Econ Behav 10(1):218–254. https://doi.org/10.1006/game.1995.1031
39. Tian Y, Pei K, Jana S, Ray B (2018) Deeptest: automated testing of deep-neural-network-driven autonomous cars. In: Proceedings of the 40th international conference on software engineering, ICSE '18. ACM, New York, NY, USA, pp 303–314. https://doi.org/10.1145/3180155.3180220
40. Voulodimos A, Doulamis N, Doulamis A, Protopapadakis E (2018) Deep learning for computer vision: a brief review. Comput Intell Neurosci 2018(7068349):1–13. https://doi.org/10.1155/2018/7068349
41. Zhang H, Xu T, Li H, Zhang S, Wang X, Huang X, Metaxas DN (2017) Stackgan: text to photo-realistic image synthesis with stacked generative adversarial networks. In: The IEEE international conference on computer vision (ICCV)
42. Zhang M, Zhang Y, Zhang L, Liu C, Khurshid S (2018) Deeproad: gan-based metamorphic testing and input validation framework for autonomous driving systems. In: Proceedings of the 33rd ACM/IEEE international conference on automated software engineering, ASE 2018. ACM, New York, NY, USA, pp 132–142. https://doi.org/10.1145/3238147.3238187

43. Zheng B, Liang H, Zhu Q, Yu H, Lin C (2016) Next generation automotive architecture modeling and exploration for autonomous driving. In: 2016 IEEE computer society annual symposium on VLSI (ISVLSI), pp 53–58. https://doi.org/10.1109/ISVLSI.2016.126
44. Zheng B, Lin C, Yu H, Liang H, Zhu Q (2016) Convince: a cross-layer modeling, exploration and validation framework for next-generation connected vehicles. In: 2016 IEEE/ACM international conference on computer-aided design (ICCAD), pp 1–8. https://doi.org/10.1145/2966986.2980078
45. Zofka MR, Klemm S, Kuhnt F, Schamm T, Zöllner JM (2016) Testing and validating high level components for automated driving: simulation framework for traffic scenarios. In: 2016 IEEE intelligent vehicles symposium (IV), pp 144–150. https://doi.org/10.1109/IVS.2016.7535378
46. Zofka MR, Kuhnt F, Kohlhaas R, Zöllner JM (2016) Simulation framework for the development of autonomous small scale vehicles. In: 2016 IEEE international conference on simulation, modeling, and programming for autonomous robots (SIMPAR), pp 318–324. https://doi.org/10.1109/SIMPAR.2016.7862413

Case Studies in Smart Cities

Big Data for Smart Cities: A Case Study of NEOM City, Saudi Arabia

Tanweer Alam, Mohammad Ayoub Khan, Natheer K. Gharaibeh, and Malik Khlaif Gharaibeh

Abstract Several countries are planning to adopt the idea of smart city in their towns and enforcing large-scale data projects that promotes smart city features to achieve the recommended level of sustainable development to enhance the quality of life. The smart city tends to mean an advanced technological area that would be capable of understanding the environment by examining the data so it can improve the livability. The smart cities allow different kinds of wireless sensors to gather massive amounts, full speed and a broad range of city data allow the use of big data technology to learn important observations to control resources, services, and infrastructure effectively. The smart cities use numerous innovations to increase the efficiency of fitness, travel, resources, education and public infrastructure, resulting in a higher comfort level for their residents. Big data analytics is one of the most recent technologies with great potential to improve smart city facilities. Big data is a huge amount of data collected from different sensors, smart objects or smart devices connected with the Internet of Things (IoT). Smart cities allow regional officials to communicate directly to both society and city services as well as to track what might be happening in the city or how the city is progressing. This chapter presents a comprehensive discussion on the fundamental of data analytics, challenges, and a case study of NEOM smart city in Saudi Arabia.

T. Alam (✉)
Faculty of Computer and Information Systems, Islamic University of Madinah, Madinah, Saudi Arabia
e-mail: tanweer03@iu.edu.sa

M. A. Khan
College of Computer and Information Technology, University of Bisha, Bisha, Saudi Arabia

N. K. Gharaibeh
College of Computer Science and Engineering, Taibah University, Yanbu, Saudi Arabia

M. K. Gharaibeh
Independent Researcher, Amman, Jordan

M. Khan et al. (eds.), *Smart Cities: A Data Analytics Perspective*,
Lecture Notes in Intelligent Transportation and Infrastructure,
https://doi.org/10.1007/978-3-030-60922-1_11

1 Introduction

The smart city is an innovative idea aimed at easing the problems generated by the rapid growth of urbanism in the cities. In order to tackle the problem policy decision-makers are funding smart cities initiatives aimed at sustainable development and improving quality of life for residents as well as tourists. Communication and Information Technologies (ICT), wireless sensor network (WSN) and Internet of things (IoT) are the underlying technologies for smart cities. The comprehensive definition of the smart city also needs to be defined to incorporate and assess smart cities in action [1].

Smart cities innovation has given people the ability to limit the size of urban development problems and/or solve them. Over the last 10 years, community services are becoming more modern and knowledge-based, but there has been a profound shift in the living environment of people and the governance of communities. Infrastructure, community, transportation, entertainment and other facets of communities have been intimately linked to smart cities, and the network is becoming a big component of resident's daily routines. The numerous achievements of digitization of the town's records not only provide the community with everyday convenience. Smart city planning work has contributed to a comprehensive understanding of the innovations required to build a smart community. Owing to the diversity of concepts suggested by various parties, there is still a wide range of smart city technologies throughout the research. Many technologies demonstrate that smart cities are powered and allowed by interdisciplinary technologies. The set of revolutionary information technologies, including cloud services, big data, information vitalization, the IoT and mobile computing, have been commonly used in the smart city model [2–5]. The general concept of a smart city is to make a city (government, healthcare, transport, and so on) smarter and more effective.

The big data has become a terminology that represents a huge amount of data, including organized or unorganized, which intermixes a business regularly. However, the volume of data is not significant. Because that is what the companies are doing for the data depends. The big data could be mined through observations that contribute to better decision-making and effective business decisions. Big data has to be so massive that is hard to manage through conventional data repository or software methods. In most business cases, the data amount would be too high and flows too quickly and exceeding the existing computing power.

The big data analytics are the use of sophisticated analytical methods toward extremely broad, complex data sets that contain organized, semi-organized and unorganized data from various channels and also in different sizes. The big data is a paradigm used for databases whose size or complexity is beyond the capacity of conventional relational databases to store, handle or analyze low-latency information. The big data includes one or more of the following criteria: massive volume, higher speeds or large range. Machine intelligence, web, media and the IoT generate data sophistication across new techniques and metrics [6, 7].

Big data analytics is one of the latest innovations with a tremendous ability to improve smart city infrastructures. Because of the digitization is becoming an essential part of our daily lives, the data processing has contributed to the creation of large volumes of data that could be used in several advantageous technology environments. Optimal data collection and management is a major factor for performance in several business and infrastructure domains, such as the smart city framework. Big data technology for smart cities allows reliable data collection and retrieval to obtain information that would optimize various smart city resources. The big data allows policy-makers to prepare for the development of smart cities resources and infrastructures. Increasingly reliable data sets, advancements in the IoT [8] and Big data technology, and the development of a broad range of machine learning techniques provide development opportunities for delivering predictive assistance to people and decision-makers for the public. Furthermore, there is a difference in the combination of the existing state-of-the-art through an advanced system that will help to minimize construction costs as well as allow new types of programs. This chapter demonstrates the Big data analytical system for smart cities.

The rest of the chapter is organized as follows, Sect. 2 presents the literature review, Sect. 3 presents big data analytics for smart cities. Section 4 presents the case study of NEOM smart city. Section 5 presents the conclusion and future scope.

2 Literature Review

In 2011, Martin Strohbach et al. was published an article entitled "Towards a Big data Analytics Framework for IoT and Smart City Applications". In this article, they presented the big data analytics for IoT and smart cities. Their contributions to this research are as follows. Firstly, they presented the relationship between Big data and IoT technologies, secondly, they have presented a case study in the smart grid domain that illustrates the high-level requirements towards such analytical Big data [3].

In 2012, Batty et.al. were published an article entitled "Smart cities of the future". In this article, they described the basics of forms a smart city that we identify as a community wherein ICT is combined with conventional systems, organized and incorporated with emerging mobile technologies. Firstly, they outlined seven goals that concern: launching a new interpretation of environmental concerns, efficient and practicable ways of integrating digital innovations, frameworks, and techniques for using public information throughout spatial measurements, creating new communication and distribution technologies, developing modes of urban leadership and organization, and identifying critical issues [9].

In 2014, Rob Kitchin has published an article entitled "The real-time city? Big data and smart urbanism", he was emphasized on the previous and, based on a variety of scenarios, discusses how cities are being fitted to smart devices and services that deliver "big data". This data, claim smart city activists, allows real-time monitoring of life in the city, different forms of the city administration, and offers the raw resources for more effective, efficient, sustainable, profitable, accessible and responsive communities [10].

In 2015, Tang, Bo, et al. were published an article entitled "A hierarchical distributed fog computing architecture for big data analysis in smart cities", they were presented the hierarchical architecture to support the integration of a huge number of infrastructure components and services [11].

In 2016, Yunchuan Sun et al. presented the idea of the IoT and Big data Analytics for smart cities in the article [12]. The smart and connected cities are emerging from the idea of new technologies. It is designed to tackle synergistic need to consider the history (conservation and redevelopment), the need to live in the current (quality of living) with the need to prepare for the future (sustainable development). A perception is therefore to raise the standard of living, conservation, restoration and sustainable development of the city. One purpose of constructing the smart city is to live in the present, to create a better future, and to consider the history of the city [12].

In 2017, Anthopoulos, was published an article entitled "The Smart City in Practice", Throughout fact, smart cities, having operated on four contrasting dimensions of historical precedent: first, to discuss and briefly describe the smart innovations that smart cities use currently according to their meaning and nature, supported by literature facts. It instead attempted to analyze and outline the digital services provided by an efficient number of smart cities on their official sites. Then, it discussed current smart city principles to understand how emerging solutions can work together to create smart city goals. Ultimately, knowing from experience comes to the end of this analysis: 13 smart city studies are examined and briefly discussed [13].

In 2018, Alfa, Attahiru Sule et al. were published an article entitled "The Role of 5G and IoT in smart cities", they have presented the construction of each smart system as a self-contained system that would communicate efficiently with other systems. It will involve improving the output of each system to ensure that the available resources are being used effectively with reduced latency for rapid central information collection and decision-making. • Construct a communications network that links and transmits information and details to a variety of smart platforms. The correspondence network can interact with the other participants of the networks by transferring messages and information to other smart platforms [14].

In 2019, Allam and Dhunny, have presented a paper on big data, artificial intelligence, and smart cities, they were discussed the strategic capability of AI and the initiative for a new system of integrating AI technologies and smart cities, although maintaining the convergence of key aspects of community, mobility, and leadership; that is considered to be crucial to the effective adoption of smart cities following the Sustainability and the new Urban Strategy. Such an article is targeted at decision-makers, information scientists searching to promote the integration of artificial intelligence and big data in smart cities with the ultimate objective of increasing the sustainability of the urban landscape and at the same time boosting growth in the economy and possibilities [15].

In 2020, Kulkarni and Akhilesh, have presented the big data analytics as an enabler in smart governance for the future smart cities, they were discussed about effective governance that has been one of the areas of the study of the different facets as to how smart government can be transformed into good governance. Big Data Analytics has been utilized by companies and organizations to enhance their decision-making by gaining insight into their market challenges, increasing productivity and providing customer support. In its existing implementations, Big data Analytics is promising to serve as a catalyst to expand the contexts in which it can be implemented. Good governance is needed to create effective sustainable communities, and big data analytics will play a key role in solving the complex day-to-day challenges that urban cities experience [16].

3 Big Data Analytics for Smart Cities

Big Analytics isn't the latest technology that has been around for decades in the form of business intelligence and data mining applications. Throughout the years, big data has evolved significantly so that it would accommodate even greater information sizes, execute requests faster, and conduct technically sophisticated simulations. Smart cities are urban center which uses various kinds of wireless IoT sensors to gather information and use the observations from the whole data to handle resources, services, and infrastructure successfully. Owing to increasing residential immigration levels and urbanization, people of big cities present several challenges. Across the other side, new problems emerge as a result of the evolution of society and the well-being of residents.

3.1 Role of Big Data in Smart Cities

The smart city is a modern idea that aims to resolve certain challenges and opportunities. Throughout smart cities, attempts are being designed to meet the expectations of technology-based people. Although addressing community issues, residents need to have access to public services everywhere and everywhere. Big data Analytics for smart cities puts together some of the most promising recent technologies in the field of incorporating sophisticated data analytics frameworks through smart cities including cloud computing, simulation and information gathering for intelligent transportation infrastructure situation evaluation, interactive simulation, cyber-physical networks, and intelligent building innovations [17].

Big data has been in use for several companies now realize that if they collect all the data that flows through the operations, they can implement analytics and obtain significant value from all of this. However, the latest advantages that big data analytics offers are efficiency and performance. Even though a few years earlier a business might have obtained information, operated analyses and discovered knowledge and could be used for future policy, nowadays the enterprise can find insight for urgent decision making. Having the ability to operate quicker and remain agile gives companies a strategic advantage they haven't had in [18].

3.2 Categories of Big Data Analytics

Big data analytics can be divided into four categories.

3.2.1 Concise Analytics

Those methods inform businesses of what has occurred. These produce clear statements or simulations which demonstrate all that has happened at a specific point in time or over some time. These would be the least developed analytical tools available.

3.2.2 Diagnosis of Data Analysis

These methods clarify how something has occurred. Rather sophisticated than simple analysis tools, these enable observers to look deeper into the information and identify core issues for a particular circumstance.

3.2.3 Quantitative Analytics

One of the most famous big data analytics available tools now, these tools use sophisticated techniques to determine what could occur soon. Such methods also allow the use of automation and machine intelligence tools.

3.2.4 Legalistic Analytics

The stage above analytics, legalistic analytics inform institutions on how to accomplish the goal. Most such techniques involve quite sophisticated machine learning functionality, and very few strategies on the market provide legalistic strengths.

3.3 Big Data Analytics Uses

The following are the big data advantages in smart cities.

1. The big data is valuable to run a warehouse because you didn't have to change pieces of infrastructure depending on the number of months or even years that they have been on the operation. It is expensive and inefficient since different elements break at various levels [19]. Big data helps you to track defective devices and determine when they will be replaced.

2. The big data is critical in the healthcare sector, which is one of the very few enterprises yet struggling with a generic, traditional method. Like a scenario, when

someone has the disease, they will have a treatment, but if it doesn't succeed, the specialist will prescribe something. The big data makes it possible for a patient to get medications that are produced based on diseases.

3. It would help businesses produce goods and services that respond to the consumers, and help them build potential opportunities for increasing revenue.

4. It can enhance customer support. The companies also use broad data analytics to analyze social networking sites, technical support, marketing, and business statistics.

5. It will allow companies to accurately assess consumer sentiment and react to consumers dynamically.

6. Improved Information Security protection is another important field for big data analytics. The security software generates an incredible amount of user information. Through implementing broad big data strategies towards this information, companies could often detect and prevent cyber threats that may have gone undetected.

7. The big data is important to keep the data secure. Data analytics applications enable developers to analyze the business data environment, that allows evaluating organizational risks. For an instance, you should know whether or not your confidential data is covered. Another more typical example is that you should be able to investigate 16-digit address or processing details.

8. The big data enables businesses to innovate the sources of income. assessing analytics will send it real data which might bring you closer up with a consistent source of income.

3.4 Big Data Analytics Challenges

The implementation of a big data analytics approach is not as simple as the businesses expect it is. Additionally, several studies show which the number of enterprises achieving tangible financial benefits through big data analytics delays under the variety of groups adopting big data [20]. Many specific challenges that make it very difficult to obtain the opportunities offered by big data providers:

3.4.1 Information Rise

The problem of big data analytics would be the exponential amount of data produced. As per IDC, the volume of data on servers around the world is increasing yearly. Such repositories are expected to store 175 zettabytes of electronic data by 2025 [21]. The big data analytics applications should be capable of performing well for a network if they want to be of interest to businesses.

3.4.2 Unconventional Information

The amount of data contained in an organizational network does not exist in organized warehouses. Additionally, there are unorganized records, like emails, photographs, documents, music files, videos, and other forms of information. Distributors are continually upgrading the Big Data Analytics technology to make themselves well off.

3.4.3 Information Siloes

Business information is generated by either a diverse range of technologies, like enterprise resources planning (ERP) systems, customer relationship management (CRM) approaches, supply chain management technology, e-commerce approaches, office suite systems, etc. Incorporating information out of all these varieties of sources is one of the most challenging tasks for any big data analytics experiment.

3.4.4 Social Issues

While big data analytics has become widespread, business culture has not penetrated all over the world.

3.5 *Trends of Big Data Analytics*

The following are the trends of big data analytics.

3.5.1 Public Access

While big data analytics improves its popularity, the emphasis is on free-source techniques that help ease down and analyze information. The participants are Hadoop, Spark, and NoSQL servers. Particularly specialized methods still integrate or encourage emerging opensource innovations. This seems impossible to improve in the coming years.

3.5.2 Business Strategy

Lots of overall-purpose big data analytics solutions have reached the business sector, however, predict a little more to appear, focusing on unique market segments, like safety, advertisement, CRM, network performance management and recruiting. Big data resources are now incorporated into current enterprise applications at a rapid speed.

3.5.3 Machine Intelligence and Learning

While involvement in Artificial intelligence has grown, businesses have raced to integrate machine learning and computational technologies into big data analytics systems. As per Gartner, every new technology service, especially big data analytics, would integrate Artificial Intelligence developments through 2020 [22]. Through 2022, more than half of the new big business applications will integrate integrated intelligence using real-time background data to enhance decision-making [22].

3.5.4 Universally Applicable Analytics

Because of this moving to Artificial intelligence, encourage businesses to become more involved in universally applicable big data. Shown as perhaps the "strongest" form of big data analytics, such techniques would not only be designed to forecast the possibilities but will also be able to recommend actions that could lead towards positive outcomes for businesses. However before certain kinds of technologies may become popular, entrepreneurs would need to make advances in the development tools.

3.5.5 Reorienting Along the Individual Decision-Making Process

When big data progress and becoming a characteristic of market scales in analytics frameworks, and not be too disappointed unless the human factor is briefly underestimated once it returns to full swing. When Big Data Analytics is becoming more popular, it's going to be more like a lot of previous innovations. The big Data Analytics has become another technique. How you're going to do with it, though, is what concerns.

4 NEOM City: A Case Study

NOEM city is among future smart cities of the world. With its economic diversification and modernization Programme, Saudi Arabia has received attention from a major project. The crown prince Mohammed bin Salman announced in October 2017 that it was in the country that plans to build NEOM, the latest smart city in the world [23]. It is constructed in an area of 10,230 square meters covering Saudi Arabia, Jordan and Egypt as shown in Fig. 1. The data about the NOEM project is presented in Table 1 [23–25].

Fig. 1 NEOM city in Saudi Arabia

Table 1 Metrics about NEOM city

Project Announcement	Oct 24 2017
Objective	To make a global hub for trade, innovation, and knowledge
Trust area	Transportation, Food Processing, Healthcare, internet of things, Nature programming, Genetic Modification
Demographic information	26,500 km^2 with Vacant land with 500 km of beachfront
Stake Holders	Crown Prince Mohammed bin Salman, Chair Masayoshi Son, chairman and CEO of the SoftBank Group Stephen A. Schwartzman, chairman, and co-founder of the Blackstone Group Marc Raibert, CEO of Boston Dynamics Klaus Kleinfeld, NEOM's CEO
Project Worth	$100 Billion project by 2030

The NEOM city has many vertical to make it smarter. Major vertical includes community, technology, nature, livability as shown in Fig. 2. A details list is presented as follows:

1. Energy
2. Water
3. Mobility
4. Bio technologies
5. Food
6. Manufacturing

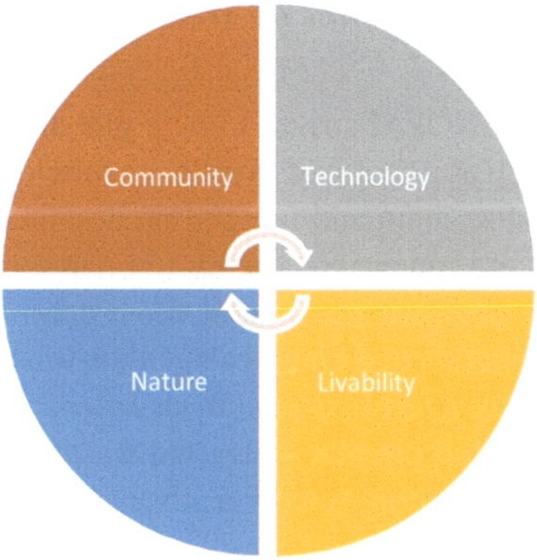

Fig. 2 Verticals of NEOM City

7. Media
8. Entertainment and culture
9. Tourism
10. Sports
11. Health and Well Being
12. Education
13. Livability

NOEM city will have many more features as follows [23–25]:

(1) The NOEM will have flying taxis. People can hire flying taxis from their roof-top to travel anywhere in the city.
(2) The NEOM will have giant artificial moon to illuminate the city.
(3) The security system of the NOEM will be based on drones, security cameras, and facial recognition technology that will enable continuous tracking for the citizen.
(4) There will be the concept of Cloud seeding technology to produce technology made clouds to produce higher rain the which can occur naturally.
(5) A new dimension of virtual called holographic teacher will be used in school for teaching. Even there will be robotic dinosaurs in the park.

4.1 Future Energy

NEOM will offer unrivalled increased solar and wind capacity by competitively priced renewable energy. The NEOM city will build a new business and push the next energy transition surge by producing green hydrogen. Using low-cost clean energy and a carbon-free energy ecosystem from scratch, NEOM plans to develop energy-intensive industries that can lead globally. The NOEM will have smart transmission and distribution network with state-of-the-art and advanced technologies [23]. NEOM also aims to be the breeding ground for start-ups and create an innovative ecosystem by combining R&D, education and innovation.

4.2 Internet of Water (IoW)

The NOEM strives to be a global powerhouse of water production and storage, as well as a centre of excellence for global water technologies, fully integrated along the value chain and focused on desalination wate. The water will be 100% desalinated using renewable energy to guarantee zero CO2 emissions, and zero brine effluent to preserve the natural ecosystem. Complete seawater refining will contain brine minerals and chemicals. Other wastewater in NEOM will be processed to generate energy, fertilizer, and fresh water for irrigation [23–25]. This will promote zero waste and full circular water and wastewater system. The water distribution network will be fully connected via advanced IoW technology to reduce water loss.

4.3 Smart Mobility

NEOM will have new future mobility definitions including international, regional and urban links via sea, air and land. There is planning to build a port that catalyses international trade with NEOM, open an intermodal doorway with the surrounding area of NEOM. A global technological benchmark in an integrated customer-cantered travel experience will be set by the best possible air connection between NEOM and the rest of the world [23–25]. The architecturally iconic airport is biometrically processed to remove bottlenecks via a seamless shutdown experience. NEOM is a global reference in manned and unmentioned air traffic management for health, safety and environmental policy. In order to allow the seamless management of goods for the residents and companies of NEOM will develop the first intelligence and autonomy supply chain in the world. NEOM will orchestrate all of this with artificial intelligence and will become the world's main trading hub.

4.4 Biotech Advancements

The NEOM will be home to both aspirating and established biotech companies by establishing a global biotech innovation hub and ecosystem for rapid integration of technologies into health practice. The planet should seek NEOM to change its understanding of human health and disease and develop diagnoses, therapies and diagnosis of the next generation. Biotechnology research is a quest for the future prosperity of the world. NEO will be a global hub for the future of biotech through the appeal of leading and emerging research, cooperation, development and application of new technologies [23–25].

4.5 Future Entertainment

At NEOM, people will be able to get through high-tech experiences, from taking out a book to shopping in the library of the future and watch a movie or in a game and play alongside amazing animals [23–25]. All these educational, cultural and fashionable possibilities created by future technologies will add up to a rich and unprecedented entertainment.

4.6 Projects 2.0

NEOM, one of the world's largest project aims to transform how we conceptualize, architecture, plan, implement, use and reuse capital projects. This is achieved by creating an ecosystem of unmatched performance and efficiency outcomes through pace of implementation, sustainable environment, talent retention, and conservation. NEOM should benefit from the latest technology that is built in this environment by a new generation of creative design and construction firms [23].

4.7 Advancement in Healthcare Technologies

There is another focus on the integrated health and well-being ecosystem in the NOEM city. NEOM will lead the world in health and welfare, creating an environment for healthy lifestyles that is constructive and personalized for world-class disease care based on machine learning and IoT [24, 26–28]. The artificial intelligence together with genetic and intelligent technologies will provide real-time assessments for the patients. In addition to the comprehensive wellness services offered at the health resorts, a network of global experts and distinctive centres of excellence shall rapidly follow best practice offering high-end medical treatment.

4.8 Livability

The prosperity and future success of NEOM will be based on liveability. For the next wave of designers and innovators, we seek to create the most advanced and livable urban ecosystem in the world. Living is essential to improve quality of life, develop inclusive communities, enhance urban sustainability and ensure the best possible utilization of digital technologies to provide world class citizen services. NEOM's social, economic and environmental outlook ensures that the next generation of creativity and spirit is generated in the most advanced quality of life [23–25].

4.9 *Robotics and Future Technologies*

This very interesting to know that Saudi Arabia has grant citizenship to "Sophia", the first humanoid robot. This shows that how much Saudi Arabia is very open and aggressive to adopt advance technologies. The vision of the Saudi leadership is to make NEOM a town of robots the first town with more robots than the people, which brands the city as futuristic city. The robots will provide security, delivery of items, and assistance in healthcare, etc.

Smart cities across the world are getting more efficient. These are introducing policies to foster a greener and healthier urban areas, clean water and air, greater accessibility and effective social resources. Certain programs are enabled by innovative technology, like the IoT and Communications technologies which provide a technological structure for the implementation of smart city initiatives. The smart cities initiatives require big data analytics to operate. IoT creates large databases that need to be evaluated and stored for the implementation of smart city operations. The Big data networks, building in the smart city infrastructure, should organize, interpret and filter information obtained from the IoT sensors and devices. Smart cities government incorporates communication technologies to connect government programs, while at the same moment involving citizens in public government and thereby encouraging collaboration.

The smart city is information-driven through its concept. The big data analytics perform an increasingly significant role in strategic planning, with certain cities now acting as Community Information Executives in the government. An Integration of big data analytics and smart cities solutions help cities improve the management in critical segments such as big cities have more challenges to manage power usage effectively. Smart city introduction helps city authorities to track energy usages dynamically. Through big data, smart cities can forecast times of high usage and schedule the allocation of power appropriately.

5 Conclusion and Future Scope

Big data performs a significant role in the analysis of information gathered by the IoT sensors in smart cities, therefore, more studies could be done to identify trends and requirements throughout the locality. IoT devices placed in the city produce large amounts of information, even if they are would use efficiently, several changes could be made. The smart cities' innovations and frameworks enable cities to transition to a purer, cleaner and more efficiently designed urban community. This is no wonder, therefore, that the smart city concept is proven to be popular with more than 100 smart cities worldwide using emerging technology to boost the standard of living of their citizens. The cities like NEOM in Saudi Arabia are formed the basis of the intelligent framework that would continue to improve, with the emerging technologies industries producing innovative solutions and alternatives to build the world smart.

References

1. Yin CT, Xiong Z, Chen H, Wang JY, Cooper D, David B (2015) A literature survey on smart cities. Science China Inf Sci 58(10):1–18
2. Alam T (2019) IoT-Fog: a communication framework using Blockchain in the internet of things. Int J Recent Technol Eng (IJRTE) 7(6)
3. Strohbach M, Ziekow H, Gazis V, Akiva N (2011) Towards a big data analytics framework for IoT and smart city applications. Modeling and processing for next-generation big-data technologies, pp 257–282. Springer, Cham
4. Mohammed A, Alam T (2015) An algorithm for accessing traffic data-base using wireless technologies. In: Computational Intelligence and Computing Research (ICCIC), 2015 IEEE International Conference on, pp 1–4. IEEE. https://doi.org/10.1109/iccic.2015.7435818
5. Alam T (2017a) Fuzzy control based mobility framework for evaluating mobility models in MANET of smart devices. ARPN J Eng Appl Sci 12(15):4526–4538
6. Alam T, Benaida M. The role of cloud-MANET framework in the internet of things (IoT). Int J Online Eng (iJOE) 14(12):97–111. https://doi.org/10.3991/ijoe.v14i12.8338
7. Alam T (2017b) Middleware implementation in cloud-MANET mobility model for internet of smart devices. Int J Comput Sci Netw Secur 17(5):86–94
8. Alam T, Benaida M (2018) CICS: cloud–internet communication security framework for the internet of smart devices. Int J Interact Mob Technol (iJIM). Nov 1;12(6):74–84
9. Batty M, Axhausen KW, Giannotti F, Pozdnoukhov A, Bazzani A, Wachowicz M, Ouzounis G, Portugali Y (2012) Smart cities of the future. Europ Phys J Spec Topics 214(1):481–518
10. Kitchin R (2014) The real-time city? big data and smart urbanism. GeoJournal 79(1):1–14
11. Tang B, Chen Z, Hefferman G, Wei T, He H, Yang Q (2015) A hierarchical distributed fog computing architecture for big data analysis in smart cities. In Proceedings of the ASE bigdata & socialinformatics, pp 1–6
12. Sun Y, Song H, Jara AJ, Bie R (2016) Internet of things and big data analytics for smart and connected communities. IEEE Access 4:766–773
13. Anthopoulos LG (2017) The smart city in practice. In: understanding smart cities: a tool for smart government or an industrial trick?, pp 47–185. Springer, Cham
14. Alfa AS, Maharaj BT, Abu Ghazaleh H, Awoyemi B (2018) The role of 5G and IoT in smart cities. In: Handbook of smart cities, pp 31–54. Springer, Cham
15. Allam Z, Dhunny ZA (2019) On big data, artificial intelligence and smart cities. Cities 89:80–91
16. Kulkarni P, Akhilesh KB (2020) Big data analytics as an enabler in smart governance for the future smart cities. In: smart technologies, pp 53–65. Springer, Singapore
17. Wenge R, Zhang X, Dave C, Chao Li, Hao S (2014) Smart city architecture: A technology guide for implementation and design challenges. China Commun 11(3):56–69
18. Javadzadeh G, Rahmani AM (2020) Fog computing applications in smart cities: a systematic survey. Wireless Netw 26(2):1433–1457
19. Ghazal A, Rabl T, Hu M, Raab F, Poess M, Crolotte A, Jacobsen H-A (2013) BigBench: towards an industry standard benchmark for big data analytics. In: Proceedings of the 2013 ACM SIGMOD international conference on Management of data, pp 1197–1208
20. Marjani M, Nasaruddin F, Gani A, Karim A, Hashem IAT, Siddiqa A, Yaqoob I (2017) Big IoT data analytics: architecture, opportunities, and open research challenges. IEEE Access 5:5247–5261
21. Patrizio A (2018) IDC: Expect 175 zettabytes of data worldwide by 2025. Network World
22. Moore S (2019) Gartner identifies top 10 data and analytics technology trends for 2019. Accessed 21 April 2019
23. Liam D (2017) Saudi Arabia's Neom: Oasis or Sand Castle? Bloomberg. https://www.bloomberg.com/news/articles/2017-10-24/saudi-arabia-s-neom-oasis-or-sand-castle. Accessed 8 Aug 2020
24. Farag A (2019) The story of NEOM city: opportunities and challenges. https://doi.org/10.1007/978-3-319-77875-4_3

25. NEOM Official Web Site. https://www.neom.com/. Accessed 10 Aug 2020
26. Quasim MT, Khan MA, Abdullah M, Meraj M, Singh SP, Johri P (2019)Internet of things for smart healthcare: a hardware perspective. In: 2019 first international conference of intelligent computing and engineering (ICOICE), Hadhramout, Yemen, 2019, pp 1-5. https://doi.org/10.1109/ICOICE48418.2019.9035175
27. Khan MA, Quasim MT, Alghamdi NS, Khan MY (2020) A secure framework for authentication and encryption using improved ECC for IoT-based medical sensor data. IEEE Access 8:52018–52027. https://doi.org/10.1109/ACCESS.2020.2980739
28. Khan MA (2020) An IoT framework for heart disease prediction based on MDCNN classifier. IEEE Access 8:34717–34727. https://doi.org/10.1109/ACCESS.2020.2974687

Smart Cities Pilot Projects: An IoT Perspective

Syed Rameem Zahra and Mohammad Ahsan Chishti

Abstract The idea of smart cities originated as a strategy for addressing the unparalleled challenges of rapid urbanization, rising population density and, at the same time, giving residents and visitors a better quality of life. A smart city consists of smart components such as smart health, smart buildings, smart transport and smart industry, which form various city domains where the meaning of the "smart" label has different connotations in each domain. Internet of Things (IoT) along with advanced communication technologies is essential means for realizing smartness in any of smart city domains. IoT saw its journey from the unthinkable and the impossible to sustainable and tenable. Its pace of expansion from agriculture to athletics; personal safety to intelligent traffic detection; baby monitors to high-end military systems is incredible, and unforeseeable. It is to say that the future implementation fields of IoT are numerous and complex, impacting all aspects of human life. The European Internet of Things Research Cluster (IERC) defines and explains key IoT technologies that cover various constituting domains of Smart City viz. smart health, smart transport, smart building, and smart industry. This chapter identifies the major pilot projects in each of these domains that were launched to turn the catchphrase 'Smart City' into reality. It summarizes the outcomes of these projects and also recognizes the major technologies, tools, challenges and future opportunities in these domains.

Keywords Internet of things · Projects · Smart city · Smart health · Smart transport · Smart building · Smart industry

S. R. Zahra (✉)
Department of Computer Science and Engineering, National Institute of Technology Srinagar, Srinagar, India
e-mail: rameemzahra_36phd17@nitsri.net

M. A. Chishti
Department of Information Technology, Central University of Kashmir, Tulmulla, Ganderbal, India

M. Khan et al. (eds.), *Smart Cities: A Data Analytics Perspective*,
Lecture Notes in Intelligent Transportation and Infrastructure,
https://doi.org/10.1007/978-3-030-60922-1_12

1 Introduction

In 1999, Kevin Ashton, a great technologist gave out the term Internet of Things and said that “IoT has the potential to change the world, just like the Internet did and maybe even more so.” It was the moment when the internet technologies were not entirely demystified, cloud was not born, and the ability to communicate wirelessly was missing in dissimilar artifacts. Around the moment, this notion seemed ridiculous and contradictory but today, everyone would vouch for the statement’s legitimacy. Today, IoT enables technology across a broad variety of fields from handling objects of insignificant significance such as thermostats to controlling life-saving surgical implants. The application scope of IoT ranges from tracking field dampness, auditing of the movement of products across a manufacturing line, to the centralized control of patients with terminal diseases, and monitoring their medical devices . Through linking daily items (plants, animals, land, and people) to the internet, IoT aims to make life a little more fun . IoT, for example, turns the refrigerator’s concept of calling the store on its own when it runs low on milk into fact, which the users find fascinating.

The IoT concept takes advantage of several ubiquitous services to enable Smart City deployments worldwide. IoT provides new opportunities, such as the ability to remotely track and manage devices, analyze and take action based on information obtained from different sources of real-time traffic data. As a result, IoT products are transforming cities by enhancing infrastructure, creating more reliable and cost-effective public services, improving transport services by reducing congestion on the roads and improving the safety of people. As such, it is important to understand which smart city projects turned IoT from Kevin’s imagination to reality. This chapter talks about the major pilot projects in the key smart city domains, executed in various countries. It also highlights the major challenges faced in the actual realization of IoT in these domains. The future directions chalked out in each domain can help the researchers in gaining an idea of the research gaps. The key contributions of this study for pilot projects in various domains of smart city, is shown in Fig. 1.

2 Pilot Projects

While any single definition of a smart city remains in conflict, its common contemporary understanding brings us to the following features of smart city: smart health, smart buildings, smart transport and smart industry. Through holding an eye on all the big processes, the vast sources of tremendous data could allow for smarter decisions. For example, better energy management is obtained when home appliances such as refrigerators and washing machines are controlled by IoT. In the simplest terms, the smart city can be described as a city planning approach to monitor and subsequently integrate and optimize the conditions and use of city lifelines such as roads, bridges,

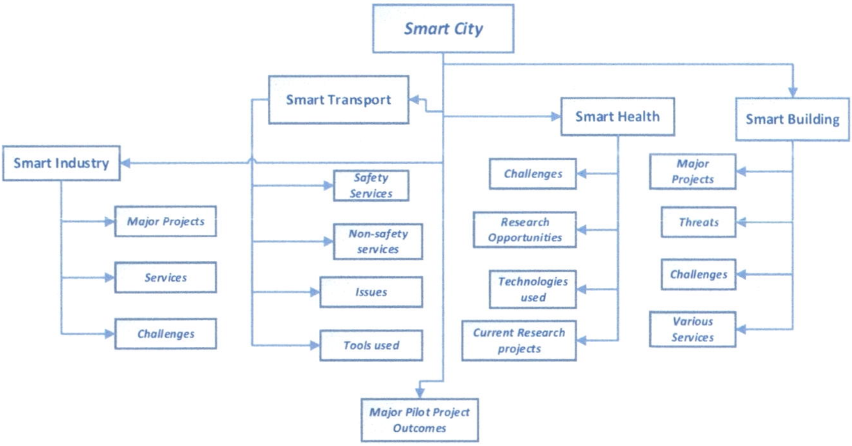

Fig. 1 Blueprint of the chapter

tunnels, railway lines, seaports, airports, electricity, water, communication, etc. Table 1 highlights the outcomes of major pilot smart city projects.

Following identifies the major projects executed in smart city domains including smart health, smart buildings, smart transportation, and smart industry.

2.1 Smart Health

Today the proportions of the aging population are witnessing a significant rise. As such, IoT Health Management Systems (HMS) have been developed to offer a feasible solution to conventional methods for dealing with healthcare. HMS seeks to offer inexpensive remote healthcare to those who need it, while retaining their privacy and preventing hectic and skyrocketing issues [24]. Following describes the various aspects of the smart health along with the challenges, future work and research opportunities:

2.1.1 Activity of Daily Living (ADL) [25]

This includes grooming activities like brushing teeth, face washing, making hair, eating, dressing, sleeping, toileting, etc. ADL is realized using technologies like home automation and convenience implants. Automation of home [26] helps the dependent people to easily operate various appliances, knobs, switches, etc. at home. While as, convenience Implants are used for detection of movements, identification of people, automation, etc. Research project in this category includes Health Smart Home [27], which uses off-the-rack, minimal cost equipments to issue alerts to

Table 1 Smart city pilot projects

Smart city pilot project	Smart city pilot project outcomes
Amsterdam, Netherlands [6, 7]	Proper flow of traffic with less jams, better security and efficient energy management
Groening, Netherlands [8]	Provides exact bus location to users by using GPS data i.e. upgraded public transportation
Vienna, Austria [9, 10]	Carbon footprint reduced, enhancement in the protection of climate, energy efficiency noted
Smart Santander, Spain [11]	Uplifted irrigation techniques for lawns and grounds, proper waste administration, better traffic management, parking area administration, track of environmental noise, temperature and light
Malaga, Spain smart city project [10, 12]	With the installation of more than 17,000 smart meters, this project desires to curtail carbon emissions by more than 6000 tons per annum
Barcelona, Spain smart city project [13–16]	Better waste collection strategies, smart street lighting, and better design of bus networks and accomplishment of smart traffic
Norfolk, England [17]	Betterment in data accumulation, analysis and delivery services
Santa Cruz, USA [18]	Breaking down the criminal activities to foresee the necessities of police and to calculate the most extreme presence of police in the required locales
Fujisawa Sustainable Smart Town (SST), Japan [19]	Aims to slash carbon emissions by 70%, pervasive sensing
PlanIT Valley, Portugal [20]	Pervasive sensing, aims to deploy 10 crore sensors
Songdo, Korea [21, 22]	Pervasive Sensing, Building automation, Smart street lights based on flow of pedestrians, Tele-presence and Smart meters
Stockholm, Sweden [23]	Aims to provide optic fiber networks throughout the city of Stockholm

physically disabled people thereby allowing them to stay safe at home. Securing medical implants [28] and detecting anomalies in people and in statistically relevant groups [29] are the challenges that need to be overcome. Moreover, transition from classic IoT sensors to Internet of Medical Things needs to be looked upon. The assimilation of sensors that are precisely interfaced with the brain will be a rolling area of research in future [30].

2.1.2 Instrumental Activity of Daily Living (IADL) [24]

IADL includes activities like preparation of meals, laundry, use of medicines, housekeeping, shopping, etc. Augmented Reality has a crucial role to play in IADL. It gives an augmented view of the reality by combining the real world with computer produced data and pictures. It can be used to help the patients who have lost the voluntary ability of movement (akinesia) to move [31]. Cao et al. [46] designed a Care-O-Bot 3 (COB) hardware and Robotic Operating System (ROS) software that offers a self-learning robotic solution for doing various household chores like providing prescribed medicine on trays to disabled people. Mining large quantities of data that wearable devices generate is an impediment that requires prompt solution [29]. Moreover, in IoT, a plethora of diverse systems communicate and thus may interfere with each other [32]. To slash these interferences, it is essential to characterize the propagation channel in every scenario. Ray Launching [33] will provide the best precision and least computational cost and thus will be extensively studied in future. In healthcare scenario, it can be used in emergency rooms when multiple beds communicate with interfering sources such as MRI systems [34].

2.1.3 Ambulatory Activity of Daily Living (AADL)

This includes static activities like lying, standing and sitting, dynamic activities like walking, running, jogging, bike riding and etc., transitional activities like standing to sitting, sitting to standing, standing to walking, etc. Medical Implants are used for monitoring such activities. These implants include programmable, configurable and low power demanding devices used for scrutinizing body vitals [35]. Coiera et al. [31] tried to establish the mobility of akinesia patients by anticipating virtual objects onto the patient's physical world to give them the feeling that they were walking through them. However, it suffers from side effects like motion sickness, equipment failure, fatigue of patients, etc. Kevin et al. [36] identifies people by surgically introducing RFID devices into their bodies that tracks their movements and allows the automated surrounding to react to their presence. Securing medical implants [28] and detecting anomalies in people and in statistically relevant groups [29] is highly crucial in this domain. Scalable SQL [37], Elastic Stack [38] and MapReduce [39] will gain importance in the future of healthcare information.

2.1.4 Monitoring of Mental Functions (MF)

MF includes memory, judgment, understanding, sense of direction, etc. Convenience implants [29] and medical Implants are used for monitoring such functions. Apple i-watch and Kardia Band [40] have the ability to detect atrial defibrillation that leads to strokes. High complexity [41] and security of medical implants [28] are the factors that need to be looked upon. In future, Recommender Systems (RS) will be developed quite fast and will play a crucial role in health applications as RS [42] assist in separating noise from useful information. However, RS in healthcare is still in its infancy. Most famous RS is Collaborative Filtering [43] and encompasses a large number of recommendation methods. The assimilation of sensors that are precisely interfaced with the brain will be a rolling area of research in future [30].

2.1.5 Physiological Activities

This includes monitoring of heart, brain and muscle working. Wearable sensors [44, 45] play a crucial role for monitoring such activities. They have the ability to revolutionize our life. These sensors keep a continuous check on the person and offer instant results for extended periods of time, thereby easing the way people communicate, exercise and get entertained. However, lack of trust [46, 47] and security of medical implants [28] are the hindrances that need to be overcome. Deep Learning (DL) has a massive potential in this domain. It consists of a number of Machine learning (ML) techniques to decipher various patterns in data and is based on multilayer Artificial Neural Networks (ANN). DL will advance the smart healthcare systems by allowing improved analysis of data in medical area such as medical imaging and translational omics [48].

2.1.6 Social Activities of Daily Living (SADL) [29]

SADL includes get together with family and friends, making phone calls, video calls, etc. Robotics and Personal Assistants [29] and micro-sensors are used for the automation of such activities: Robotics and Personal Assistants help people viably in their homes, and have the capacities of self-learning, versatility, pro-activeness and consistent reasoning. Micro-Sensors refer to body patches like tattoos, stamps, stickers and etc. that are less than 1 mm thick. The primary benefit of using micro sensors is that they could be powered by internal sources such as body heat [49] or pressure [50]. Smart Labels [51] have the ability to read temperature of the environment. Perry et al. [52] developed a micro-sensor that is worn on forearm and measures the level of oxygen. Electric Skin [53] has the ability to feel any external stimulus just like human skin. This has been achieved by using various kinds of pressure sensors. Lack of consideration of user profile (people differ in terms of age, profession, cultural beliefs, mental statuses, societal needs, daily routine and they also change with time so a smart system working today may not work tomorrow) [54], making of customized and context-aware sensors [55], and heterogeneous data integration [55] are the impediments in this domain that require urgent solutions. A promising way for mining of data and predictive process monitoring will be Long-Short Term Memory (LSTM) neural networks [56].

Various aspects of smart health are tabulated in Table 2.

Table 2 Various aspects of smart health

Smart health monitoring activity	Technologies used	Research projects	Challenges to overcome	Future Research Opportunities
Activity of daily living	Home automation, convenience implants	Health smart home [27]	Securing medical implants, anomaly detection	Transition from classic IoT sensors to Internet of Medical Things
Instrumental activity of daily living	Augmented reality	Care-O-Bot 3 (COB) hardware and Robotic Operating System (ROS) software [46]	Mining large quantities of data	Ray Launching
Ambulatory activity of daily living	Medical implants	Mobility of akinesia patients [31], movement tracking [36]	Securing medical implants, anomaly detection	Scalable SQL, Elastic Stack and MapReduce will gain importance in the future of healthcare information
Monitoring of mental functions	Convenience implants, medical implants	Apple i-watch and Kardia Band [40]	High complexity, securing medical implants	Recommender systems and Collaborative Filters
Physiological activities	Wearable sensors, medical implants	Apple i-watch and Kardia Band [40]	Lack of trust, securing medical implants	Deep Learning will advance the smart healthcare systems by allowing improved analysis of data in medical area such as medical imaging and translational omics
Social activities of daily living	Robotics and personal assistants, micro-sensors	Smart Labels [51], oxygen level measurement using fore-arm sensor [52], electric skin [53]	Lack of consideration of user profile, Making of customized and context-aware sensors, heterogeneous data integration	A promising way for mining of data and predictive process monitoring will be Long-Short Term Memory

2.2 *Smart Building*

One of the main building blocks of a smart city, a Smart Building, is the one in which all the service structures are automatically managed and interconnected, working together to maximize resource usage and improve the savings of the invested capital and running costs, reliability and efficiency. Following highlights the services provided by smart buildings, their integrations, various projects and challenges that exist:

2.2.1 Energy Management [57]

The major system for energy management in a smart building is Building Automation System (BAS) [58]. It utilizes BACnet, LonWorks standardized protocols for communication and offers Direct Digital Control (DDC) thereby offering better flexibility, definite control, enhanced comfort of residents and less energy costs as compared to electronic and pneumatic control systems. BAS employs a huge no. of hardware units like operator workstations, control units for network, sensor systems, etc. It can be integrated with facility management. Random change of data that may lead to faulty functioning of various devices of smart building set-up is a challenge that needs to be resolved [59]. The usefulness of smart building depends on its integration and potential to evolve with the changing lifestyle design and the environment. Moreover, it should not lead to redundancy. Such integration is challenging [60].

2.2.2 Communication Management

HVAC Systems [61, 62] in smart building consist of air handling equipment controller, air volume system controller, heating and cooling components that are spread throughout the occupancy regions of the floor, distributed controller, condenser systems and various sensors. Moreover it consists of software programs like Chillers optimum start-stop program, unoccupied period program, unoccupied night purge program, load reset program, zero energy band program, heating cooling plant efficiency program, duty cycle program, etc. Central BAS can remotely talk to the dial-up system using a modem and hence monitor the building remotely. Data leakage because of weak security postures installed on these devices remains an impediment that needs an urgent solution [59]. An administrative challenge to realize smart buildings is that every owner needs to have some kind of technical know-how of the working of things. This reduces the complexity level of services provided by a smart building [60].

2.2.3 Office Automation [62]

The major smart building system in this category is smart lighting system [63]. The software employed in smart lighting system consists of temporal lighting system programs, light control programs for detecting occupancy, etc. and hardware employed consists of Organic Light Emitting Diodes (OLED), Solid State Light (SLL) sources, Charge-Coupled Device (CCD) cameras, motion detectors, intelligent lighting controllers (ILC), touch switches, etc. BAS can interconnect fire alarm system with other frameworks like HVAC, lighting and security systems. When a fire breaks out, HVAC can open exhaust dampers and close outside air intake dampers to stop smoke from spreading. Gain of information from malicious sources which may lead to erroneous and trouble causing behaviors [59] and the interoperability of the devices from various manufacturers using different communication protocols are the issues that need to be resolved [64].

2.2.4 HVAC Control System

The major building system in this category is Fire Protection System [65]. It consists of fire detectors, fire alarms, intelligent fire controller (IFC) and typical safety software programs for detection and protection against fire. When alarm situations are created, interfacing of alarm systems with security system can allow the opening of security locked doors to allow faster evacuation. However, faulty internet function can upset the working of devices that massively rely on services of the cloud that is accessed via internet and remote controlling will be badly influenced [59]. Moreover, reliability is a challenge because errors in smart building services like fire detection alarms and security systems cannot be tolerated. Since data from unreliable sources can come, wrong interpretation of data can happen thereby making reliability an issue.

2.2.5 Light Control System

The major smart building system in this category is vertical transportation systems. It consists of passenger detectors, lift sensors, CCD cameras, various controllers, programs for monitoring and lift operations. The integration of security system with lighting system and HVAC can allow the glowing of specific lighting paths to guide people into specific locations for protection. In advertent failure of communication channel will result in the loss of connectivity among various devices that will eventually lead to the loss of total functionality [59].

2.2.6 Fire Protection System

This includes security system [65] that consists of CCTV surveillance, intruder detection alarms, presence detectors, Intelligent Access Controllers (IAC) and Programs for implementing safety and security systems. By integrating vertical transportation systems with security systems/fire systems, it is possible to calculate the count of elevators that are required.

2.2.7 Vertical Transportation System

This includes communication system that consists of classical phone systems, transmission media, mixers, aerials, repeaters, amplifiers, attenuators, Local Area Network (LAN), Private Automatic Branch Exchange (PABX), Integrated Service Digital Network (ISDN), etc. [65]. Combination of computer vision systems with HVAC can allow the counting and distribution of people in an AC space. Dependence on third party devices for any purpose may put security at stake and make the service unavailable for the time the faulty third party device is replaced [59].

2.2.8 Security System

This includes CCTV surveillance, Intruder detection alarms, presence detectors, Intelligent Access Controllers (IAC) and programs for implementing safety and security systems [59]. Masquerading threats causing the exploitation of building services by an intruder is a challenge that needs to be resolved [59].

Various aspects of smart building are tabulated in Table 3.

2.3 Smart Transport

Intelligent Transport System (ITS) technology has given direction for the construction of smart vehicles, smart motorcycles, smart buses and trains by arming them with different types of sensors and actuators, including radars, Global Positioning System (GPS), Event Data Recorders (EDRs), cameras, omni-directional antennas, electronic chassis numbers, electronic license plates, etc. Various aspects of smart transport are summarized below.

2.3.1 Public Transport Administration

This includes services like public travel security, in-transit, personalized and public transit information. Moreover, electronic payment of toll charges, highway charges, bus charges, etc. is facilitated in ITS. Payments are made with smart cards that are

Table 3 Smart building

Smart building service	Major smart building system	Threats and challenges faced
Energy management	Building automation system	Random change of data that may lead to faulty functioning of various devices of smart building set-up
Communication management	HVAC systems	Data leakage because of weak security postures installed on these devices
Office automation	Smart lighting system	Gain of information from malicious sources which may lead to erroneous and trouble causing behaviors
HVAC control system	Fire protection system	Faulty internet function can upset the working of devices that massively rely on services of the cloud that is accessed via internet. Remote controlling will be badly influenced
Light control system	Vertical transportation systems	Inadvertent failure of communication channel will result in the loss of connectivity among various devices that will eventually lead to the loss of total functionality
Fire protection system	Security system	Device failures, Increasing cost of the technology
Vertical transportation system	Communication system	Dependence on third party devices for any purpose may put security at stake and make the service unavailable for the time the faulty third party device is replaced
Security system	Security system	Masquerading threats causing the exploitation of building services by an intruder

detected by automated systems at toll plaza. However, security remains an important concern in this domain. The topologies in which smart vehicles move are extremely dynamic and change abruptly and continuously. As such the authenticity maintenance of vehicles is different and the environment is vulnerable to attacks [66]. This eventually leads to less efficiency and less safety. An important in-vehicle software named Carwall has been developed for security of cars against cyber-attacks. It works by making the ECU software quite strict to prevent attacks. The external device is barred from placing any unwanted codes into the car system.

2.3.2 Cooperative Collision Warning

If a car gets out of control or is over-speeding, it informs other cars in the vicinity about the danger using V2V communication to avoid collision. This requires an efficient broadcasting algorithm with very small latency. Travel and Traffic Management is a non-safety service in ITS. It encompasses pre-tour travel information, information about driver, route guidance, trip co-ordination and reservation, voyager service information, traffic control, travel request administration. Attacks like malicious code propagation and data integrity attacks could be launched. A company called Visual Threat produced a Cyber Security protection framework called as FUSE. FUSE is an acronym for Firewall, Umbrella, Security over the Air (SOTA) and Event Intelligence. Firewall boosts protection on vehicle, umbrella provides a security policy framework, SOTA encourages instant avoidance of car hackings and Event Intelligence performs event analytics and threat correlation among diverse attack vectors.

2.3.3 Lane Change Warning

If a vehicle wants to change the lane and wants an empty room for the same, warning is sent to all the other vehicles using V2V communication. Coupling/Decoupling is a non-safety service in ITS that allows joining two or more buses or trains together to reduce travelling time and head way distance, data transfer among vehicles and traffic information exchange. The amount of data generated by the transport industry is huge and to efficiently collect, store and process the same is an issue. To address the issue cloud and fog based architecture could be used. CANtact, a tool for knowing the happenings in the vehicle has been developed. It represents an open-source Controller Area Unit (CAN) to Universal Serial Bus (USB) interface for computers. It has the ability to connect to any CAN-enabled car by employing the OBD-II cable. When connected to OBD-II port, the happenings inside the vehicle can be found.

2.3.4 Intersection Collision Warning

On the blind corner intersections of roads, the cars remain out of vision from each other and thus the chances of accidents are elevated. As such V2I safety application is used and warnings are broadcasted among the vehicles to avoid high speeds and sharp turns on the curves. To keep a check on vehicles used for commercial purposes and include services like: electronic clearance of such vehicles, safety inspection of roads, security check of OBU of vehicles [66], administrative check on vehicles, response to accidents and freight movements is a non-safety service in ITS. Latency requirements of various applications require more research in the direction of V2I communication, infrastructure and protocols. CANard is a toolkit based on python that has been developed for intersection collision warning. It performs functions like sending/receiving messages and encoding/decoding data, hardware abstraction, protocol implementation and information sharing with ease.

2.3.5 Approaching Emergency Vehicles

When the emergency vehicles like ambulance, police cars or fire brigade approach they warn other vehicles using V2V communication to clear the lanes till they pass. Information Management Services is the non-safety requirement for realizing this service. Addressing of moving vehicles becomes a challenge for packet delivery. Extension of IP addressing with geo-addressing continues to be a challenge for the actual realization of smart transport system. A tool called as Rolljam has been devised that has the ability to catch and store keyless entry thereby allowing the owner of the vehicle an uninterrupted access to it.

2.3.6 Safety from Hit and Run Cases

The EDRs of smart vehicles store everything that happens in the surrounding. Hence if a car hits and runs away, the EDRs of the neighboring vehicles would record the Electronic chassis number as well as catch the number from Electronic License plate. The police can then have a shortlist of suspect vehicles. Economic simulation platforms need to be developed that have the ability to blend events, traffic and applications together.

2.3.7 Emergency Management

This deals with services like emergency notifications, disaster management and evacuation. Infotainment services improve the experience of travel and include audio, video streaming services. Vehicles have vulnerable components like Electronic Control Unit (ECU), CAN, Global Navigation Satellite System (GNSS) and software.

2.3.8 Roll Over Warning

The roadside units arranged at the critical bends, communicate the message to the vehicle about street conditions and bend points with the goal that vehicle can alter its speed according to the condition of the road. This requires location based services that boost the driving experience and prevents the fatigue of driving. The services included are navigation, tourist spot advising, etc.

2.3.9 Work Zone Warning

An RSU can be confined at the work zone to caution the approaching vehicles about the work in advance so that the vehicle can change the course or back off the speed as need be. Traffic Management was designed to reduce the driving time and fuel consumption by informing the driver about the conditions of road and shortest route in that particular direction.

Various aspects of smart transport are tabulated in Table 4.

Table 4 Various aspects of smart transport

Smart transport public safety services	Smart transport public non-safety services	Issues in smart transport	Tools used in Smart Transport
Public transport administration	Electronic payment Services	Security	Carwall
Cooperative collision warning	Travel and traffic management	Attacks like malicious code propagation and data integrity attacks could be launched	FUSE
Lane change warning	Coupling/decoupling	Voluminous magnitude of data	CANtact
Intersection collision warning	Management of commercial vehicles	Latency requirements of various applications require more research in the direction of V2I communication, infrastructure and protocols	CANard
Approaching emergency vehicles	Information management services	Addressing of moving vehicles becomes a challenge for packet delivery	Rolljam
Safety from hit and run cases	Maintenance and construction management	Economic simulation platforms need to be developed that have the ability to blend events, traffic and applications together	–
Emergency management	Infotainment services	Vehicles have vulnerable components like Electronic Control Unit (ECU), CAN, Global Navigation Satellite System (GNSS) and software	–
Roll over warning	Location based services	Security and privacy	–
Work zone warning	Traffic management	Security and privacy	–

2.4 *Smart Industry*

The manufacturing sector will soon be witnessing a revolution as its production mode will shift from digital to smart. This is due to the rapid improvement of electrical and electronic innovations, manufacturing technology, and the Information Technology. For example, Industry 4.0 Strategy, China Manufacturing Plan, USA Reindustrialization, Europe 2020 Strategy, and Manufacturing reflow Strategy.

Following illustrates the various services offered by the smart industry as well as the challenges which cripple this development.

2.4.1 Water Monitoring

This includes checking of clean and waste water levels and measuring the quantities of piped clean water and bottled clean water. Moreover, the collection and treatment of waste water is managed [67]. However, designing an efficient plan for asset management is highly required [68].

2.4.2 Transport Assessment

By installing sensors inside vehicles, a check is kept on all the forms of transportation i.e. air, rail, maritime, road and off-road to make early predictions about malfunctions so that failures could be avoided [69].

2.4.3 Manufacturing

Handles everything about manufacturing starting from smart designing and planning, efficient making of items, their quick delivery and service [69]. Loss of assets is a challenge that needs to be looked upon [68].

2.4.4 Retail

Monitors everything from storage to the point of sale by optimization of logistics, supervision of levels in silos, tanks and cisterns and quick movement of vehicles between the production and demand sites [69]. Creating an efficient plan for maintenance of assets is highly crucial [68].

2.4.5 Electricity Monitoring

This includes generation of all the forms of electricity viz. fossil fuels, nuclear sources and renewable sources, their transmission as well as the distribution gets monitored [69]. Advanced technologies are needed to detect and isolate faults in distributed, dynamic and interconnected environments like power generators [70].

2.4.6 Gas and Oil Monitoring

The exploration, extraction, refining and distribution of gas and oil are carried out smartly through the use of advanced technologies [69]. Fault diagnostic techniques that focus on non-linear modeling, non-linear state estimation, non-linear dynamics and non-parametric state estimation are needed [71].

2.4.7 Agricultural Services

Remote control over the operational routines at aquaculture and fisheries sites, crop farming sites, forestry and livestock sites through the use of sensors, cameras and wireless communication [69]. Data preparation in heterogeneous environments is a time consuming process. New methods of data cleansing, accounting, grouping and conversion are required for presenting data in a homogenous fashion [70].

2.4.8 Worker Safety Services

By installing meters that can monitor the levels of oxygen, ozone and toxic gases inside chemical plants, safety of workers is assured [69].

2.4.9 Location Services

By using active tags like Zigbee, UWB and passive tags like RFID and NFC, location of individuals and assets can be calculated [69]. However, lack of talented and skilled workers remains a challenge [71].

Some of the important projects related to smart industries are tabulated in Table 5.

Figure 2 presents a plot representing the cost involved (in Million Euros) and the number of partners in each smart industry project described in Table 5. The plot shows that Digital Lifecycle Twins for Predictive Maintenance (DayTiMe) [75] project involves highest cost while as Social Local Mobile indoor shopping experience (SOLOMON) [82] project involves the lowest cost among the mentioned projects. Moreover, CyberFactory [74] project has maximum number of partners

Table 5 Smart industry projects

Project title	Project head	Aim	Cost (k€)	Participating countries	Partners
Autonomous DataCenters for long term deployment (AutoDC) [72]	Tor Bjorn Minde from Sweden working for Ericsson	Enable the autonomous data centers to perform self-healing and other tasks without worrying about the contextual interference e.g. correcting over-heating in a factory without human intercession	7,305	Austria, Canada, Finland, Republic of Korea and Sweden	27
Cognitive Services for IoT based Scenarios (COSIBAS) [73]	Karl Waedt from Germany working for Framatome GmbH	To mix the cognitive AI and semantic technologies so as to extract better decisions, greater value of information and better forecasts with enhanced flexibility of the underlying system	13,986	France, Germany, Portugal, Spain and Turkey	28
CyberFactory [74]	Adrien Phillippe Becue from France working for Cassidian cyber security organization	It goes for planning, creating, coordinating and demonstrating a cluster of key empowering capacities to encourage optimization and flexibility of the Factories of Future (FoF) Uses electronics, machine and automobile manufacturing industries as use cases to develop procedures to deal with safety and security concerns of FoF	24,605	Finland, France, Germany, Portugal, Spain, Turkey and United Kingdom	39

(continued)

Table 5 (continued)

Project title	Project head	Aim	Cost (k€)	Participating countries	Partners
Digital lifecycle Twins for Predictive Maintenance (DayTiMe) [75]	Ad de Beer from Netherlands working for Philips electronics Nederland BV	Using a Generic value chain model, it aims to offer solutions for issues encountered in smart Manufacturing Predictive Maintenance (MPM)	29,131	Belgium, Canada, Finland, Netherlands, Norway, Spain, Turkey and United Kingdom	38
Intelligent IoT based Port Artefacts communication, Administration and Maintenance (I2PANEMA) [76]	Franz- Josef Stewing from Germany working for Materna GmbH	It plans to convey the intensity of IoT to enhance port activities, making them more productive and sustainable Aims to pave the way for a network of smart ports	20,081	Finland, France, Germany, Spain and Turkey	36
Predictive and Prescriptive Automation in Smart Manufacturing (PIANISM) [77]	Ahmet Cagatey Talay from Turkey working for Koç´Sistem	To create norms for empowering more adaptable and applicable solutions for manufacturers, PIANISM aims to develop standards and provide related missing investigation systems and algorithms, present new data identification, integration and modeling processes	6,167	Canada, Portugal, Spain and Turkey	15

(continued)

Table 5 (continued)

Project title	Project head	Aim	Cost (k€)	Participating countries	Partners
Smart Additive Manufacturing (SAMUEL) [78]	Alain Coulombe from Canada working for 3D Semantix	To develop an Additive Manufacturing (AM) knowledge base from various machine learning, data mining and advanced analysis ideas for helping the business people in all major phases of AM	6,092	Belgium, Canada and Spain	16
SMART-PDM [79]	Baris Bulut from Turkey working for Enforma Information and communication Technologies A.S	To lower the maintenance costs, improvement in throughput and quality Provide diagnostic and prognostic information while maintaining the fiscal feasibility of the system	12,195	Finland, Germany, Portugal, Romania, Spain and Turkey	29
Optimized Industrial IoT and distributed control platform for Manufacturing and Material handling (OPTIMUM) [80]	Anja Maria Fischer from Germany working for Demag Cranes and Components GmbH	To utilize the capacities of IoT technologies in industrial needs, improve the control and backing applications using contextual awareness and common-model oriented 3D engineering and supervision	12,571	Germany, Republic of Korea, Romania, Spain, Turkey and United Kingdom	22

(continued)

Table 5 (continued)

Project title	Project head	Aim	Cost (k€)	Participating countries	Partners
A new interface standard for integrated Virtual Material Modeling in Manufacturing Industry (VMAP) [81]	Klaus Wolf from Germany working for Fraunhofer SCAI	Expects to pick up a typical understanding and interoperable definition for virtual material models in Computer Aided Engineering (CAE) To set up an open and seller unbiased material exchange interface standard community that will work for alleviating various standardization issues	15,884	Austria, Belgium, Canada, Netherlands and Switzerland	29
Social Local Mobile indoor shopping experience (SOLOMON) [82]	Martin Johannes Treiber from Autria working for IKANGAI GmbH	To provide a "shop operation and experience" framework that will connect various technologies and information sources into smooth and coherent services for communicating with retailers, shop keepers and customers	2,481	Austria, Romania, Spain and Turkey	7
Industrial Enterprise Asset Value Enablers (INVALUE) [83]	Arda Gureller from Turkey working for Ericsson Ar-Ge	To support recent paradigms of smart industry by designing and displaying an open and shared integrated service platform	4,715	Belgium, Portugal, Spain and Turkey	15

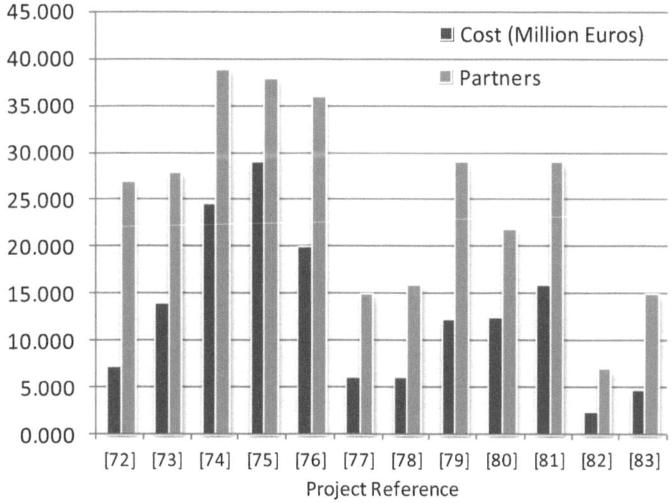

Fig. 2 Cost involved and number of partners in each project

and Social Local Mobile indoor shopping experience (SOLOMON) [82] project has minimum number of partners among the mentioned projects.

3 Conclusion

Smart cities provide an opportunity to connect people and places using innovative technologies that contribute to better urban planning and administration. The striking advances in device miniaturization, cloud growth and wireless connectivity paved the way for the realization of smart city concept. At the core of smart cities are the collection, management, analysis and visualisation of massive magnitude of data that is produced every minute in an urban environment through social, anthropogenic or natural environmental events or other activities. Smart cities data can be collected directly from a variety of sensors, smartphones, residents and incorporated with city data repositories to conduct analytical reasoning and generate required information or new decision-making knowledge for better urban governance. ICT technologies provide the ability to monitor and process smart city data, and provide local stakeholders with accurate and appropriate information for decision-making. In this chapter, we tried to bring a clear understanding of which projects led to the actual realization of the smart city concept. The most important pilot projects in the constituting domains of smart city viz. smart health, smart buildings, smart transport and smart industry were looked upon. Moreover, in each of these domains, major challenges and future research directions were sketched out to help researchers in identifying the research gaps.

References

1. Bandyopadhyay D, Sen J (2011) Internet of things: applications and challenges in technology and standardization. Wirel Pers Commun 58:49–69
2. Gubbi J, Buyya R, Marusic S, Palaniswami M (2013) Internet of things (IoT): a vision, architectural elements, and future directions. Future Gener Comput Syst 29:1645–1660
3. Showkat S, Qureshi S (2020) Securing the Internet of things using Blockchain. In: 10th international conference on cloud computing, data science & engineering (confluence). IEEE
4. Khan Z, Anjum A, Kiani S (2013) Cloud based big data analytics for smart future cities. In: Proceedings of the 2013 IEEE/ACM 6th international conference on utility and cloud computing. IEEE Computer Society, pp 381–386
5. Kitchin R (2014) The real-time city? Big data and smart urbanism. GeoJournal 79:1–14
6. Amsterdam Smart City (2019). https://amsterdamsmartcity.com/. Accessed August 2019
7. The Internet of Everything for Cities (2019). https://pie.pascalobservatory.org/sites/default/files/ioe-smart-city_pov.pdf. Accessed August 2019
8. Smart Cities Pilot—Public Transport Planner—Groningen (2019). https://www.smartcities.info/smart-cities-pilot-public-transport-planner-groningen/. Accessed August 2019
9. Smart City Wien (2019). https://eu-smartcities.eu/blog/project-smart-city-wien/. Accessed August 2019
10. Gerhard P, Hancke et al (2013) The role of advanced sensing in smart cities. In: Sensors
11. Santander Summary (2019). https://smartsantander.eu/wiki/index.php/Testbeds/Santander/. Accessed August 2019
12. Malaga Living Lab. Available online: https://www.openlivinglabs.eu/livinglab/m%C3%A1lagaliving-lab/ (accessed August 2019).
13. Smart City|Servei de Premsa|El Web de la Ciutat de Barcelona (2019). https://ajuntament.barcelona.cat/premsa/tag/smart-city/. Accessed August 2019
14. Laursen L (2019) City saves money, attracts businesses with smart city strategy. https://www.technologyreview.com/s/532511/barcelonas-smart-city-ecosystem/. Accessed August 2019
15. TMBAPP (Metro Bus Barcelona)|Apps|iTunes|apps4BCN|All the Apps You Need for Barcelona! (2019). https://apps4bcn.cat/en/app/tmbapp-metro-bus-barcelona/111. Accessed September 2019
16. Transport & Traffic|IOS|The Best App Selection for Barcelona|Apps4bcn|All the Apps You Need for Barcelona! (2019). https://apps4bcn.cat/en/apps/index/Category:transport-i-tr-nsit. Accessed September 2019
17. Smart Cities Pilot—Norfolk Customer Insight—Norfolk County Council (2019). https://www.smartcities.info/smart-cities-pilot-norfolk-customer-insight-norfolk-county-council/. Accessed September 2019
18. Writer SB-S (2019) Staff modest gains in first six months of Santa Cruz's predictive police program. https://www.santacruzsentinel.com/article/zz/20120226/NEWS/120227300 (accessed September 2019
19. Fujisawa SST (2019). https://panasonic.net/es/fujisawasst/. Accessed September 2019
20. PlanIT Valley (2019). https://planitvalley.org/. Accessed September 2019
21. Strickland E (2011) Cisco bets on South Korean smart city. In: IEEE Spectrum
22. Songdo (2019). https://www.songdo.com/. Accessed October 2019
23. ICT Regulation Toolkit (2019). https://www.ictregulationtoolkit.org/practice_note?practice_note_id=3244. Accessed October 2019
24. Zahra S, Chishti M (2020) A collaborative edge-cloud internet of things based framework for securing the indian healthcare system. Int J Sens Wirel Commun Control. https://doi.org/10.2174/2210327910666191218144157
25. Katz S, Ford A, Moskowitz R, Jackson B, Jaffe M (1963) Studies of illness in the aged, the index of adl: a standardized measure of biological and psychosocial function. J Am Med Assoc 185:914–919
26. Harper R (2003) Inside the smart home. Springer, Berlin

27. Freitas D, Marcondes T (2015) A health smart home system to report incidents for disabled people. In: Proceedings of the 2015 international conference on distributed computing in sensor systems, Fortaleza, Brazil, pp 210–211
28. Radcliffe J (2011) Hacking medical devices for fun and insulin: breaking the human SCADA system. In: Proceedings of the black hat conference presentation slides, Las Vegas
29. Bennett J, Rokas O, Chen L (2017) Healthcare in the smart home: a study of past, present and future. In: Sustainability
30. Strickland E (2017) DARPA wants brain implants that record from 1 million neurons. In: IEEE Spectrum
31. Coiera E (1996) The internet's challenge to health care provision. BMJ 312:3–4
32. Aguirre E, Flores M, Azpilicueta L, López-Iturri P, Falcone F, Ramos V, Solanas A (2014) Implementing context aware scenarios to enable smart health in complex urban environments. In: IEEE international symposium on medical measurements and applications, pp 508–511
33. Casino F, Azpilicueta L, Lopez-Iturri P, Aguirre E, Falcone F, Solanas A (2016) Optimised wireless channel characterisation in large complex environments by hybrid ray launching-collaborative filtering approach. IEEE Antennas Wirel Propag Lett PP(99):1–4
34. Casino F, Azpilicueta L, Lopez-Iturri P, Aguirre E, Falcone F, Solanas A (2015) Hybrid-based optimization of wireless channel characterization for health services in medical complex environments. In: 6th international conference on information, intelligence, systems and applications (IISA)
35. Zhang X, Jiang H, Zhu B, Chen X, Zhang C, Wang Z (2010) A low-power remotely-programmable MCU for implantable medical devices. In: Proceedings of the IEEE Asia Pacific conference on circuits and systems (APCCAS), Kuala Lumpur, Malaysia, pp 28–31
36. Warwick K, Gasson M (2006) A question of identity. In: Proceedings of the wireless sensor networks conference, London, UK, 4 December 2006, pp 1–5
37. Cattell R (2011) Scalable SQL and NOSQL data stores. SIGMOD Rec 39(4):12–27
38. Gormley C, Tong Z (2015) Elasticsearch: the definitive guide, 1st edn. O'Reilly Media, Inc.
39. Dean J, Ghemawat S (2008) MapReduce: simplified data processing on large clusters. Commun ACM 51(1):107–113
40. Alivecore. https://www.alivecor.com/en/press/press_release/new-kardia/. Accessed October 2019
41. Sahni Y, Cao J, Shen J (2018) Challenges and opportunities in designing smart spaces. In: Internet of everything, internet of things. Springer Nature
42. Schafer J, Frankowski D, Herlocker J, Sen S (2007) Collaborative filtering recommender systems. In: The adaptive web. Springer, pp 291–324
43. Goldberg D, Nichols D, Oki B, Terry D (1992) Using collaborative filtering to weave an information tapestry. Commun ACM 35(12):61–70
44. Mukhopadhyay S (2014) Wearable sensors for human activity monitoring: a review. IEEE Sens J 15:1321–1330
45. Bonato P (2003) Wearable sensors/systems and their impact on biomedical engineering. IEEE Eng Med Biol 22:18–20
46. Cao T, Li D (2013) Human interaction based Robot Self-Learning approach for generic skill learning in domestic environment. In: Proceedings of the IEEE international conference on robotics and biomimetics (ROBIO), Shenzhen, China, pp 203–208
47. Cook D (2012) How smart is your home? Science 335:1579–1581
48. Rav D, Wong C, Deligianni F, Berthelot M, Andreu-Perez J, Lo B, Yang G (2017) Deep learning for health informatics. IEEE J Biomed Health Inform 21(1):4–21
49. Kumar L, Vigneswaran C (2015) Electronics in textiles and clothing: design, products and applications, 1st edn. CRC Press, Boca Raton
50. Fan F-R, Lin L, Zhu G, Wu W, Zhang R, Wang ZL (2012) Transparent tribo electric Nano generators and self-powered pressure sensors based on micropatterned plastic films. Nano Lett
51. Boden R (2019) Thinfilm shows printed NFC smart label with temperature sensor. https://www.nfcworld.com/2014/05/28/329390/thinfilm-shows-printed-nfc-smartlabel-temperature-sensor/. Accessed October 2019

52. Mitsubayashi K, Wakabayashi Y, Murotomi D, Yamada T, Kawase T, Iwagaki S, Karube I (2003) Wearable and flexible oxygen sensor for transcutaneous oxygen monitoring. Sens Actuators B Chem 95:373–377
53. Someya T, Sekitani T, Iba S, Kato Y, Kawaguchi H, Sakura T (2004) A large-area, flexible pressure sensor matrix with organic field-effect transistors for artificial skin applications. Proc Natl Acad Sci USA 101:9966–9970
54. Mennicken S, Vermeulen J, Huang E (2014) From today's augmented houses to tomorrow's smart homes: New directions for home automation research. In: Proceedings of the 2014 ACM international joint conference on pervasive and ubiquitous computing. ACM, pp 105–115
55. Solanas A, Casino F, Batista E, Rallo R (2017) Trends and challenges in smart healthcare research: a journey from data to wisdom. In: IEEE
56. Tax N, Verenich I, Rosa M, Dumas M (2017) Predictive business process monitoring with LSTM neural networks. Springer International Publishing, pp 477–492
57. Martirano L, A smart lighting control to save energy. In: Proceedings of the 6th IEEE international
58. Bushby S (1997) BACnet: a standard communication infrastructure for intelligent buildings, automation. Construction 6:529–540
59. Anwar M, Nazir M, Mustafa K (2017) Security threats taxonomy: smart-home perspective. In: IEEE
60. Ozkan N, Davidson R, Bicket M, Whitmarsh L (2013) Social barriers to the adoption of smart homes. Elsevier Energy Policy 63:363–374
61. Ashrae handbook HVAC Systems and Equipment, American Society of Heating, Refrigerating and Air-Conditioning Engineers, Inc (2004)
62. Wong J, Li H, Wang S (2004) Intelligent building research: a review. In: Automation in construction. Elsevier
63. Castro M, Jara M, Antonio F, Skarmeta G (2013) Smart lighting solutions for smart cities. In: 27th international conference on advanced information networking and applications workshops. IEEE
64. Fitzpatrick G, Stringer M, Harris E (2006) Lessons for the future: experiences with the installation and use of today's domestic sensors and technologies. In: Pervasive computing: proceedings of the 4th international conference. Springer, pp 383–399
65. Chan W (1999) A.T.P. So. In: Intelligent building systems. Kluwer Academic Publishers, Boston
66. Zahra S (2018) MNP: malicious node prevention in vehicular ad hoc networks. IJCNA 5(2):9–21
67. Beecham Research (2014) M2M Sector Map. https://www.beechamresearch.com/download.aspx?id=18
68. Smart industry (2019). https://www.softwebsolutions.com/resources/IoT-solutions-for-smart-industry.html. Accessed October 2019
69. Boyes H, Hallaq B, Cunningham J, Watson T (2018) The industrial internet of things (IIoT): an analysis framework. In: Computers in industry, vol 101. Elsevier, pp 1–12
70. Serpanos D, Wolf M (2018) Industrial internet of things. Springer International Publishing AG
71. Sadiku M, Wang Y, Cui S, Musa S Industrial internet of things. Int J Adv Sci Res Eng
72. AutoDC (2019). https://itea3.org/project/autodc.html. Accessed December 2019
73. Cosibas (2019). https://itea3.org/project/cosibas.html. Accessed December 2019
74. CyberFactory (2019). https://itea3.org/project/cyberfactory-1.html. Accessed December 2019
75. Daytime (2019). https://itea3.org/project/daytime.html. Accessed December 2019
76. I2panema (2020). https://itea3.org/project/i2panema.html. Accessed January 2020
77. Pianism (2020). https://itea3.org/project/pianism.html. Accessed January 2020
78. Samuel (2020). https://itea3.org/project/samuel.html. Accessed January 2020
79. Smart-pdm (2020). https://itea3.org/project/smart-pdm.html. Accessed January 2020
80. OPTIMUM (2020). https://itea3.org/project/optimum.html. Accessed February 2020

81. Vmap (2020). https://itea3.org/project/vmap.html. Accessed February 2020
82. Solomon (2020). https://itea3.org/project/solomon.html. Accessed March 2020
83. inValue (2020). https://itea3.org/project/invalue.html. Accessed March 2020

Performance of Smart Cities Concerning the Use of Internet of Things: A Case Study of Four Indian Himalayan Cities

Mahfuzuar Rahman Barbhuiya, Navneet Munoth, and Ranjeet Singh Rajput

Abstract The rapid urbanization of Indian cities is leaving the available resources very limited for the city dwellers. The problems related to cities need special attention else it may never be solved. The best example is the Dharavi slums in Mumbai, where even after decades, proper infrastructural facilities could not be provided [1]. Similar fear is growing among Urban planners and policymakers to transform cities in such a way that it can address the unforeseen problems efficiently without disturbing the regular city routine. The efficiency in managing a city's available resources can be brought with the help of the Internet of Things (IoT). The present study tries to investigate the performance of the smart city initiative, which is adopted by 100 Indian cities and investigates how IoT can be used efficiently. The results from the four selected Himalayan cites shows that policymakers ignored the need of the city dwellers and prioritized tourism and aesthetic appeal of the cities. The lack of attention to the management of primary resources can be seen by the percentage of funds that is allocated to each City. The study recommends allocating funds equally in the seven identified categories. The study also suggests and discusses the strategies that can be taken up as a top priority. Building trust and citizens participation were found to be the most critical factors for the implementation of a development project in a city.

Keywords Internet of things · Smart city · Citizen participation · Trust building · Himalayan cities · India

M. R. Barbhuiya (✉) · R. S. Rajput
Industrial and Management Engineering Department, Indian Institute of Technology Kanpur, Kanpur, Uttar Pradesh, India
e-mail: oyemehfuz@gmail.com

R. S. Rajput
e-mail: ranjeetsme@gmail.com

N. Munoth
Department of Architecture and Planning, Maulana Azad National Institute of Technology Bhopal, Bhopal, Madhya Pradesh, India
e-mail: navneet.munoth@gmail.com

M. Khan et al. (eds.), *Smart Cities: A Data Analytics Perspective*,
Lecture Notes in Intelligent Transportation and Infrastructure,
https://doi.org/10.1007/978-3-030-60922-1_13

1 Introduction

In India, more than 30% of the population lives in cities, and it is expected to exceed 50% by 2050 [2]. The management of such a vast population is a big task for a country like India. Many countries have adopted the smart city initiative and are further using the Internet of Thing (IoT) to manage cities. The concept of 'smart city' and 'smart architecture' in the age of IoT are complementary and should be used for better planning and designing of buildings and cities [3]. The Internet of things is instrumental and widely used in some countries, but for a country like India, this concept is relatively new [4]. With the help of IoT, a city can be smarter and can make decisions concerning the City, quickly and appropriately using the right kind of information [5]. Sembroiz et al. [6] observed that the development of both software and hardware systems have tremendously led to the Smart city revolution using IoT based Wireless Sensor Networks [6]. The smart City tries to manage and control heterogeneous domains such as healthcare, waste and traffic management intelligently and provides smart solutions [7, 8].

Presently the cities are consuming more resources and have become more demanding. IoT, Big data analytics and cloud computing are the three vital pillars of the smart city. IoT works on smart sensors and intelligent networks and further generates big data from these smart and intelligent devices which are finally stored in the cloud for computing, analysis, monitoring and privacy and security purposes [9]. Young et al. [10] noted that the IoT revolution has made the Smart city concept possible [10]. City planners and developers are focused on working on the data extracted from the innumerable sensors across the City for the benefit of the city dwellers. Although the City becomes more secure, energy-conscious and mobile, the implementation of IoT, along with big data analysis, is still very costly [11].

Thirumalai et al. [12] noted that with the adoption of IoT in our daily life, our efficiency could be improved considerably [12]. IoT enabled transportation network, water supply, drainage systems and renewable energy sources are some of the functions which can help in making a city dweller more efficient [12]. India has implemented the smart city initiative in 2015, which aims to transform 100 cities across India. With the massive advancement in the field of IoT, this study aims to investigate the performance of a few selected smart cities. The research questions that will be addressed are as follows:

RQ1. Tounderstand the extent to which the different factors that are associated with smart cities help in making the cities smarter.
RQ2. Toevaluate the existing literature in the Indian context and find gaps in theoretical and practical understanding of the smart city initiative in India.
RQ3. To understand the performance of the selected cities with the help of data.

To understand each question thoroughly, first, we look at the reports and guidelines issued by the government at three levels; namely, the centre, state and City, were studied. Secondly, a thorough literature review was undertaken. Thirdly, the present status of the smart city initiative was studied, and the practical challenges and expectations were outlined. Finally, the performance of the City was evaluated based on the fund allocation to each category, as well as the number of new and technologically advanced measures were used. It is worth mentioning that the study does not rank the cities; it only evaluates the selected cities. This is because all the cities, although similar, differ in terms of location, area and population.

2 Research Motivation and Methodology

The 'Smart City Mission' was launched by the Government of India on June 25, 2015. Since then, there has been a tremendous push to improve the quality of life for city dwellers by providing better infrastructure, a clean and sustainable environment and by making cities sustainable and inclusive. An exhaustive and detailed process was adopted to select cities for development. All the states were invited by the Ministry of Urban Development to send a list of the cities that they want to develop.

The present study has chosen four Himalayan cities to see where the Indian cities stand presently after five years from the launch of the initiative. The study also investigates the use of the Internet of Things (IoT) in the smart city initiative. The study also tries to find gaps and suggest strategies to a better approach in meeting the objectives of the smart city initiative. The cities chosen are Dharamshala, Namchi, Shillong and Pasighat from the states of Himachal Pradesh, Sikkim, Meghalaya and Arunachal Pradesh respectively. These cities are exclusive to their states; thus, they do not have the issue of divided attention as has been the case with many other Indian cities where two or more than two cities were selected.

Furthermore, these cities lie in hilly regions and need a similar kind of attention from their respective city authorities. The gaps and current status were identified from the literature review and by studying the reports published by the City, state and centre. A detailed strategy is suggested for smooth implementation by the city authority. To develop the conceptual framework on factors that will help in building a smart city with the help of the Internet of Things, an extensive review of the literature was conducted. This literature review was done to have some idea of the literature available in the field of Smart Cities in the Indian context. This study was conducted between February 2020 and May 2020. Articles published in the English language were used in the present study. Only peer-reviewed journal articles, book sections, reports from government and international bodies, and articles presented in the conference of international repute were considered for this literature review. Other sources from which literature can be gathered, such as working papers, book reviews, theses, abstracts, internet columns and blogs were excluded from this study.

The databases used in this literature review are Google Scholar and Scopus. The keyword searching technique was used for finding relevant articles. Articles were considered only if the keywords' Internet of Things', 'IoT', 'Smart City', 'City', 'Architecture', 'Planning', or 'India' were found in the title or abstract of the paper.

Articles published between 2000 and 2020 were considered. The base year was taken as 2000, as the Internet became increasingly popular around that time, and cyber law came into effect from that year. A few crucial citations mentioned in the journal articles were added if they met the required criteria. A total of 214 articles were found in the database. After carefully reading abstracts and conclusions, 94 papers were kept. 48 of them were either not matching the interest of this research or out of scope, and all these articles were removed. Later, after going through methodology and findings, 46 articles were found to be relevant to the present study. 6 articles were included externally after reading the references of selected articles.

The selected articles were first separated out into the categories mentioned by the Government of India, and those which did not fall under any of these categories were grouped together under the heading ‘Sustainable Factors’. Utmost care was taken to keep the selection of articles fair and honest. However, there might have been possible bias during selection as a human error, which could be a limitation to this study. In addition, literature in languages other than English, which discuss the aspect of sustainability and resilience, must exist, but this literature review only considers articles published in the English language.

Finally, the selected cities are compared and ranked according to their progress and the initiatives that they have taken. The ranking is done using two criteria: First, how the funds were distributed in the seven identified categories, and recommendations were made accordingly; Second, the number of IoT based aspects used by each City to make them smart. Finally, the study suggests a few stepwise strategies for cities.

3 Literature Review

The literature review is broadly classified into categories like those given by the Ministry of Urban Development, Government of India [13]. A new category named ‘Sustainable Factors’ was added to accommodate the factors that are missing from the list provided by the government.

The six categories into which smart solutions were divided in India are:

1. **E-governance and Citizen Service**

   i Public information and grievance redressal.
   ii Electronic service delivery.
   iii Citizen engagement.
   iv Citizens – City’s eyes and ears.
   v Video crime monitoring.

2. **Waste Management**

   i Waste to energy and fuel.
   ii Waste to compost.
   iii Wastewater to be treated.
   iv Recycling and reduction of C&D Waste.

3. **Water Management**
   - i Smart Meters and management.
   - ii Leakage identification, preventive maintenance.
   - iii Water quality monitoring.
4. **Energy Management**
   - i Smart Meters and Management.
   - ii Renewable sources of energy.
   - iii Energy-efficient and green buildings.
5. **Urban Mobility**
   - i Smart parking.
   - ii Intelligent traffic management.
   - iii Integrated multimodal transport.
6. **Others**
   - i Tele-medicine and tele-education.
   - ii Incubation/trade facilitation centres.
   - iii Skill Development Centres.

A few factors, such as Industries and Social Media are not included in any of the categories within this literature review as the technological advancements in these fields are still in progress. Industries need more openness and acceptance towards IoT apart from factors such as industrial infrastructure and expertise, industrial acceptance and compatibility, security, cost-effectiveness and management support, for the successful implementation of Industrial IoT [14]. The use of Social Media as a source of IoT sensors is tricky, as it can be manipulated very easily. Until a proper framework is developed, Social Media cannot be used as a reliable source for data collection, even though it is used for rapid information transfer from authorities to citizens [15]. The following subsections are based on similar lines as proved in the Indian smart city Mission, with an added subsection for Sustainable Factors.

3.1 E-Governance and Citizen Service

This section lists the factors that help in making citizens' lives better and achieving better governance. Security is one of the essential factors in this category, and many researchers have opined its importance equally. Citizens want help from the government in every possible way to get a sense of security. The use of IoT can help the government in the smooth running of the City by assuring a sense of security. Narwane et al. [4] found that factors such as big data analytics, trust, customer–supplier intention, perceived security risk, manufacturing infrastructure, security and privacy, big data analytics, top management support, knowledge sharing, perceived ease of use,

IT support, trust, and external environment, all support finding the mediating role of IoT in governance and business [4]. Similarly, Asthana and Dwivedi [16] suggested using more user-friendly ways to incorporate IoT techniques into third party logistics to tackle the growing needs of a country like India [16]. Jiang et al. [5] observed that the reason behind the success of smooth governance and management largely depends on cloud-based system factors such as cloud security, scalability, on-demand self-service and resource visualization [5].

The process involved in incorporating IoT technology into city governance should be designed and implemented after proper security checks. Otherwise, the privacy of citizens will be at risk [17, 18]. Availability, appearance, and power grid access are essential factors when deciding the hackability of a city, and each City should consider these factors with utmost importance [18]. Since the concept of IoT is very new, the design should be universal so that the various models, applications, configurations and makes do not go out of use too soon, as a considerable cost of installation is also involved [19]. With proper care and judicious use of resources, the trust and transparency of information can be attained [4, 20].

3.2 Waste Management

Managing city waste is one of the most critical concerns, but it is not given the amount of attention it needs. For a city to provide a good quality of life, waste management and maintenance of infrastructure are equally important. Vamsi et al. [21] suggested using sensors in sewage systems to figure out the exact location where maintenance is required [21]. Singhvi et al. [22] studied city dumping grounds and dustbins, only to find that in India, dustbins do not give real-time information, and thus, sometimes, there is an overflow of toxic materials as municipal corporations have fixed cleaning times per day [22]. The use of smart dustbins with IoT based sensors can help municipal corporations to efficiently manage garbage and prevent an overflow of toxic materials, as well as skip trips to locations where daily cleaning of dustbins is not required [23]. This would help in saving lots of energy and human resources [22].

Panjagal and Ramaiah [24] noted that with rapid urbanization, municipal waste dumping has become the most common challenge in India, exposing citizens to the risk of being more vulnerable to numerous health issues apart from air, soil and water pollution [24]. Open dumping also leads to the emission of dangerous greenhouse gases such as carbon monoxide, carbon dioxide, volatile organic compounds and methane [24, 25]. An IoT based sensor to detect the concentration of poisonous gases can be a possible sustainable solution [24].

3.3 Water Management

It is said that the third world war will be fought for safe and potable drinking water. Water management has become even more critical, so much so, that it can be found directly or indirectly in five out of seventeen Sustainable Development Goals, 2015 [26]. To manage the safe and continuous supply of water, sensors can be used in plumbing systems with the help of IoT technology. By using seasonal and daily water usage data, water can be redirected to parts of the City which need it more [21, 27]. Irrigation and construction activities consume a considerable amount of water, so with the integration of IoT systems, operational cost and time can be saved, apart from the visible reduction of energy consumption [9]. Use of machine learning techniques will help in better management of data and accurate predictions of the quality and quantity of water required for different uses such as domestic, field and industrial in a city [28].

3.4 Energy Management

With the world moving towards technological advancements, the use of electricity consumption cannot be reduced but can be made more sustainable. Yadagani et al. [29] suggested the implementation of Smart Grid systems using IoT in residential sectors, and dynamic pricing systems to meet the variance in demand during the day [29]. Yadagani et al. [29] further concluded that a city needs to have sound IoT systems, communication techniques and advanced metering infrastructures for the efficient running of smart grid systems at home [29]. Using green, sustainable and intelligent transmission systems can help in the reduction of loss of energy during transmission as compared to the traditional method [30]. The use of IoT enabled devices in the form of smart meters can help in leading a sustainable lifestyle and can forecast energy consumption requirements, thus reducing wastage of energy in industrial, commercial and densely populated cities, which consume energy at an accelerated rate [31]. Pires et al. [32] suggested the use of an advanced metering system to measure electrical energy consumption by using real-time analysis of data from various consumption sources, which will help the City in managing electric energy efficiency and maintaining a balance between consumption and demand [32].

3.5 Urban Mobility

Traffic lights used in smart cities are static, time-based and pre-configured. This makes them costly and time-consuming. Metro cities in India waste more than 40,000 L of fuel each, only because of improper traffic management, and more than 400 traffic-related deaths are reported per day in India [33]. The use of IoT in traffic signals can help with having better user-friendly traffic control systems, thus saving time and energy [34]. A better-organized traffic system using IoT and remote sensing technology will help in reducing the accident rate, fuel consumption and traffic waiting time [33]. Also, an IoT system can be used to detect blind spots on the road, such as diversions on highways and crossings. Early warnings can be used to alert drivers about these blind spots [35]. Further, the technology can be used to analyze real-time data and manage traffic signals, lights and delays in traffic [7, 34, 36].

The use of Radio Frequency Identification (RFID) for smart parking is very convenient, efficient, secure and cost-effective [8]. Using RFID techniques can help in identifying and grouping vehicles in public places according to the need of the vehicles. For example, an electric car can be parked in a location which has free charging ports [8]. Tiwari et al. [8] also suggested that data from smart parking locations be analyzed in real-time and stored in the cloud using IoT techniques, thus giving users real-time data about the availability of parking spaces available [8].

The traditional streetlight systems have been one of the critical infrastructural facilities of the City. With the advancement of technology and the introduction of IoT, streetlights can be used as nodes for IoT sensors, thus solving the problem of connection and maintenance issues [37].

3.6 Others

Any city can adopt smart techniques, but to maintain, sustain, and finally evolve is the most challenging part. The investors are the government, and the private players want their money to be utilized efficiently, and after some period, they must get their money back with or without profit. The City can only generate revenue if it minimizes waste and optimizes its available resources [29, 38]. Implementation of smart grid systems for all types of services availed by the City's residents, and estimating economic benefits based on the return on investments is a way forward for every smart City [29, 38].

The importance and use of IoT in the healthcare system cannot be overlooked. Murthy [39] suggested a management solution for emergency facilities: Bed and blood booking using IoT with better integration between hospitals, blood banks and patients [39]. The study has found that delays caused in doing this for emergency cases like accidents can be reduced drastically [39].

3.7 Sustainable Factors

The study has identified a few more factors which are very much needed for the City to be categorized as a smart city. These factors are being used in many cities around the world and should be included in the Indian context as well.

a. **Big Data**
Data is fundamental, and it needs to be stored and analyzed in real-time. The depth of expertise with regards to usage skill, knowledge sharing, awareness and experience in IoT significantly determines the success of a city [5]. There is an urgent need to use urban-social data to make the City's infrastructure IoT enabled; this can be done using many mobile applications and collecting real-time data in a city dashboard [40]. Nayak and Joshi [25] suggested the implementation of a big data server in every proposed City from where the collected data will be mined and simplified [25]. Data will then be sent to different agencies such as planners, engineers, medical staff, city police and district authorities in order to make the City a fully automated one. There are many innovative ways to collect data, like the City of Rio de Janeiro collects and processes traffic data by using sensors attached to the city buses [41]. Furthermore, Young et al. [10] suggested connecting the IoT enabled vehicles to the server so that the data can be processed and analyzed in real-time and used smartly [10].
b. **Smart buildings**
More than 90% of a human being's life is spent inside a building [42]. Smart buildings and smart societies are critical aspects of a smart city, as these buildings and communities make up the smart City [43]. To study the smartness of a building, its adaptive behaviour needs to be looked at in detail, to understand the information flows and temporal behaviour in a building [44]. Before using any device, which is IoT based, it should be appropriately tested. Human behaviour should be noted to understand and improve the performance of the devices [6, 42]. The two most important aspects of designing a building are planning and operation. Finding the optimum location of the sensor and the occupant behaviour will help in developing the building management system (BMS) [6]. Rao et al. [45] suggested the integration of all household appliances to IoT sensors to make energy consumption more sustainable, thus making a house smarter [45]. Villegas-Ch et al. [11] studied a small university campus to see how the integration of IoT by using big data analysis would help in making a traditional campus smart, and concluded that the smart city initiative should start with small societies and then be developed in a better manner, rather than focusing on the whole City [11].

4 Data Used for Measuring the Smart City Performance

A thorough investigation was undertaken to get an idea of actual ground-level data for measuring the performance of the selected Cities. The data used were taken from the respective City's latest annual report on smart city initiative. Each City has divided funds according to its need and requirements into different categories. The sub-heads under which the funds were distributed was also mostly according to the needs of the City. Table 1 gives the idea of the funds allocated in different categories by the city administrators. The percentage of funds allocated under each category and the best performing city according to the fund allocation is discussed in this section.

Table 1 The table below shows the allocation of funds (in rupees crores) in different categories

Categories	Dharamshala	Namchi	Pasighat	Shillong
E-governance and citizen service	79.68	42.5	52.1	92.51
Waste management	166.04	111.75	38.17	114.45
Water management	253.07	22	15.81	98.09
Energy management	0	67.36	161.92	104.33
Urban mobility	551.86	260.95	222.68	411.35
Others	81.01	179	147.24	30
Sustainable factors	978.47	237	848.03	181.34
Total	2110.1	920.56	1485.95	1032.07

E-governance and Citizen Service is one of the most important categories for the smooth running of a city. Many cities, especially European cities, have taken different innovative strategies for making the city smart using IoT technology and implementing e-governance. The four selected cities somewhat lacked in allocating adequate funds in this category which is very visible with the kinds of initiatives undertaken by the cities which are mostly the installations of fibre networks, CCTV camera installation and data centres. Though these initiatives are essential, it's not looking at the future. Rupees 266.79 crores was allocated by four cities out of an available 8905 crore rupees, which is nearly 2.75% shows the improper distribution of funds.

The Shillong city focused mostly on the collection and metering of waste with funding nearly 97 crore rupees on them out of allocated 114.45 crores. The remaining funds were allocated for waste management and using smart techniques. Other three cities also saw an uneven fund allocation and not opting for advanced techniques for improving the present conditions. Overall, all four cities combined allocated 430.41 crore rupees for waste management, and only 82.63 crore rupees was awarded for smart and IoT based techniques. Water management was given some attention regarding metering and billing techniques. In contrast, the funding related to the implementation of leakage proof IoT based sensors or getting water to all the corners of a city lacked very much. All four towns combined allocated 388.16 crore rupees for water management out of the total allocated 8905.12 crore rupees. This allocation is just 4.35% of the total funds in this category. Dharamshala completely missed on allocating any funds for energy management or smart metering. Other three cities awarded a nearly equal amount of money; however, the highest allocation of funds was by Pasighat, which was almost 5.5% of its fund for energy management. The total combined allocated funds were least among all categories. It was only 3.75% of the total allocated funds.

Urban Mobility is one of the most acknowledged categories mainly because of the kind of problems faced by the city dwellers, mainly because of the steep terrain and harsh weather conditions. This category saw generous funding from each City. Shillong, Pasighat, Namchi and Dharamshala allocated 411.35, 222.68, 260.95 and

551.86 crores of rupees respectively for Urban Mobility. The category Others and sustainable factors are significant not only for the present but also for sustainable development. The funds' allocation completely ignored the immediate requirement of human beings, i.e. the homes and environments. Though these two categories were handsomely funded but mostly for tourism development and somewhat for infrastructure development for the public buildings like schools and hospitals. The combined funds allocated were 2682.09 crore rupees. With such a considerable investment, care should have been taken for developing smart homes and protecting the cultural identity of the cities.

5 Study Area and the Smart City Performance

This section first discusses briefly the profile of the City taken. It then presents the level of work undertaken by each City and where the work is undertaken—Whether pan-city or only for area-based development. Almost all the work has already started, but none of the work has been finished yet. The amount of funding dispersed is different for different cities. To check the sincerity and progress, this study investigates the list of work the City intends to do or is doing, and then compares and ranks the cities accordingly. Each City has proposed a list of work that it is willing to do with the fund allocated by the Government of India. All the seven identified categories are equally important, and the City's rank in this study will depend on the equal distribution of the fund among each category. A radar-based statistical evaluation is carried out for each City. In addition, this study makes a list of the works in which IoT technology can be used or is being used. Here, the study ranks the cities according to their technological advancements. Finally, the ranking of the cities is separately done in accordance with the number of initiatives that are taken, which need technological advancements. The initiatives that need the usage of Internet of Things, Big Data, Cloud Computing and Advanced Sensors will be counted. The motive is to design and develop the City for the next two–three centuries in such a way that they can adapt to new changes quickly rather than adopting measures which are already decades old.

Figure 1 shows the location of the selected cities. It can be noted from the map that only Pasighat City has somewhat straight roads while all other cities have curvy roads. This is mainly due to the terrain of the City, with Pasighat being situated at an elevation of 152 m whereas all other cities are at more than 1300 m elevation. The map also shows essential features like Police Stations, Banks, Hospitals and Markets; all these public infrastructures are almost evenly distributed within the cities, which is a unique advantage for them. Since the cities are in hilly terrain, the width of the roads is not enough for emergency purposes. Also, although most of every City is accessible, there is an urgent requirement for all-weather roads. Table 2 gives the cities' vital information in one place.

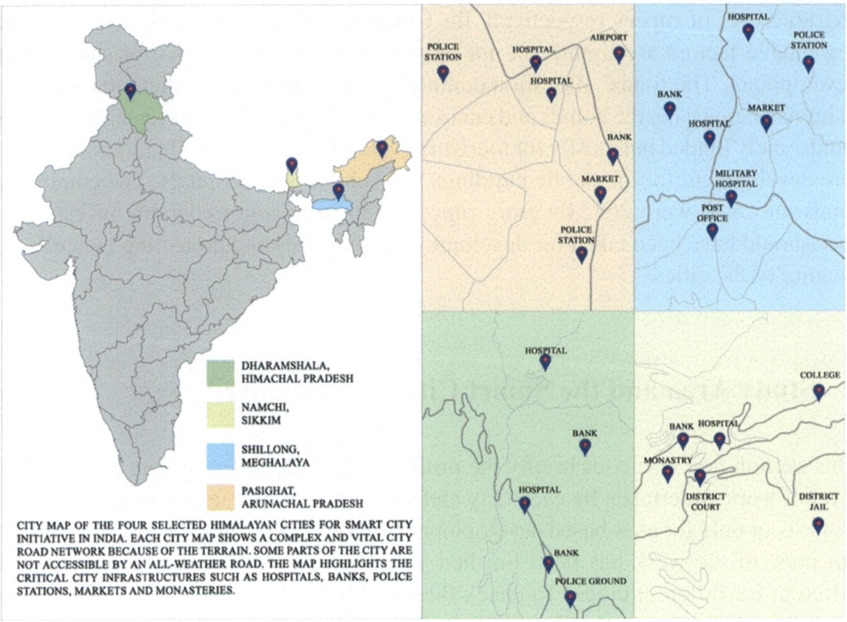

Fig. 1 The city map of the selected himalayan cities with important roads and public places

Table 2 The table below shows the city at Glance

City and state	Elevation	Population	Literacy (%)	Sex-ratio
Pasighat, Arunachal Pradesh	152	24,656	79.60	975
Shillong, Meghalaya	1730	1,43,229	92.81	1042
Dharamshala, Himachal Pradesh	1457	30,764	77	824
Namchi, Sikkim	1,315	12,190	91.30	977

5.1 Dharamshala

Dharamshala is in the Kangra district of Himachal Pradesh. It is the only third City in India which has the distinction of being the second capital of the state. The elevation is about 1457 m and 77% of the 30,764 residents are literate. Further, the City has a sex-ratio of 824 [46]. The City is committed to working on seventy-one aspects of the smart city project and has allocated money for different aspects of it. Surprisingly, the City has completely excluded the Energy Management aspect of the smart City. Urban mobility and Sustainable factors have got nearly three-fourth of the total budget, while the other five categories combined receive one-fourth of the total fund available. Dharamshala has mostly focused on conserving culture and heritage along with the development of tourism, which is good, but they have somehow ignored the basic needs of the people residing in the City. Figure 2 shows the difference in

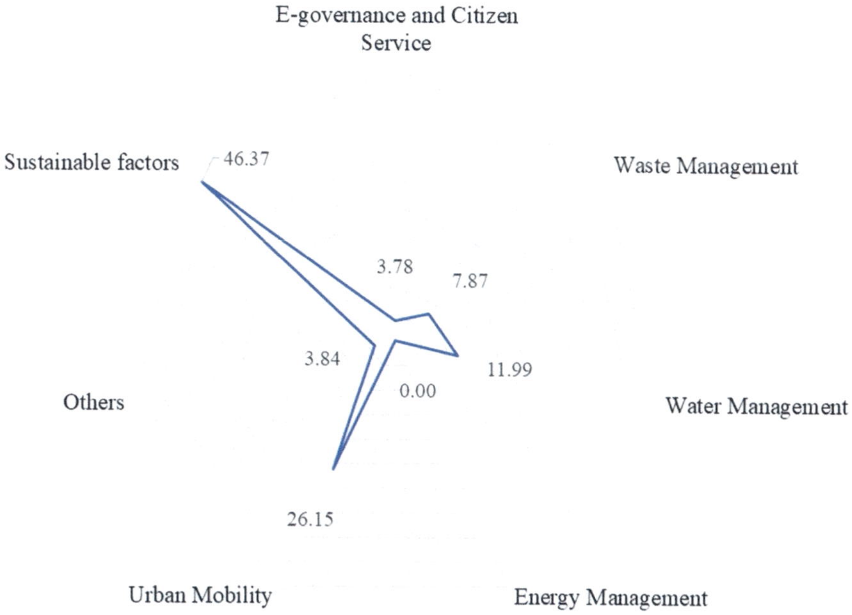

Fig. 2 The rader map of Dharamshala

fund allocation in categories by the city administration is 46.37% [47]. Out of the seventy-one aspects the City has committed to, thirty-seven aspects were focused on the use of the Internet of Things. Thus, 52% of the smart city aspects taken up by city administrators require IoT based development.

5.2 *Namchi*

Namchi is in the south district of the Sikkim state with an elevation of 1315 m. This small municipal council city has a population of only 12,190 out of which 91.30% are literate, and 977 females exist per1000 males [48]. Namchi city has divided the funds in a better way with none of the categories getting more than 30% of the total funds. The City has committed to working on Fifty-nine aspects of the smart city project, but it is worth noting that it has not allocated any funds for a few of the water management related aspects. Here, tourism and identity conservation were given more importance than the necessary facilities in the City. Figure 3 shows the category' e-governance and the citizen service again lack funding. The fund allocation difference here is around 25.96% [49]. Namchi city has taken twenty-three out of fifty-nine initiatives related to implementing IoT technology. Thus, nearly 39% of the initiative needs advanced technology.

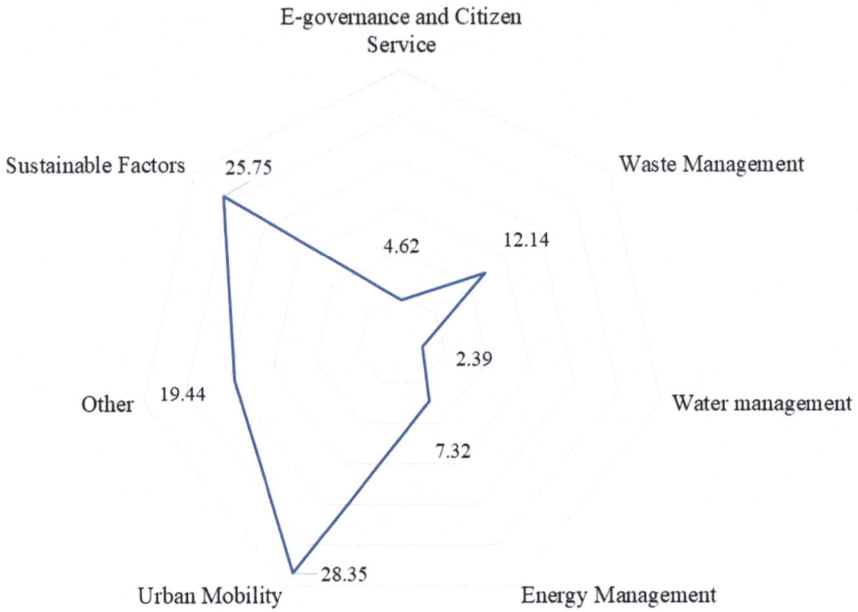

Fig. 3 The rader map of Namchi

5.3 *Pasighat*

Pasighat is in the East Siang district of Arunachal Pradesh. This small city is in the valley, situated along the Brahmaputra river with an elevation of 152 m. According to the 2011 population census, the City has a population of 24, 656 with a sex ratio of 975 and a literacy rate of 79.60% [48]. The City of Pasighat saw a very uneven distribution of the allocated funds. The City has committed to working on eighty-one aspects, but more focus was on sustainable factors, where importance was given to developing tourism and public buildings such as hospitals and slums. The City's waste management needs a quick review because of the proximity of the City to the river Brahmaputra and possible contamination if left unchecked. Figure 4 shows that the water management was given the least priority, with only 1.06% of the total fund allocated; and the City had the most uneven fund allocation with a difference of nearly 56.01% [50]. In Pasighat, out of eighty-one aspects, thirty aspects mentioned the need for IoT-based features. Thus, 37% of the aspects need IoT technology.

5.4 *Shillong*

Shillong is the capital city of the Meghalaya state. The City is located at an average elevation of 1730 m from the mean sea level. The City's population, according to

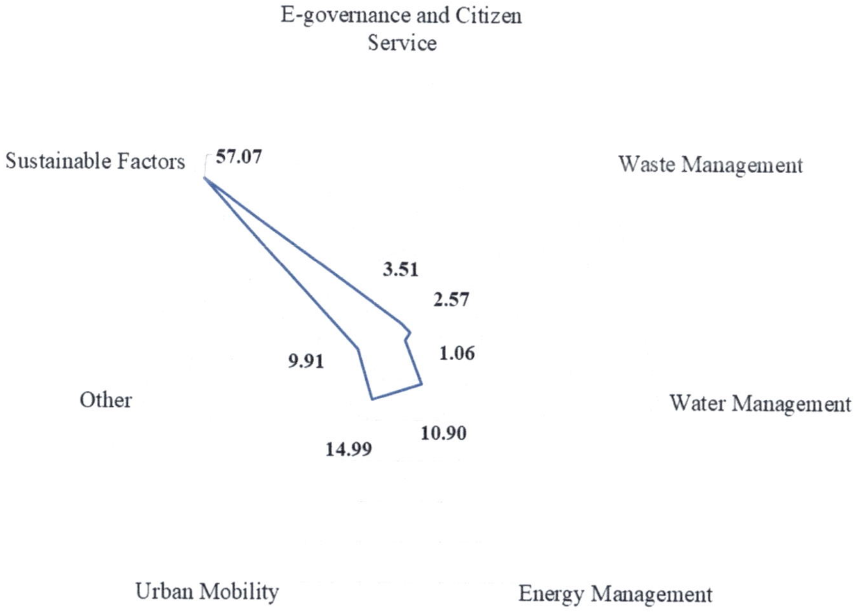

Fig. 4 The rader map of Pasighat

the census, was 143,229 with a literacy rate of 92.81% and sex ratio of 1042 [51]. The list of work that the City has compiled is only thirty-four. It has focused more on urban mobility, followed by sustainability factors. As per Fig. 5, 'Others' category, which investigates skill development, funding and investments, was given the least share of funding. The difference between the funds in different categories was 36.95% [52]. Fourteen out of thirty-four items in the Shillong city need technological advancements in the form of IoT; which is around 41% of the smart city initiative in the Shillong city.

6 Findings from the Smart City Performance

It has been observed in all the cities that the focus was on the development of tourism, as the City's economy is mostly dependent on it. Shillong focused more on urban mobility as, being the state capital, tourism was bound to happen, but to move people in and out of the City, it needs infrastructure. The three significant aspects of any city are water, waste and energy, and all these three categories were given the least share of funding in each City. Urban mobility was mostly given the second level of priority in each City, as it is evident from Fig. 6 itself that the cities need better road infrastructure due to terrain, and because of the heavy rainfall and harsh winter conditions that they experience. The best performing city according to the fund

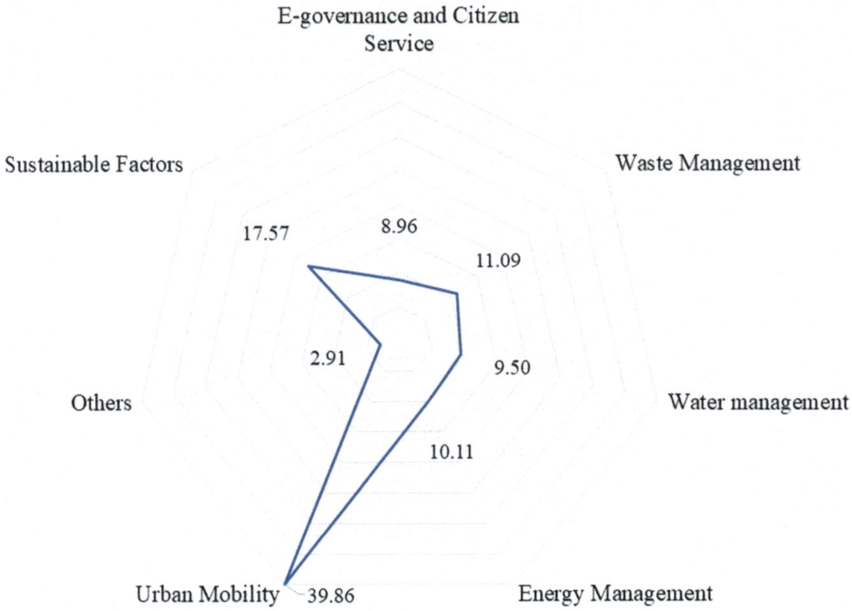

Fig. 5 The rader map of Shillong

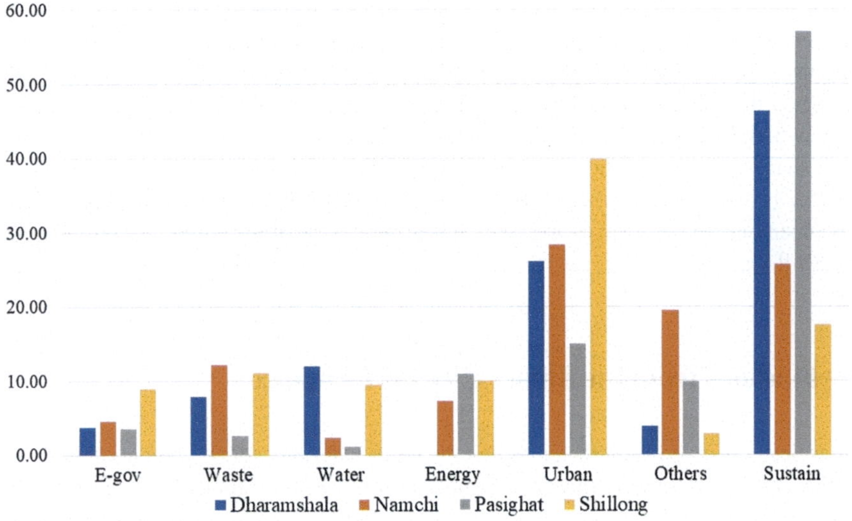

Fig. 6 The City-wise percentage of funds allocated in different categories

allocation mechanism is Namchi city, while Dharamshala and Shillong have done equally badly in the skewed funding allocation to different categories.

In terms of the incorporation of the Internet of Things based aspects in the smart city initiative, Dharamshala is the only city with more than half of the initiative being advanced. Although all the aspects are smart, the focus is on the technologically advanced aspects. Shillong, Namchi and Pasighat have respectively used 41, 39 and 37% advanced aspects from the total committed aspects by the City.

It is worth mentioning that both the relatively bigger cities, Dharamshala and Shillong, have focused on technological integration, as well as on developing their roads, tourism, and maintaining the identity of the City for which they are known. The Namchi and Pasighat city is relatively small, and the city administration is more focused on uplifting the quality of life of the residents. The cities have an identity of their own. Still, the character is not unique in a big country like India. Thus, the cities have smartly decided to improve the necessary amenities for their city-dwellers rather than going for branding. Overall, the City of Namchi has shown the most honest and sincere approach in the smart city initiative, and both the big cities need to do much to catch up in their approach.

7 Role of IoT in Improving the Performance of a City

The technological advancements have made everyone feel that the IoT can be used in every aspect of life, be it water, waste, energy, or mobility. This powerful IoT technology can be used by the government in the smooth running of City using e-governance methods and other sustainable approaches. From the literature review and the study of the selected cities, we can safely say that some factors affect at individual levels and some at community levels, as shown in Fig. 7. The identified factors that use IoT and impacts at individual levels are lifestyle activities and buildings; and the identified factors that use IoT and affects at community levels are transportation, environment, infrastructures and public utilities. All the above-identified factors collect data using different sensors and which can be analysed for better future decision making as well as for real-time analysis.

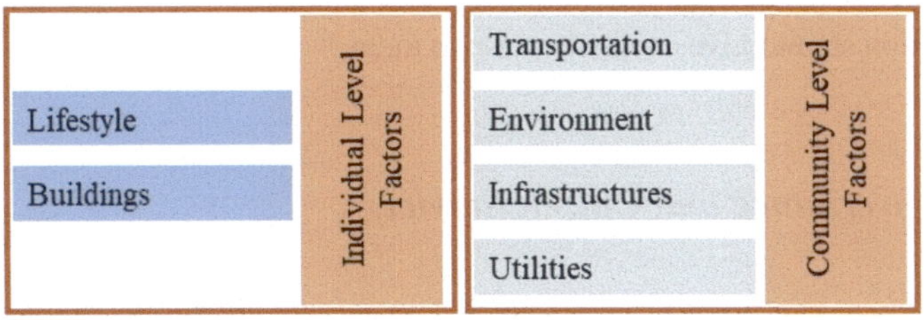

Fig. 7 The identified factors that uses IoT in improving the performance of a city

Transportation factor includes electric transport facilities, smart traffic controls and smart, fast lanes for emergency purposes. All these activities require data and sensors, which can easily be controlled and analyzed for decision making. Environmental factors include green buildings, landslide and forest fire prevention measures, early earthquake detection techniques, wind turbines as energy sources, solar energy systems and air pollution control techniques. The data used in the environmental factor will not only help in maintaining systems but also the prevention of disasters and thus significantly saving lives and property. The third identified community-level factor is infrastructures, which includes management of waste and detecting the release of poisonous gases, smart lighting across the City and checking structural health of all the City's physical buildings. The last identified factor is the public utility which includes aspects like water leakage detection, monitoring of the quality of potable water, leakages from the factories and industries and smart energy grid. All the aspects mentioned above the whole City and are mostly governed by the city administrators.

The buildings affect an individual directly, and the aspects of the buildings are the fire safety sensors enabled with IoT technology, overall building management systems used, safety sensors and smart metering techniques used. All the smart aspects in a building that used IoT technology and gives real-time data helps in the conscious use of resources by the occupants. Finally, the lifestyle activities that use IoT technology are Wi-Fi connections, smart parking and real-time city updates. As we can see, there is a vast use of data driven IoT in the smooth running of a city. Proper management, utilization and analysis of the data will help in smooth running and better performance of the City.

8 Discussions and Strategies

In the Indian context, the selected cities for the smart city's initiative have taken a positive step towards becoming more technologically advanced. The factors and aspects of technology integration are not structured enough. Through this section, the study aims to provide a stepwise strategy to approach the smart city aspects. All the aspects are broadly divided into the seven categories as found in the 'Sect. 3, literature review' of the study. All the categories are equally important, and the city administration should give equal attention to them.

8.1 E-governance and Citizen Service

The government can bring in the use of IoT in serving citizens better. Area-based development is the best way to check the efficiency of smart devices that are being used. Citizen registration and identification can be moved from 'Aadhar Cards' to a more secure system where the location and profile can be updated regularly. Also, the benefits that a citizen requires can be updated and directly implemented with the use of smart governance methods. A smart women safety tracker can be developed which, when turned on will work the heart rate data, and thus, give signals to the concerned authorities in case of unusual data. The e-governance part is very crucial as the performance of this category will instil confidence among the citizens to use devices that will track almost everything about the users.

8.2 Waste Management

Knowing the exact demand for water will help city managers to have the required amount of stock for the demand that may arise. It will help in reducing wastage. Collection, segregation and dumping of waste, although sounds straightforward, needs a lot of human resources, who would, however, be exposed to dangers; with the introduction of IoT based sensors, management of waste can be done more efficiently.

8.3 Water Management

Poisonous gases and unhealthy conditions are detrimental for both human lives and the environment. With smart waste management, the level of toxicity can be measured, and the corrective measures can be taken immediately. Also, sensors can help in identifying the areas which need immediate attention and the corrective measures that are required. Data-driven water management should be considered so that water can be saved and, if needed, diverted to the areas which require more water.

8.4 Energy Management

In India, smart metering and smart grid electric power supply are not a very common thing, thus leading to a massive loss of power. The smart meters can help in saving much energy which otherwise gets lost in the traditional method of power transmission. The City should also push for the usage of smart appliances in homes, which in turn will not only help in saving energy but will also alert users of any faulty connections.

8.5 *Urban Mobility*

The smart metering system should be used in all kinds of transportation mode both within and for domestic travel between cities. This will help tremendously in forecasting demand, thus making proper provisions for supply. The smart street lighting system will help in providing the lighting requirements intelligently and will also act as the nodes for collecting information and sending it to cloud servers.

8.6 *Others*

Smart initiatives will attract investors to pump in money for more development related works. With the explicit assurance that the money that is being invested will benefit the investors, more money can be brought into the City and quality of life can be improved. Health and safety of the workers are very crucial, whether the employees are working in risk zones or in closed spaces; with IoT sensors, the behaviour of the employees can be tracked, and recommendations can be made if they need rest or a break from their present work.

8.7 *Sustainable Factors*

Tracking and monitoring of the data are vital in any system that uses IoT technology. The focus should be on making these tracking activities more robust and safer. One of the best solutions can be to use smart heating, ventilation and air-conditioning systems, which can cover and collect all the data from a house. A smart and adaptive home should be a priority for every City. If every home is smart, naturally, the society and the City becomes smart. Proper care should be taken in handling the data generated from each home, and the multiple data sources should be combined to avoid leakages of data.

9 Conclusion

The development of a nation depends on the quality of life that its citizens enjoy. With ever-increasing load on the City's infrastructure, there is an urgent need to make cities smart so that the limited resources can be used judiciously. The general guidelines for the smart city mission lack in many aspects. This is the reason for not having any well-known Indian smart cities. The guidelines have suggested six broad categories for smart solutions but have entirely missed the few fundamental aspects like conservation of heritage and culture and also the tourism sector, which contributes to nearly 10% of Indian GDP [53]. Heritage plays a significant role in giving identity to a city, and proper care should be taken [54]. Also, the guidelines have lacked proper strategies for public infrastructures like hospitals, schools, colleges, etc. Data management and securing data is one of the most crucial aspects of using IoT in smart cities, and this aspect was also not adequately addressed.

The present study on the four cities shows that there is a considerable gap in the way city managers proceed with the funds allocated to them and what should be the sustainable way of moving forward. The cities have tried to give more focus to tourism and its related services rather than taking up necessary services and improving the quality of life of the citizens. The hilly location of the cities needs special attention as it has unique requirements and should be dealt with accordingly. The underlying focus should be to make cities self-reliant and to minimize the use of resources from outside the City [55]. The three essential and crucial categories, namely, water management, waste management and energy management, should be given the utmost importance. These three categories are to be used very intelligently as they take most of the natural resources available. One of the primary categories is the usage of IoT in the e-governance and citizenship services. The use of IoT can help citizens to get information electronically in real-time and have a sense of security. Crime and unusual activities can also be tracked using data generated from the IoT sensors. To instil confidence among the citizens, every city authority should take up e-governance first so that they can demonstrate to citizens the benefits of using IoT based sensors. It has also been noted that none of the cities has a well-maintained smart city dashboard.

Moreover, in cities like Namchi and Pasighat, a dedicated website is altogether missing. To make a city smart, the city administrators should take smart and calculated decisions on the way to approach the simple-looking complex problem. More involvement of the resident community and gradually building trust will help in getting more cooperation from the locals in managing and implementing plans and schemes more efficiently.

10 Study Limitations and Scope for Future Research

A more granular level of data can help in getting a clear picture of the level of development in a city. The data, in general, are mostly old, and thus, it cannot be used to forecast any development related project efficiently. The citizens perspective and their behavioural analysis can be studied to know the expectations of the citizens in a city development project and more specifically, in projects that are connected to the smart city initiative. A framework may be developed so that it can have policy implications on city development.

References

1. Kshetrimayum B, Bardhan R, Kubota T (2020) Factors affecting residential satisfaction in slum rehabilitation housing in Mumbai. Sustainability 12(6)
2. Bank W (2017) World population report. Washington DC
3. Figueiredo SM, Krishnamurthy S, Schroeder T (2020) What About Smartness? Archit Cult [Internet] 1–15. Available from https://doi.org/10.1080/20507828.2019.1694232
4. Narwane VS, Raut RD, Mangla SK, Gardas BB, Narkhede BE, Awasthi A et al (2020) Mediating role of cloud of things in improving performance of small and medium enterprises in the Indian context. Ann Oper Res [Internet]. (0123456789). Available from https://doi.org/10.1007/s10479-019-03502-w
5. Jiang N, Tian E, Daneshmand Malayeri F, Balali A (2019) A new model for investigating the impact of urban knowledge, urban intelligent transportation systems and IT infrastructures on the success of SCM systems in the distributed organizations. Kybernetes
6. Sembroiz D, Careglio D, Ricciardi S, Fiore U (2019) Planning and operational energy optimization solutions for smart buildings. Inf Sci (Ny) [Internet] 476:439–452. Available from https://doi.org/10.1016/j.ins.2018.06.003
7. Pradeep Pai T, Shashikala KL (2019) Smart city services—challenges and approach. In: International conference on machine learning, big data, cloud and parallel computing: trends, prespectives and prospects. IEEE, pp 553–558
8. Tiwari R, Sharma HK, Upadhyay S, Sachan S, Sharma A (2019) Automated parking system-cloud and IoT based technique. Int J Eng Adv Technol 8(4C):116–123
9. Radhika D, Aruna KD (2019) The smart triad: Big data analytics, cloud computing and internet of things to shape the smart home, smart city, smart business and smart country. Int J Recent Technol Eng 8(2):3594–3600
10. Young R, Fallon S, Jacob P, Deyer DO (2019) A flow based architecture for efficient distribution of vehicular information in smart cities. In: 2019 sixth international conference on internet of things: systems, management and security, pp 93–98
11. Villegas-Ch W, Palacios-Pacheco X, Luján-Mora S (2019) Application of a smart city model to a traditional university campus with a big data architecture: a sustainable smart campus. Sustainability 11(10)
12. Thirumalai R, Babu SS, Sharan KT, Shunmugam K, Siva S (2019) Smart City and Security Issues—IoT Prespective. In: International conference on computing methodologies and communication. IEEE, pp 905–911
13. MoUD (2015) Smart lcities: mission statement and guidelines. ministry of urban deveopment. New Delhi
14. Sivathanu B (2019) Adoption of industrial IoT (IIoT) in auto-component manufacturing SMEs in India. Inf Resour Manag J 32(2):52–75
15. Chatterjee S (2019) Internet of things and social platforms: an empirical analysis from indian consumer behavioural perspective. Behav Inf Technol [Internet] 39(2):133–149. Available from https://doi.org/10.1080/0144929X.2019.1587001
16. Asthana S, Dwivedi A (2019) Performance measurement of India-based third party logistics sector: an empirical study of user versus provider perspectives. Prod Plan Control [Internet] 31(2–3):259–272. Available from https://doi.org/10.1080/09537287.2019.1631467
17. Komninos N (2011) Intelligent cities. Electronic Government, pp 4205–4212
18. Vestergaard LS, Kasenburg N, Jorgensen MS (2019) Implications of conducting internet of things experimentation in Urban environments. In: Global IoT Summit
19. Maloney M, Reilly E, Siegel M, Falco G (2019) Cyber physical iot device management using a lightweight agent. In: IEEE International Conference on Internet of Things, pp 1009–1014
20. French T, Bessis N (2012) Towards a context-aware and adaptable room system for intelligent "trusted" office-spaces in smart cities. In: Proceedings—6th international conference on innovative mobile and internet services in ubiquitous computing. IMIS 2012, pp 148–154
21. Vamsi, Manikanta, Anandhi T (2019) Survey on enhancing drainage maintenance system using IOT. Int J Recent Technol Eng 7(5C):130–135

22. Singhvi RK, Lohar RL, Kumar A, Sharma R, Sharma LD, Saraswat RK (2019) IoT basedsmart waste management system: india prospective. In: 4th international conference on internet of things: smart innovation and usages. IEEE, pp 1–6
23. Mokale P (2019) Smart waste management under smart city mission—its implementation and ground realities. Int J Innov Technol Explor Eng 8(12):3095–3103
24. Panjagal SB, Ramaiah GNK (2019) Design of landfill site monitoring system to study the health effects of greenhouse gases using autonomous sensing nodes. Int J Recent Technol Eng 8(1):2521–2525
25. Nayak SS, Joshi DA (2019) Ingenious techniques for creation of smart cities by big data technology and urban modelling simulation by matsim. Int J Recent Technol Eng 8(2):1922–1927
26. Saikia U, Hosgelen M, Dasvarma G, Chalmers J (2018) National human development report. New York
27. Saravanan K, Anusuya E, Kumar R, Son LH (2018) Real-time water quality monitoring using Internet of Things in SCADA. Environ Monit Assess 190(9)
28. Ponraj AS, Vigneswaran T (2019) Machine learning approach for agricultural IoT. Int J Recent Technol Eng 7(6):383–392
29. Yadagani J, Balakrishna P, Srinivasulu G (2019) An effective home energy management system considering solar PV generation. In: 1st IEEE conference on susinternationaltainable energy technologies and systems. IEEE, pp 57–62
30. Sodhro AH, Pirbhulal S, Luo Z, de Albuquerque VHC (2019) Towards an optimal resource management for IoT based Green and sustainable smart cities. J Clean Prod [Internet] 220:1167–1179. Available from https://doi.org/10.1016/j.jclepro.2019.01.188
31. Sehovac L, Nesen C, Grolinger K (2019) Forecasting building energy consumption with deep learning: a sequence to sequence approach. In: IEEE International Congress on Internet of Things. IEEE, pp 108–116
32. Pires FM, Leon Quinonez L, De Souza Mendes L (2019) A cloud-based system architecture for advanced metering in smart cities. IEEE 10th Annual Information Technology, Electronics and Mobile Communication Conference, pp 1087–1091.
33. Manjunath MN, Meghana SK, Rajini S (2019) IoT for intelligent traffic management. In: 3rd international conference on computing methodologies and communication. IEEE, pp 779–781.
34. Zitouni R, Petit J, Djoudi A, George L. Iot-based urban traffic-light control: Modelling, prototyping and evaluation of mqtt protocol. In: 2019 IEEE International Congress on Cybermatics. IEEE; 2019. p. 182–9
35. Vinoj J, Gavaskar S (2020) Finding on roads risk and blind spot warning system IoT techniques and Hbrp algorithm. Int J Sci Technol Res 9(1):4216–4219
36. Velmani R, Srinivasa Reddy A, Ranjith Reddy M (2019) Design of its through IoT. Int J Eng Adv Technol 8(5C):1559–1561
37. Sun B, Dong M, Yang K, Bi S (2019) A design of high-performance streetlights management system. In: International conference on dependable, autonomic and secure computing. IEEE, pp 1–6
38. Park S, Lee S, Park S, Park S. AI-based physical and virtual platform with 5-layered architecture for sustainable smart energy city development. Sustain. 2019;11(16).
39. Murthy GNK (2019) Smart emergency management with iot for integration of emergency centers and patients. Int J Recent Technol Eng 8(4):3979–3983
40. Raetzsch C, Pereira G, Vestergaard LS, Brynskov M (2019) Weaving seams with data: Conceptualizing city APIs as elements of infrastructures. Big Data Soc. 6(1):1–14
41. Meslin A, Rodriguez N, Endler M (2018) A scalable multilayer middleware for distributed monitoring and complex event processing for smart cities. Int Smart Cities Conf
42. Mariouli M El, Laassiri J (2019) Using IoT and social networks for enhanced healthy practices in buildings. In: Information systems and technologies to support learning [Internet]. Springer Nature Switzerland, pp 471–481. Available from https://doi.org/10.1007/978-3-030-03577-8
43. Mamatnabiyev Z, Suliyev R (2019) Development of house automation system controlled using IoT technologies. In: 14th international conference on electronics computer and computation. IEEE; pp 206–212.

44. Urquhart L, Schnädelbach H, Jäger N (2019) Adaptive architecture: regulating human buArchitecture: regulating human building interaction. Int Rev Law, Comput Technol [Internet] 33(1):3–33. Available from https://doi.org/10.1080/13600869.2019.1562605
45. Rao PR, Srinivas S, Ramesh E (2019) A report on designing of wireless sensor networks for IoT applications. Int J Eng Adv Technol. 8(6 Special Issue 3):2005–2009
46. Pradesh H (2011) District census handbook shimla. Shimla
47. Corporation DM (2016) Dharamshala: smart city mission [Internet]. Dharamshala. Available from: https://repositorio.unan.edu.ni/2986/1/5624.pdf
48. India G (2011) Census 2011. New Delhi
49. Namchi Municipal (2016) Namchi: smart city mission. Namchi
50. Municipality P (2016) Pasighat: smart city mission. Pasighat
51. Commissioner C, Delhi N (2011) Primary census abstract for slum. Census of India, pp 1–40
52. Corporation SM (2016) Shillong: Smart city mission. Shillong
53. Group TWB (2017) Tourism for development. Washington
54. Dadwal A, Barbhuiya MR (2013) Preserving ecosystem by conserving traditional buildings. J Indian Inst Archit. 78(12):30–34
55. Barbhuiya MR, Garg YK, Munoth N (2015) Developing guidelines for a smart city in India; Using Garden city theory as a tool. In: The 9° Study Day of INUNaples, Italy, pp 5–12

Sustainability and Smart Cities: A Case Study of Internet Radio

Haifaa Omar Elayyan

Abstract Distinguishing between smart solutions and sustainability is always a problem when we discuss about the vision of "Smart Cities". This chapter is written and thoughtfully designed to bring into reader's consideration new dimension of smart cities and how to start thinking about them. The chapter promotes the smart city vision and presents smart solutions, with the revolution of technology we witness, as building blocks of sustainability that required to create and sustain smart city visions and goals. We do also present a customized smart solution called "Hybrid radio" technology to bridge the gap of the digital divide problem which we have not only in developing countries but also the rural areas of developed countries.

Keywords Smart cities · Smart solutions · Sustainability · Digital divide · Hybrid radio · Rural areas

1 Introduction

A smart city as a frequently used term includes phalanges more than just being an axiomatic and straightforward terminology. The formalization of a "Smart city" goes behind being a designating municipality founded incorporating the smart technology and information-driven interest to enhance the quality and performance of Urban services such as transportation and utilities increasing operational efficiency while condensing the resource consumption. Nevertheless, Smart cities should propose convoluted models stationed on the integration of smart band-aid solutions to excogitate sustainability into these models. Notwithstanding, there is an anomalous interpretation of smart cities because of the breadth of the technology and industrial science that can be assimilated into a city that takes the advantage of information and communication technologies (ICT). The referred transpire is to accost the demands of its citizens and address the community involvement in the processes which are

H. O. Elayyan (✉)
Middle East University-MEU, Amman, Jordan
e-mail: haelayyan@gmail.com

M. Khan et al. (eds.), *Smart Cities: A Data Analytics Perspective*,
Lecture Notes in Intelligent Transportation and Infrastructure,
https://doi.org/10.1007/978-3-030-60922-1_14

considered prime points to construe and exchange views on thoroughly. Apprehensively, many obligations are ascending, these days, in relational to the manifestation of smart embodies or spaces, and also in terms of taking relational communication, sensual experiences, and perception into consideration and analysis practices [1]. Yet, "smart" solutions have transformational power that we need to enlighten and emerge into a subsystem that is easier to embed into current running systems to enhance the outcome and reach the point of stability.

The twenty-first century has been witnessing and practicing a deviation from sustainability assessments and expected measurements and control to promote the vision of a smart city and promote goals [2]. This means that city assessment tools have transformed and developed toward defining smart targets and apply methodologies to promote cities "in a smart way" to function in smart frameworks. Smart solutions have auto-generated smart processes to integrate, interconnect, and interact to form in its turn, the scheme of sustainability. This ordered structure demonstrates the fact of sustainability being substantially more than just thinking and creating a simple but smart solution. Sustainability is a complex terminology that we need to consider but as a sperate entity from smart solutions as each has a different concept and methodology to be applied. Setting the two concepts apart is beneficial to give also more space to discuss the problem of the digital divide and trying to suggest new methodologies to bridge this gap. This will be a step forward for smart cities to enjoy the success of smart initiatives that depend upon the capability of integrating the advantages of ICTs and empowering citizens to make significant contributions, share, and interact coming over such obstacles. This obstacle, the digital divide, can simply be a barrier that limits urban revitalization and the full realization of a "smart city vision. The digital divide is now a digital chasm as Smart Cities World mentioned in their study and published a report on April 18, 2017.

> "The Huawei Global Connectivity Index 2017 of nations warns of an ICT version of the 'Matthew Effect' whereby the rich get richer and the poor get poorer. Digitally – developed economies around the globe are continuing to progress due to larger investments and adoptions in information communication technology (ICT). Meanwhile, digitally-developing economies have also started to accelerate their growth by investing strategically in ICT capabilities and their digital transformation journeys yet the gap continues to grow", said Smart Cities World news team, "Digital divide is now a digital chasm" – Smart Cities World, was released (18 Apr 2017).

The mentioned study presents facts about how far countries are going forward with the potentiality of digital transformation counting on only on technology enablers such as broadband, IoT, cloud and big data without any precontrive efforts or considerations to achieve sustainability. "Nations must prioritize ICT infrastructure development," says Huawei. Bear in mind that the digital divide at its start had divided the world to have and not have, which is the gulf between those who have stable and consistent access to computers and the internet and those who don't have. The major cause of the digital divide is the access based mainly on reliable ICT infrastructure that can't be replaced with any smart solutions. Moreover, crucial contributors to the basket of the digital divide such as the cost of technology, access for the disabled, lack of skills and experts, lack of knowledge-based education, lack of information to solve problems as well as the lower performance computers have been delaying the realization of sustainability that each smart city is trying to reach [3]. Important to understand that the digital divide is not reliable on infrastructure only neither when you are staying in rural areas as you might think of. It is a problem you can find while you are staying in a nice, well-defined city. For many US cities, the digital divide is an obstacle that goes beyond lacking a proper infrastructure as broadband access problems are not limited to rural areas. The fact is that the digital divide is currently defined as a barrier to urban revitalization too which limits the full realization of a "smart city" vision as mentioned before. In other words, and come to the fact, "smart cities can't exist while there's a digital divide".

It is well known; the digital divide concept is also related to what is known Digitally developed and digitally developing economies where both are continuing to expedite as a result of broad endowments and adoption in the information communication technology (ICT) field. Both economies evoked the race to stimulate prosperity by investing strategically in ICT competences, but the contention is harsh for the developing countries as their digital transformation efforts are not yet bridging the digital divide gap. Nevertheless, the contest is not offhand for developed countries as well. However, "It is sometimes said that developing countries are richer than developed ones, because of their human and natural resources, but to exploit these resources, technology is necessary and this technology is controlled by the developed world" [3].

According to Huawei Global Connectivity Index (GCI) (2017), the fourth annual study presented five technology enablers as main pillars in the digital transformation progress which are: big data, data centers, clouding platforms, well- structure broadband architecture, and Internet of things (IoT). The more investments are paid toward these enablers, the easiest digitizing economies can go. As you can tell, that the mentioned enablers are the platforms of smart solutions invading the smart city vision. Yet, no clear view on how sustainability can be merged into these digital concepts unless we are characterizing the sustainability with "smart solutions "theme. In other words, with many hardships' economics have especially in developing countries, the hope is more related to evolve smart solutions other than overloading expenses with sustainability solutions. Smart cities are sustainable when they can serve smart citizens with smart solutions that make their instant needs are met and obstacles are overcome at the spot. The vision of a smart city is to look for sustainable solutions such as green energy and infrastructure, but still, when it is expensive, smart solutions have the right to overwrite.

This chapter will help to distinguish between the two terminologies "Smart solution" and "Sustainability" related to the smart city vision. The chapter will discuss two case studies about two smart cities: detailing the most important models and schemes. Later, the chapter suggests a starting model for " a smart city" in the developing countries, showing smart solutions are the best building blocks to frame the vision. We do also, at the end of the chapter presenting a new smart solution trying to bride the digital gap enhancing communication in the transportation sector.

2 A Smart Solution

2.1 Definition

Most likely any one of us has used a smart solution or parts of smart solution. This utilization didn't require the knowledge to know what lies behind the term" smart solution". Smart solution terminology indeed introduces other related terminologies such as Smart city, smart technology, smart systems, and smart economic, and of course, these are examples not for listing. If we try to define what exactly "smart solution" and present the smart knowledge for all to get to know how these terms are connected and integrated, it might not be a beneficial or easy step to take. This section will help a lot of people who have no previous idea or tried before to acquire the knowledge behind the smart solution, to understand and take this as a piece of cake for further activities in their lives. Interchangeably, we can say that smart solution is a net package of inventiveness to present the technology as platform benchmarking advanced technologies, patents, and services models enabling automation and analytics that take our daily life and small details to new levels of performance by maximizing production, lowering costs while being away from irrelative details that require deep knowledge we don't have to acquire to enjoy this luxury. Dealing with smart objects, smart technologies, smart systems, and smart devices which are connecting and interacting with each other to optimize machine productivity, mine our needs and anticipate our service requirements is called "Smart aesthetics" [1].

Smart aesthetic conceptualization of smart solutions that bring comfort and flexibility into our daily routines. It is about the implementations of smart thinking in functional objects of common use leading to the definition of a new branch of intelligent technology, not only a smart. This new definition requires to integrate all relevant objects with the characterization of being "intelligent", "clever "and "technological into a connotation of including continuous quality processes to manage and control these technologies transform and inflammation. So, if we try to present a new definition of smart solutions, we would say that a " smart solution" is a technology-based representation of our intellectual needs and requirements. It is an interest-driven system that comes with architecture and structure that takes a place into the digital word promoting a life quality and high-performance personalized theme works according to available resources and infrastructure. The referred smart-systems interact and incorporate smart functions of collecting data, sensing, reasoning, actuation, and control to describe and analyze specific situations. And then take the interests into major accounts to decide and take a course of actions that supposed to control the results for quality purposes. These solutions, systems, and sometimes sub-systems that get integrated to predict or adapt to perform smartly. Smart solutions or smart systems as they are overlapped in meaning can be attributed to autonomous operations based on a bolted circlet of control energy efficiency for quality purposes and available networking capabilities. We don't claim that smart solutions can work without infrastructure, but we do emphasize that smart solutions should come with minimum requirements of surrounding change. By taking the solution into this direction, we are avoiding the hassle and huge cost if we stick our planning to the old version of infrastructure, in other words, Smart solutions can be the other coin-face of sustainability.

2.2 *What Can SMART Solutions Bring to Us?*

Smart solutions, conferred into smart systems and devices, equip our quotidian practices with promising technical keys that expected to subsidize to form enhancement and improvement of quality, reliability and promote economic efficiency of technical products [4]. In effect, and by human fallibility, every one of use chases comfort and entertainment without expenses that overtax planned budget, therefore, we do like the smart solution that being able to sustainability and stability to our plans. They afford sustainable mobility, intelligent energy management of the building, informatics, communication technologies, and their synergy, as witnessed examples in our daily life.

We do strongly believe that smart solutions improve the quality of life we want to live, in our town, municipalities, and regions, as we all, the dream of having and living in smart cities because of these promises. Smart solutions, with smart models, are expected to control input and output with maximum savings and merest strain to our environments so, as a crow flies, the smart solution demonstrates our concerns as smart citizens and residents rather than being just technology consumers.

3 Sustainability

3.1 *Definition*

The smart cities can be described as "The sustainable urban model" which epitomizes a society of standard technology size, interconnected and sustainable, comfortable, tempting, and secure [5]. As it was mentioned before that there is no definitive explanation of smart city because of the breadth of technologies that can be incorporated, consequently, there is no definitive terminology of sustainability when it is related to smart cities in specific or to our life in general.

Let try to present a definition here to satisfy the reality behind sustainability. Sustainability, as we author may lay it out, is the ultimate result when it comes to the "price". Value in this context is not limited to the monetary charge, but it also surpasses the ordinary cases to cover customer's contributions such as time, efforts, thoughts, memories, and knowledge acquiring. It is not one object as it was the case with smart solutions, inversely, it is a layered structured model based on suggestions, trials, thinking, and finding solutions that last and serve the best. It is not a one-step procedure; it is a quality-oriented assembly defining the transition process and identifying all opportunities that guarantee the case to sustain for the longest time can be.

3.2 How Do We Understand Sustainability in Terms of Smart Cities?

Presenting some of the historical facts, we would say that "development" as progress was achieved on different layers and using different resources. therefore, Development mainly located in Europe and North America was supported in two phases: Resources and technology, which crowned as "developed economy" or developed countries. The developed economy has a less social problem and afforded fair chances as it could be with well-defined strategical planning. Population standards in developed countries are advanced as individuals have careers and accordingly, they have a stable income to fee the family [3]. This is a progression haven ever since, and we can say that they have achieved sustainability at the most important level: citizen and basic needs. The developed progress had started at very early stages in these developed countries and has been a fruitful achieving the outcome when it comes to the "value". When we discuss the case for the developing countries, we would say that as most of these countries lack strategic planning, smart city vision, and required technology to evolve the economy. Many factors would contribute to stop the development progress, not to forget also the lack of required knowledge, sensibility, and technology. Nonexistence of urban plans for the informal city, the frail urban and housing management procedure, energy and resources wastage and the inflammation of irrelative details in people's lives that consume budget, resources and overlap the crucial components to grow, like, education systems, transportation systems, and health systems. The wreckage has been there for a long time that makes acknowledgment and cognition of "sustainability" a hassle, or almost inconvenient to achieve as it will cost beyond what these countries can afford. The bill is not finite to the cost, but prospecting "sustainability" from scratch will negatively impact the economy and intercept the recovery to take place as soon as it should be.

Over the past years, sustainable development has been motivated by a comprehensive and conventional "megatrend" such as fluctuating demographic profiles, changing economic and social dynamic forces, trends, and innovations in technology. These trends are linked according to the Global sustainability challenges conferences held over the years by the united nations. The link among these trends and the associated changes in economics, social, and environmental conditions is needed as the united nation pronounced. Every united nation conference on Sustainable development since 2012 at least, has highlighted a range of interlaced challenges which all call for priority attention, including decent jobs, energy sustainable cities, sustainable agriculture that secures the quality of food, water, disaster readiness, and risk management systems. Reaching this point, we can say that progressing sustainable development requires commitment and global actions to deliver on the legitimate aspiration towards further and optimistic economic and social achievements. This will be requiring growth and employment, and at the same time strengthening environmental protection [6]. Eventually, sustainability can not be a pure self-interest driven system, it is a more global structure that requires systems to incorporate and unite together accomplishing mile-stones of fame layers to ample the model of sustainability that lasts long enough to revel in outcomes and get paid and appreciated. These layers of systems and sub-systems are technology-based smart solutions that we propose to them, in the next section, as the building blocks of sustainability in general and the functioning brickworks in the smart city illustration.

4 Smart Solution: Internet Radio

4.1 Introduction

Internet Radio was presented by Carl Malamud, 1993 when launched " Internet Talk Radio" which was the first personal computer -radio broadcasting through the internet [7]. Since then, internet radio attracted significant media and investor attention. So, what do we mean by Internet Radio? Internet radio, also known as web radio, online radio, streaming radio, and sometimes e-radio, is a digital audio service transmitted via the internet. This service can broadcast (or webcasting) on the internet without the need for the wireless means. It might be used as a stand-alone device running through the internet, or as a software running through a single computer [8].

Internet radio services are usually accessible from anywhere in the world without the availability of a constant internet connection, which means that using your smartphone you can receive this broadcasting offline. What we are trying to present here, is that internet radio is generally used to communicate and spread messages through the form of talk. To broadcast, you need a distributed wireless communication networks connected to a switch packet network, Internet in other words, via a disclosed source [9, 10].

4.2 How Internet Radio Works: Internet Radio Technology: Concept

As we mentioned that internet radio is accessible from anywhere with a suitable internet available, and it is suited to listeners with special interests allowing them to pick from a multitude of different stations and genres less commonly represented in the case of traditional radios [11]. Internet radio technology consists of three embedded technologies for listening, streaming, and simulating. In the following Fig. 1 shows a model of these layers:

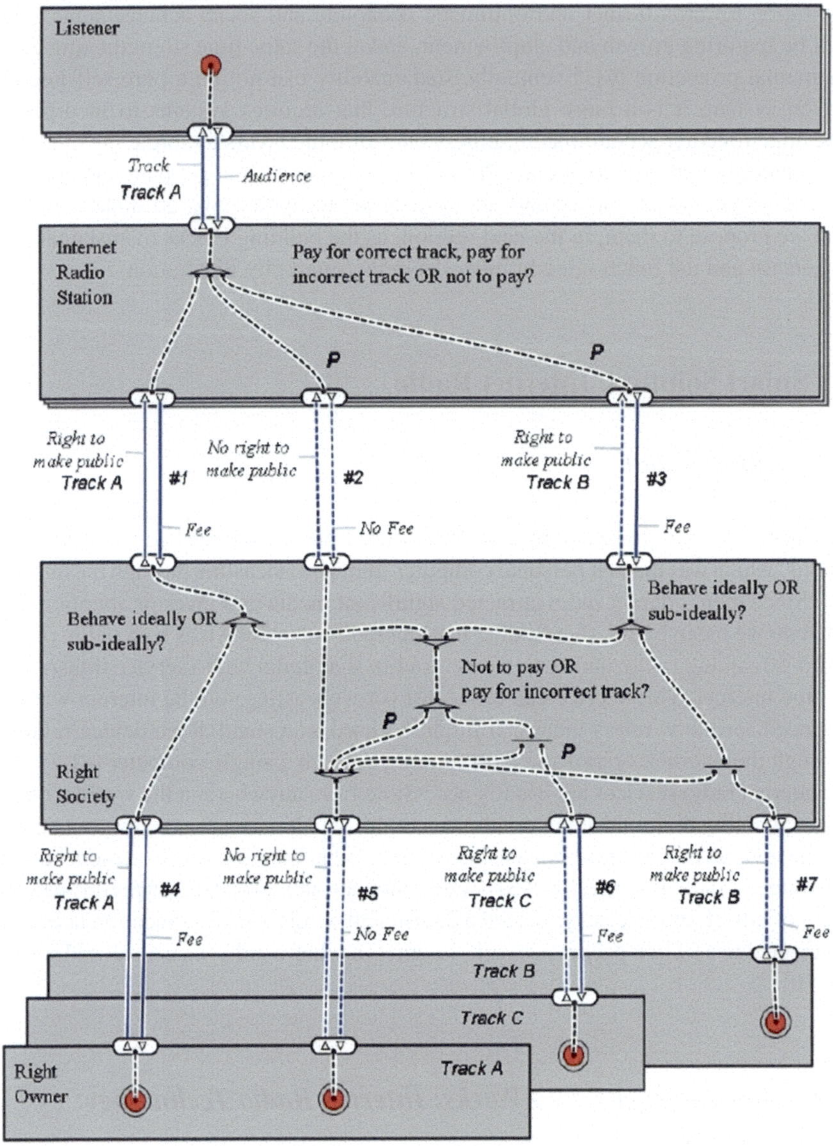

Fig. 1 Sub-ideal value model for Internet radio. *Source* Designing Control Mechanisms for Networked Enterprises: The Internet Radio Case Study—Scientific Figure on ResearchGate. Available from: https://www.researchgate.net/figure/Sub-ideal-value-model-for-Internet-radio_fig4_228956280 [accessed July 14, 2020].

1. **Listening**
   Listeners are the most important party in while broadcasting as, if you don't have listeners, no benefit of broadcasting. Listeners can access the stations' website with a suitable internet connection and they also can enjoy full of interaction with the broadcasting such as broadcasting cab be accompanied by the presentation, or animation, and they also can call-in for truly interact with the on-going discussion. They can access the web with any available smart device, such as pcs, smartphones, or the internet of things platform. Some enablers and service providers provide services and mobile applications to work as a radio receiver to give a full experience for the customer while listening to the broadcast.
2. **Streaming**
   Streaming technology is being used to distribute inter radio using a loose audio codec which is a device or computer program capable of encoding or decoding a digital data stream [12]. The streaming format includes MP3, Ogg Vorbis, Window Media Audio. Streaming means that audio data will be transmitted serially and continuously over the local network or internet in TCP or UDP packets. At the receiver side, these packets are re-assembled and played a second or two later, which known as a delay and called lag—Well known in online gaming as there is a noticeable delay between the action of players and the reaction of the server supporting the video game [13].
3. **Simulating**
   Local tuner simulation program includes all the online radio that can also be heard in the air in the city. Internet of radio is what we call it "e-content" revolution which means that you specify and decide on content, concept, or context and try to promote it and share your experience, achievement, and future work in terms of such concept. You use cyberspace to present and advertise for your ideas and thoughts. With advanced technology such as powerful computers, consistent internet and smartphones, smart applications, broadband, and inter of things, you can launch your internet radio station in no time on your own as the providers afford you the required knowledge and consistent support. You can be a broadcaster who communicate, share, and contribute to spreading the knowledge and transforming our life to more joyful according to what we desire or look forward. We would like to list a few important points to consider before you start an internet radio station from your home:

1. **Decide on your concept and context**
   It is always worth the time to consider what we want to broadcast and what kind of station we want to launch. If we give examples, some YouTubers have started with knowledge, lessons, and work experiences. Then some they came up with life-style sharing and documenting. Followers like such a new context as most they learn by example and visualizing the events. For internet radio, we do recommend also consider size, collaborators, and content to ensure consistent branding across all your programming. There are a lot of collaborators that can help to decide and launch your station.

2. **Brand your radio station**
   This means that you decide the name of your station and your program. This decision may impact the type of audience and listeners you want them to follow your station. There is always a tool where the place you stay to check for the trademark website, for example, in the USA there is a tool can be found on USA trademark (https://www.uspto.gov/trademark) while UK tool can be found on the UK version (https://www.gov.uk/search-for-trademark).
3. **Check copyrights and laws**
   Well, you will face this specifically if you decide to broadcast music rather than other talks or discussions. If you want to broadcast music then you most properly need to look into buying a license. You need to check the list at your broadcasting place.
4. **Equipment**
   All that you need to start is a microphone, pair of headphones, and a registered Airtime Pro station. Collaborating with internet radio station systems will save you the hassle of broadcasting as we will discuss next.
5. **Context to share**
   These will add a touch as they contribute to keeping your broadcasting consistent especially if your station is news-oriented. There are few online tools that you can use for this purpose such as Freedly which allows you to subscribe to different news and blog feeds, Tweetdeck also a useful tool for gathering content from Twitter and so on.

4.3 Proposed System: Bridging the Gap with Digital Divide: Hyper Internet Radio

Hyper internet radio combines Broadcast radio and the internet which means that broadcasting radio signal (FM, DAB, DAB +) continues to carry audio and some data while a radio with an internet connects (Wi-Fi, 3G, 4G, LTE) can connect back to your station for multimedia and interactivity.

Broadcasting radio is a technology to deliver radio to people across cities and reaches rural areas as infrastructure has been added a long time ago, and embedded even in cars and smartphones. It is a very economical way of achieving the choice of your preferences which means “a broadcast that you can listen, receive and enjoy from home”. With the internet technology revolution (IP), we could promote delivering multi-media and interactivity to people we want to share moments and transfer knowledge. The cost of using the internet especially in the rural areas is not yet described as “economic”, as we still face the digital divide problem not only in the developing countries but also with the rural area in the developed ones. The cost also to the broadcaster rise with demands, smart devices we need and install, where the idea of creating a hybrid Radio arises to combine broadcasting with the internet which endorses radio enhancement and development.

4.3.1 Hybrid Radio: Methods to Achieve Switching from Broadcast to IP

Hybrid radio offers a multi-staging service following the opportunity to switch listeners between IP and broadcast, where available. This means that hybrid radio combines broadcast and additional data from the internet among the features. A traveler between London and Liverpool, for example, will allow the traveler to enjoy London's local station through the journey. The mutual thinking of this will guide you to the start listening to a signal from FM, switching DAB + where available and as they leave the broadcast area, the vehicle system would switch to internet-delivered audio using smart devices available to provide internet stability. This goes beyond switching, it is an integrated smart solution that can estimate, track, and make the decision when the signal is lost. Also, a huge integration smart system when we decide to have a platform that manages the received data from the signal.

In the UK, radio player provides a hybrid radio application for the LG stylus 2 mobile phones, which has an addressable DAB + chip internally. Swapping algorithms will promptly switch to IP if the DAB + signal is unattainable. while connected to IP, it monitors the obtainable DAB + + signal. It is available, consistent, and error-free for a period (two minutes), it is acceptable to switch back. This switching algorithm minimizes the amount of switched times despite the fact of the user will still hear a drop in the transmitted audio. As a pragmatic solution, lessening the number of switches is a simple way to condense, else, the algorithm will confiscate the trembling effect of a switch. At the receiver side, audio has a hybrid receiver in the new Audio A8, as this product works with Fraunhofer IIS to create the Sonamic time-scaling algorithm that explained next.

Fraunhofer Sonic

1. **Sonamic Loudness**:
   Several in-vehicle media sources significantly increased lately. It is auto-action to play audio data from your smartphone in the car. Technologies of IoT and Bluetooth (basic solution) made transferring data an effortless process. Each audio source produces different levels of volume, which the driver must adjust while driving or being occupied with more priority issues like weather extremes or so which ends to a safety risk for all on the road. The loudness technology increases passenger comfort and safety by providing elasticity of volume level at the same time of switching between radio channels, media sources, or audio tracks.
2. **Sonamic Enhancement**
   To maximize the quality of delivered audio, audio codecs must be at adequately high bit rates. However, to reduce transmission charge when used for commissary, low bit rates are correlated which causes audible artifact. Fraunhofer's Sonamic enhancement technology reduces this deficiency as the algorithm uses all relevant audio codecs such as MP3, AC3, or codecs used within the satellite radio can be also processed. So basically, the algorithm doesn't require any additional

information about the codecs or bit rates, therefore, it doesn't matter which type of content is being played. This is expected to not affect the channel streaming in which the audio content is acknowledged.

3. **Sonamic Time Scaling**
   Hybrid radio will afford a continuous broadcast even when leaving the broadcast areas by combining a radio transmission with the accompanying web stream, bear in mid the transmission delay as transmission switch to and from IP. Sonamic TimeScalling has achieved on this front by synchronizing both signals with each other to produce a twin transition but at the same time precise. This technology overcomes the delay problem, lost or playing twice which could irritate the listeners. The switch works by having the system smarter to recognize the radio signal being interrupted soon, so the radio smart application begins to compute the existing counterpoises to the webstream and the resulting value is the required time till suspension, which will cause consequentially, the signal being inaudible to the customer. The algorithm works on compensating the time difference which indefinitely allows switching and listening to the station far beyond the transmission areas. The wonderful phase of this algorithm that it has been generated to work reversibly which allows the smart system to recognize the radio signal being strong enough for the reception. This means the system switch from web streaming to the radio signaling while preserving the mobile data for internet access to a minimum level, which is considered profitable thinking of sustainability as it could be especially when you are traveling long distances.
   To conclude, the best precise of Radio broadcasters should primarily avoid excessive postponements in their IP streams and should also have comparable dynamic processing on their IP to make the switches between diverse resources inaudible.

4.3.2 Hybrid Radio Offers

This section is designed briefly to explain what you can do with hybrid radios and we can summarize some like the following:

1. **Visualization**:
   Envisioning means that it can be that the audio, information on the screen, engaging visual content, and additional features that invite the listeners to stay tuned. This visualization comes as combining an audio broadcast system like FM and all the add-ons that internet empowers. Visualization comes through applications that let the broadcaster to deliver timely, relevant visuals alongside the broadcast itself. Content such as news, weather, and commercials will for sure, progress the experience and value of radio.
2. **Metadata**:
   Hybrid radio broadcasting and switching promote the chance of providing comprehensive information and programmed data in an open, uniform format to empower next-generation radio applications and platforms.

3. **Interaction**:
   The rule of "listeners who have a chance to express an opinion and interact, will never unfollow you" which means that the broadcaster guarantee life subscription. When the broadcaster permits listeners to explore the discography and biography of the artist that is being played, at the moments and enables them to admire the song or do personal customization to playlist, the broadcaster attained the outcome of the broadcast, merely by meeting the user's needs and desires. Interaction, in general, is a success key for any internet user to contribute, add values, share information, and experience.

5 Bridging the Gap for Smart Cities: Smart Solutions as Building Blocks of Sustainability

"Hybrid Internet Radio Smart Solution"

5.1 Mutual Design

In this section, we propose the mutual design of a mobile application that works on a hybrid radio platform to deliver interactive messages to the user/s in specific in case of being out of internet coverage. The hybrid technology can help a lot while traveling long distances being aware of environmental changes, Weather forecasting, and the new developments, or accidents happening on the same road this traveler is driving. The application is designed to interact with the driver, being single, or being on public transportation systems. A smart transportation system can play the role of the public platform for these interactive messages. What we are keen on is the type and the way of these warning messages reaching the user. In the following section we will divide the system not phases:

1. **Broadcasting**:
   Broadcasting will be through the technology of hybrid radio transmission that combines both regular radio transmission and internet radio transmission. The technology is explained above but what we are trying to innovate here is about creating a new smart solution when we deploy this technology for promoting the smart transportation system in smart cities. This includes smart devices, smartphones, broadband, IoT platforms that can work locally without internet access to outdoor, Smart information management systems that manage, visualizes, manipulate, and interacts with these transmitted messages adding knowledge, sharing experience, and interacting with others through the broadcast.
   Listeners will be able to send pre-recorded audio/multimedia messages by uploading them into the buffer that is broadcasted. This means that they can

send warning messages if they are witnessing accidents for example or whether sudden change allowing others to reply to their message and suggest solutions or alternative exits if possible. Using the technology of hybrid radio will allow two options: When the internet is available senders, can upload their messages directly from the microphones of their smart devices, or record them or attach multimedia files. If there is no internet coverage, the smartphone will switch to radio transmission by calling the number of broadcasting call systems allowing them to record their messages by voice or converting a text into speech to modulate a voice message. The application will be having a built-in option that helps to save time and hassle, as pressing on-screen, or sending an icon emotion will be translated automatically to pre-defined message in the system to warn or share a laugh or a joke, or a greeting for example. Converting audio to text and vice versa is an old technology that is used for a while in smartphones and being deployed at many phases, such as reading your inbox or send a nice greeting. The system will be for sure accessing your profile, IP, contact number, internal files, photos, and multimedia you are desiring to share.

2. **Receiving application**:
   At the receiver side, the radio application will show messages received transmitted by the radio channel or internet IP (in case available), these messages can be auto-played and replied to the relevant source as we are counting on the IP in specific, or as broadcasting for the public that all subscribers can receive. Sending private messages will take us to the side of the encrypting and decrypting algorithm that can be easily deployed in such a case.
   What we do suggest here is deploying the hybrid radio technology to a hybrid radio social network that provides a continuous receiving and listening with and without the internet connection. The proposed radio application will bring tour mind a new way of communicating with rural areas or people with no access to the internet while having smart devices that can communicate with radio frequencies to transmit audio and data that can be translated to something they understand and be aware at the right time saving their time, keeping in touch and providing a risk management systems while traveling. The application will be with three main concepts of hybrid radio: visualization, listening, sharing, and interacting in real-time (with acceptable delay slots). What we propose here is a smart solution for the transportation system that can promote the fact of being.
   Figure 2 will approximately present the module we are promoting.

6 Conclusion:

Smart solutions have proven the ability to generate technology that we need to improve our life and daily routine. The continuous generating of smart solutions will integrate efforts and meet the requirements of people while still having a “quality of life” presented in the methods and techniques these smart solutions apply and promote the technology. Solving problems is the first step of sustainability and managing

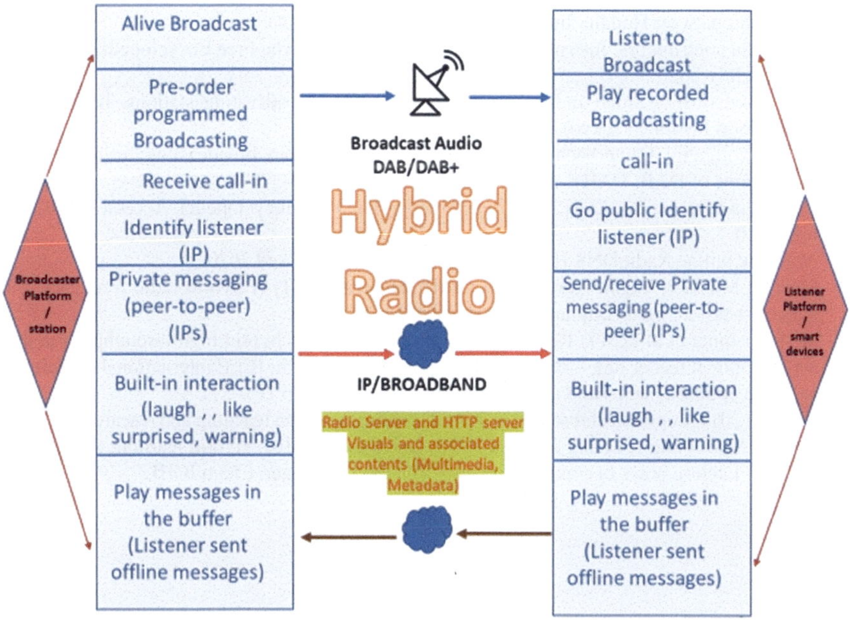

Fig. 2 Mutual promoted system

smart solutions that come with the ultimate and quick recovery will help to sustain resources. We would like to end this chapter by endorsing that playing the right and most effective role of "smart cities" we must go in parallel generating smart solutions while searching for applications of sustainability.

References

1. Cecilia Russo, Anna, Ferrara, Marinella (2017) Smart solutions, smart aesthetics?. Des J 20:S342–S353
2. Hannele Ahvenniemi, Aapo Huovila, Isabel Pinto-Seppä, Miimu Airaksinen (2017) What are the differences between sustainable and smart cities?, Cities, vol 60, pp 234–245, Part A. ISSN 0264–2751
3. Dania González Couret: developed and developing sustainable world-BVSDE, pp 1–5 (2000).
4. Tränkler, Hans-Rolf, Kanoun, Olfa (2007) Smart systems and devices: innovative key modules for engineering applications
5. George Cristian Lazaroiu, Mariacristina Rosica (2012) Definition methodology for the smart cities model. ELSEVIER 47(1):326–332
6. United Nations report: World Economic and Social Survey 2013, Sustainable Development Challenges, E/2013/50/Rev. 1, ST/ESA/344
7. Available online "Internet Talk Radio". museum.media.org. Accessed 7 Aug 2019
8. Kiraly, Jozsef (2001) Method and system for implementing an internet radio device for receiving and/or transmitting media information. Patent: US6249810B1 United States. Accessed 8 July 8 2020

9. Cerf, Vinton, Scott Huddle, Internet radio communication system
10. Wikipedia contributors: Internet radio (2019) In Wikipedia, The Free Encyclopedia. Accessed 19:45, 8 July 2020
11. Tang hoon (2013) Smart radio: cloning internet radio broadcasting stations. International Information Institute (Tokyo). Spring
12. Hoeg, Wolfgang; Lauterbach, Thomas (2009) Digital audio broadcasting: principles and applications of DAB, DAB+, and DMB. Wiley, p 26
13. Wikipedia contributors, Lag (2019) In Wikipedia, The Free Encyclopedia. Accessed 20:42, 18 July 2020
14. Avaliable online RadioDNS HYBRID RADIO (2014) Accessed 2020
15. Rahaim, Michael, Vegni, Anna Maria, Little, Thomas (2011) A hybrid radio frequency and broadcast visible light communication system, pp 792–796
16. M. Toyoshima et al (2017) Research and development on a hybrid high throughput satellite with an optical feeder link—study of a link budget analysis. In: IEEE international conference on space optical systems and applications (ICSOS), Naha, pp 267–271
17. Wang S (2014) A practical research of hybrid learning model in teaching and research activities
18. Cheung SKS, Fong J, Zhang J, Kwan R, Kwok LF (eds) (2014) Hybrid learning.: theory and practice. Lecture notes in computer science, vol 8595. Springer, Cham ICHL

An Assessment of Smart Urban Furniture Design: Istanbul Yildiz Technical University Bus Stop Case Study

Gökçen Firdevs Yücel Caymaz and Kürşat Kemal Kul

Abstract The aim of this study is to evaluate smart urban furniture design standards on the example of smart bus stop design. For this purpose, smart urban furniture design standards are gathered under seven headings including safety, form and character, function, value, maintenance, sustainability and technology. After examining the examples of limited number of smart city furniture in Istanbul, the checklist related with smart furniture design criterias was tested at Yıldız Technical University smart stop located in Beşiktaş, Istanbul. In order to explain the results more clearly, a traditional stop near the smart stop was also evaluated simultaneously. Looking at the comparative results; It has been determined that smart bus stop offers a more comfortable use environment in terms of technology and security than traditional bus stops.

Keywords Smart street furniture · Urban technology · Smart devices · Sustainable production

1 Introduction

Public spaces have had many different functions in social, political and economic life, which change according to the needs of the time. Considering the usage habits of them in the historical process, the Ancient Greek agora and the Roman forum were the main political centers of the city and the spatial focus of democracy. In the medieval urban environment, the square became the main market, assuming a strong mercantile and economic role. During the Renaissance, the city became a work of art with public spaces designed to leave a visual impression and showcase

G. F. Y. Caymaz (✉) · K. Kemal Kul
Architecture and Design Faculty, İstanbul Aydın University, İstanbul, Turkey
e-mail: gokcenfyucel@aydin.edu.tr

K. Kemal Kul
e-mail: kursatkemalkul@gmail.com

M. Khan et al. (eds.), *Smart Cities: A Data Analytics Perspective*,
Lecture Notes in Intelligent Transportation and Infrastructure,
https://doi.org/10.1007/978-3-030-60922-1_15

architecture. The factory became the heart of the city during the Industrial Revolution, and because *it* was the main activity of the community, it caused the quality of public spaces to decline. Today, the renewal of public spaces continues, whereas people tend to spend more spend more of their free time in them than in the past [1]. Particularly as technology advances, a transformation has begun to turn public spaces into sustainable active public spaces, with a whole stream of energy, materials, services and costs to support a high quality of life, defense and economic development. Intelligent urban technologies include a combination of hardware and software that provides real-time management of time usage, real-time data capture and public space interaction, storage and reuse, material adaptability, energy generation [2].

In smart cities, which are basically shaped by internet infrastructure, data received from sensors embedded in all city furniture is transferred to the internet whereas apps realized with this processed data is integrated back into the city furniture. Smart city technology in the public sphere is generally divided into four categories. These include overhead projectors, loT platforms (*which can be considered as a system for the Internet of objects*), which connect to street posts or buildings that measure various micro-climate characteristics such as carbon footprint, air quality and traffic; street infrastructure products, such as urban USB charging stations that provide a fixed solution to a problem and structures such as wifi counting/monitoring banks, ICT platforms aimed at mapping and displaying data findings and providing an interactive interface between user data; and placemetrics and parameters in which professional staff research and develop more nuanced and more understandable measurement systems (*for instance, measuring the percentage of women on the streets or the number of people using their mobile phones as an indication of the safety of the public sphere*). Each of these is developed software that works with various metrics aimed at testing and optimizing design results (*such as existing CAD software plugins*) [3].

Smart urban and living laboratory initiatives are now an almost essential feature of any city or town in the developed world, including Europe. Digital cameras, audio and other sensors, and increasingly widespread information and communication technologies (ICT), including adaptive lighting and wifi monitoring technology all combine to give cities a contemporary feel. In the smart urban/life laboratory discourse—as spoken by businesses, local governments and international organizations such as the European Union (EU)—these technologies are regarded as solutions to all urban problems caused by traffic congestion, high gas emissions and security matter [4].

Urban furniture is a small scale element in urban environments. Furniture such as traffic lights, markings, illumination elements serving different purposes, closed-circuit television (CCTV) cameras, kiosks, trash receptacles, seating elements, fountains, public toilets, trees and other vegetation, floors, grilles in urban environments [5–8] have now begun to be integrated with technology. Due to ever-changing technology and needs, solar energy charging units, walkways, charging units for electric devices and vehicles, smart mass transit stops, working units suitable for mobile devices, smart bicycle systems, smart transportation systems, smart waste collection systems like a new smart urban applications have developed and continue to develop. Smart urban furniture that run on renewable energy helps garner more attention upon public spaces by providing informatics and connectivity while providing public services for cities and society (Fig. 1) [9]. Including trash receptacles, seating elements and bus stops, the urban environment has become smarter because it has been outfitted with particulate matter equipped with several sensors which collect

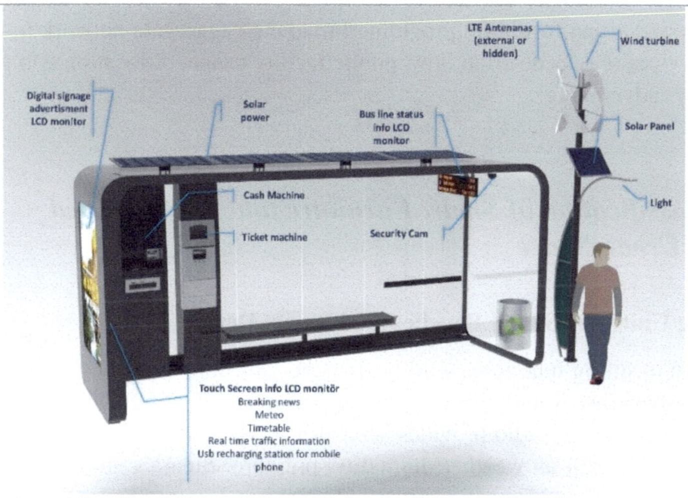

Proposal smart bus stop designs were improved and and designed by authors.

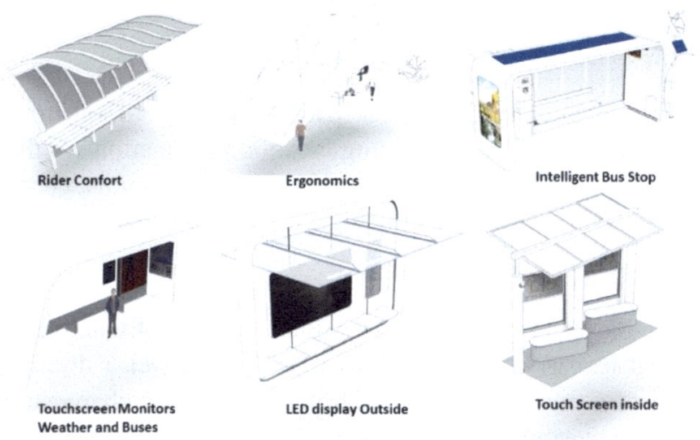

Proposal schematic bus stop designs were improved and and designed by authors.

Standard Components: Roof, Seats, Poles / Bars, Lighting, Sustainable Materials, Sidewalk Lanes, Route map on bus stop or bus signpost, direction sign for LRT station	Other Possible Components: Landscaping, Trash receptacles, Newspaper racks, Audio / Video comm, solar power, wind panels, billboards, viable canopy design suitable for LRT station -WIFI

Fig. 1 Component of standard and innovative bus stop [43–47]

information on the air pollution and environmental conditions, wireless modules, processors and microcontrollers. As a consequence, no doubt wired furniture will become a vital part of the IoT infrastructure and become one of the driving forces of such cities in the future [10].

Amongst the smart urban furniture functions are telephone/ electronic items solar-powered charging points, WIFI connection, of public services and citizen information displayed on continually updated screens; energy and therefore cost savings with solar-powered urban furniture; providing broad data aggregate network to improve public services (e.g. pedestrian flow, public facility usage, of location comparison); customary 'advertising'.

1.1 Classification of Smart Furniture that May Be Used in Urban Space

Charging Units For Smart Benches And Smart Devices

In addition to sitting functions, seating elements can store energy by means of solar panels and batteries. Smart benches possess features such as free wi-fi, smart device charging feature, rear audio provides acoustics waypoint, location assistance, environmental data such as weather, humidity, pressure. Steora's study entitled 'The Smart Bench' [10] and Forestier's work [11] and seating products that collect and store rainwater designed by Mars Architect are examples of the world's modern smart bench applications [12].

Smart Bus Stops

In addition to the bus waiting function, panels showing arrival times in minutes and which routes are operating, providing location information via the information display, air temperature information, wifi, USB charging points, charging units for electric vehicles and bicycles are technological components of smart bus stops [13].

Smart Bicycle Rental Stations

The purpose of bicycle rental systems, which provide alternative transportation without the need to carry bicycles, is to be able to travel a distance of 3-5 km without the use of a motor vehicle. Electronically tracked charges cover the time between when the bicycle is removed from the station and when it is returned to any station. The use of solar panels at such stations can also serve the function of charging bicycles [14].

Smart Lighting

It is important in the efficient use of energy in urban areas. Smart illumination includes cloud-based automation that can be controlled from a single center for the purpose of increasing visibility for night pedestrians and vehicles at night. Especially with recent advances in IoT technology, sensors and wireless solutions have replaced electrical cables in illumination control. Security systems, advertisement bilboard, environment sensors, electric vehicle (EV) Charging, asset management, emergency response, weather systems, traffic systems [15], control and monitoring of all diffused lighting from a single center, a cloud-based lighting automation, as well as the use of sodium gas bulbs can be considered components of smart illumination [15, 16].

Smart Urban Toilets

Smart toilets are systems that provide cleaning, personnel and maintenance advantages over traditional ones. They can be transported to any desired place in the urban environment. They can be put into service either for a fee or complimentary. They provide features such as 20-minute usage time, audible warning and light systems, self-cleaning with sensors. City maps can be affixed to the exterior of the toilets. Automatic toilets located at Madison Avenue and 23rd Street in Manhattan, New York [18]; 24-h free toilets [19] located at 400 locations in Paris and automatic public toilets set up in Toronto [20] are examples of their use worldwide.

Smart Waste Management Systems

The separation of non eco-friendly solid wastes, such as plastic PET bottles, aluminum cans and recycling, reducing carbon emissions by preventing the unnecessary loitering of vehicles collecting waste, measuring the fill rate of waste containers with wireless fill detection sensors and transferring it to the data cloud system with M2M (Machine to Machine) technology, creating daily optimized daily waste collection routes for these vehicles are among the features of smart solid waste collection systems. The fill rate of garbage containers and garbage trucks can be monitored instantly via wireless sensors. After processing this data collected through cloud systems, an automatic route is created for waste collection vehicles, covering containers that reach a sufficient occupancy rate, thereby saving time and fuel. Sensors dedecting fullness and odor in smart garbage cans where solar panels are used are included in features such as discharge warning at maximum fill up and some even have automatic compression [21].

Information Systems

Information and Communication Screens (ICS)

This is a system set up at the stops, that allows passengers to make live voice calls with the Call Center for information or for emergency assistance. At the same time, the line number of the stops, the scheduled route and arrival time of the bus are given. It can be uploaded online at the information communications point.

LED (Light emitting diode) Information Display

Depending on the modular structure having long-life LED technology, the LED Information Screen can display all types of graphics via apps such as effective flash, animation presentation, text and imagery through video and scrolling font displaying text-based information. LCD-based informatic systems are those that can conduct individual interactive transactions or visual information. The fundamental units that comprise the LED display are; cabin, metal support construction (holds the cabinets together), power/ signal cables, control computer (or special image transmitters) and power distribution board.

LCD (Liquid crystal display) Information Screen

LCD-based information systems are those that can partake in developing technology, make interactive individual transactions or provide visual information. While these systems provide savings to institutions and organizations, they also ensure a environments stable and efficient flow of information flow which provides convenience in work.

Fee Collection Systems

The Electronic Fare Collection System is an information management system that provides passengers with pre-payment travel rights through an electronic ticket, located at the entrance and exit points of public transportation and transit and sales points. Instant information about passengers passing through turnstiles and types of tickets used can be collected and all kinds of statistical reports can be obtained. Components of this system are; camera, kiosk—ticket/card sales and filling device, validator, driver control panel, smart ticket, digital video recorder (DVR), public transportation route panel, ABYS (Smart Ticket Management System) Mobile/Web Application and smart stop. This system renders the use of paper ticket and cash money unnecessary, enabling passengers to travel comfortably and quickly [22].

2 Literature Review

2.1 General Smart Street Furniture Design

In emphasizing the importance of the use of smart urban furniture in smart cities in her 2017 study conducted in Cairo, Hala Hassanein [23] examined technological systems that can be used in city furniture under two headings, ICT Technologies and Energy technologies, defining design principles of smart furniture as functionality, people oriented, unity, and identity. In 2018, Oihane Gómez and her colleagues [24] presented their eco-system vision of multifunctional urban furniture designed for bus stops, not only from a digital interaction standpoint, but also for sustainable use and integration into the urban environment. Nassar et al. [10] discussed the advantages of using technology in their assessment of the Present and Future Role of Smart Street Furniture in Intelligent Urban Environments. In dealing with the innovation and validation process of solar-powered furniture in research conducted in 2018, Avila and Jorge Toledo [25] emphasized the importance of using clean and sustainable energy sources with a multi-functional urban tree-type module which facilitates constant shade during the day and throughout the year with light and heat sensors. In 2019, Yao [26] underlined the contribution and importance of using Internet of Things technology in furniture to public administration.

2.1.1 Investigating the Use of Smart Urban Furniture in Istanbul

Current smart city furniture usage in Istanbul covers the range from charging units for smart benches and smart devices, to smart bus stops, smart bike rental stations, smart lighting elements, smart city toilets, smart solid waste systems, an in-formatics system as well as toll collection systems (Table 1) [27]. Most applications within the scope of the Smart City Project in Istanbul are in the area of smart transportation. Traffic Control Center, Smart Signaling, Vehicle Tracking System, Geographic Information System (GIS), Electronic Control System (EDS), which is monitored with 24/7 live cameras of Istanbul, Environmental Control Center, where waste management is performed, 'IBB (İstanbul Metropolitan Municipality) Mobile Traffic', IBBNavi Mobile applications such as 'İTaksi' and smart stops are still applicable.

2.2 *Smart Bus Stop Design*

In 2001, Kaya [30] investigated the points that need to be considered in the design of bus stops and furniture used at bus stops. Ashwin et al. [31], Kamal et al. [32], Alikhanova et al. [33], Barns [34], Wachira and Karthik [35] conducted research on the use of smart infrastructure technologies at bus stops, emphasizing the necessity and importance of technological development of bus stops (Table 2).

It was seen that the use of intelligent urban furniture rose in urban design scale, and their presence increases the spatial comfort in urban environments with heavy human traffic. This study has two main objectives; one is to define the basic principles in sustainable street furniture design, and the other is to determine what differences there are with a standard stop by trying to determine the criteria of smart bus stops.

3 Material and Method

Sustainable furniture design criteria were researched in the first stage of the study, and a checklist was created after a literature review. While creating the control list, studies on urban furniture design and standards were examined, and then smart furniture design, as well as smart and traditional stall design issues were investigated in detail (Fig. 1). The created control list was examined under six categories: security, form and character, function, value, maintenance, sustainability. In the following step, the use of smart city furniture in Istanbul was investigated to select a sample area. The control list created as a result of the research was tested on two stops located in Istanbul Beşiktaş (Fig. 2), one smart and the other standard. On-site detection and

Table 1 Smart urban furniture utilization in İstanbul

Smart Benches and Charging Units to Charge Smart Devices	
 Miro, Tünel Stop, Beyoğlu	8 device chargers, its digital screen provides city-related information, temperature, humidity and tramway schedule
Smart Bus Stops  Smart stop, Beşiktaş	**Smart Bicycle Rental Stations**  Millet Garden, Esenler
	Anatolian side of İstanbul: 380 bicycles 38 stations, European side of İstanbul: 1120 bicycles, 102 stations
Smart Lighting  Kağıthane Cendere Road, Kağıthane	**Smart Urban Toilets**  Hidiv Kasrı, Beykoz
	Disability usage, children sick, child care room, air condition, toilets
Smart Solid Waste Systems A waste recycling station recognizes PET bottles ranging between 0.5 - 1.5 l. and metal cans ranging between 200 - 500 ml. with image processing algorithm and barcode reading system. The machine also recognizes and returns full bottles and different materials [29].	 Eyüp Sultan
Information Systems	
Information Communications Monitors  Kadıköy	Information regarding travel durations trip cancellations, emergency situation messages that can be made from the center.

(continued)

Table 1 (continued)

LCD Information Monitor  Smart Stop, Beşiktaş	Led Information Monitor
Information Systems	
Information and Communications Monitors  Kadıköy	**LCD Information Monitor**  Metrobüs Entrance, Kadıköy
LCD Information System  1 in Bağcılar District, 2 in Beylikdüzü District, 3 in Fatih District, 2 in Maltepe District, 2 in Slivri District [30]	12 mm pixel pitch, Dip 346, 4,032 pixel to 3.072 pixel, 12.3863 m² (for each face) double sided led display project with 10 totem central management system. With these LED screens and central management systems, Istanbul Metropolitan Municipality provides weather, special day information, live broadcasts, traffic information messages, city event information, presentation of projects, television broadcasts, special content production, advertising campaigns, organization contents, video projects. They feature LED screens with remote management system from which it may broadcast audio and visuals in a wide range of promotional films.
Payment Collection Systems **Validator** is a smart device that controls the scope of the electronic ticket it interacts in accordance with defined parameters, collects the specified tariff amount, records the transaction result and features an improved user (passenger).	 Metrobüs Entrance, Kadıköy
Transit Control Systems These are systems that monitor personnel entries and exits to provide corporate security and time management.	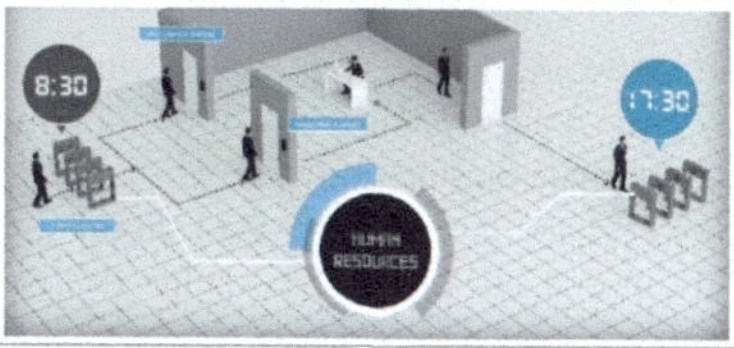

Table 2 Smart Bus Stop Furniture Literature Review

*Purpose, ** Method, *** Result
GENERAL ISSUES Imperial County Transportation Commission [36] * Provide safety and security for riders to the maximum extent practicable. **literature review *** Creating a checklist to improve the quality of bus stops and their environs [37] * Weaknesses, problems and shortcomings of existing bus stops were evaluated ** Literature review ***Use of composite materials and new technology for bus stop shelter is better than traditional ones [37] * Improving the bus waiting area ** Literature review ***the waiting stop USES solar energy to supply the lighting system of the whole bus waiting stop, and the green environment beautifies the urban environment. Smith, V., 2019 [38] * giving recommendations on ways the three bus stops could be altered to increase feelings of safety. ** literature review, safety audits and validation interviews *** safety improvements to public transit can substantially increase transit ridership and provide an impetus for more active transportation. Zhang, K. J., 2012 [39] * The waiting area at bus stops tries to improve with urban design techniques **literature review, site analysis ***The project identifies 7 primary objectives in designing a decent bus stop; integration: acoustic and thermal comfort, visual comfort, accessibility, wind protection and safety.
TECHNOLOGY ISSUES Li, C. (2020) [40] *The content of bus station landscaping based on the ecological environment concept. **Literature review ***The importance sustainable bus station development of China's urbanization infrastructure.
Ashwin, D., Mounika, V., Kommineni, M., & Swetha, K. (2019) [31] * Identifying and solving problems in urban area practices including smart environments **Site analysis *** Smart solutions using solar and turbine energy to infrastructure and services bus stop applies in rural and urban areas in order to make them better.
Kamal, M., Atif, M., Mujahid, H., Shanableh, T., Al-Ali, A. R……. (2019, June) [32] *Suggests an IoT-based eco-friendly enhanced bus stop design ** A mobile app featuring a map interface, enabling operators to monitor bus stop conditions, i.e., humidity, temperature, estimated occupancy and air pollution levels remotely *** Suggested design looks to optimise energy consumption by means of forecasting bus stop occupancy, remote air conditioning and lighting monitoring.
Alikhanova, A., Kakimzhan, A., & Mukhanov, A. (2017) [33] *this project puts forward bus stop modernizing at "Cardiac Surgery Center" bus stop by means of an optimal combination of clean-burning, conventional energy resources. **Site evaluation, passenger load survey *** Proposed sustainable bus shelter design is more efficient and economically viable.

(continued)

Table 2 (continued)

*Purpose, ** Method, *** Result
Barns, S. (2017) [34] *Possible utilization of smart infrastructure technologies for bus stop designs **Literature review ***The report scans opportunities to address the impact of climate change in Western Sydney by means of smart Technologies
Kaufman, B. (2015) [41] * The idea of combining bus stops and parklets ** Literature review. Site analysis, Photography, Questionnaire *** Findings from the analysis of this paper validates the stoplet concept feasibility.
Sungur, C., Babaoglu, I., & Sungur, A. (2015, April) [42] * Enables administers to monitor their mass transit systems effectively whereby those utilizing this system simultaneously observe information about location and status of their vehicles. ** Utilization of mini-computer based systems and digital screens *** The software system were developed
Wachira, K., & Karthik, J. (2016) [35] * A new method proposal by integrating IoT into bus shelters ** Literature study, Field research, Observation *** The importance of technological development of bus stops

observation methods were used to test the checklist. While both stops have main design components such as seating, lighting and trash receptacles; in the smart stop; information about which stop the bus will pass, the arrival time of the bus, which route to be operated, traffic density, charge station for battery powered vehicles, free wifi, ticket automat, instant communication with IETT, solar panel.

3.1 Research Question

Research Question: In conjunction with the study of smart bus stops, it is to investigate as to whether smart city furniture has advantages over traditional ones.

Subtitles of the Research Question:

What are the design criteria of urban furniture design?
What are the design criteria of traditional bus stop design?
What are the design criteria of smart bus stop design?

4 Research Findings

On security; with the smart stop, the interior lighting and environs is better than the standard stop; the perception of oncoming buses at the smart stop is worse than the

Table 3 Major Factors in the Sustainability of Site Furniture, improved by au-thors from references [25, 51–62]

CRITERIA	Smart stop			Traditional stop		
G-Good P-Poor F-Fair	G	P	F	G	P	F
1.SAFETY						

CRITERIA	Smart stop G	Smart stop P	Smart stop F	Traditional stop G	Traditional stop P	Traditional stop F
Appropriate Installation			x	√		
Periodic Inspection			x			x
Sufficient Lighting (2 - 5 Footcandles)	√				x	
Locating Bus Stops Near Existing Streetlights For Indirect Lighting	√				x	
Stability	√			√		
Durable Linkage Elements	√				x	
Good Visibility		x		√		
Avoid Planting Evergreens	√			√		
Vegetation that Stands Above 1.8-2.1 m, and Lower than 60-90 cm	√			√		
Locate Stops At Highly Visible Sites	√			√		
2. FORM AND CHARACTER						

CRITERIA	Smart stop G	Smart stop P	Smart stop F	Traditional stop G	Traditional stop P	Traditional stop F
Design That Fits the Surroundings			x	√		

(continued)

Table 3 (continued)

3.FUNCTION

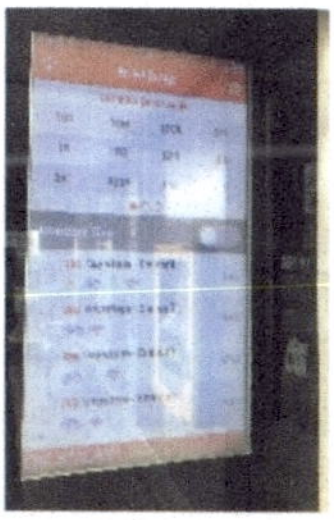

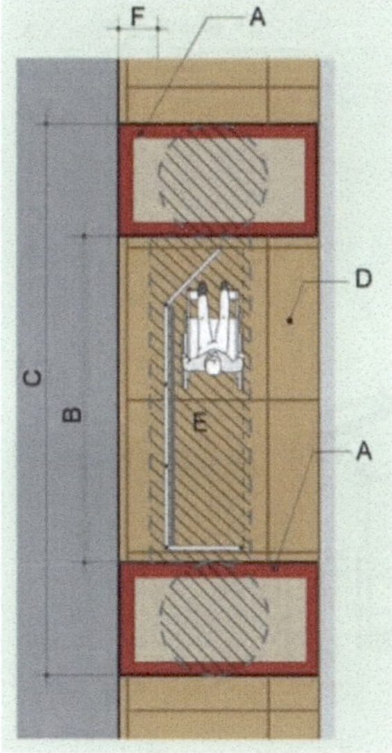

Being Functional	√				x
Create Area For Bus Loading And Layover	√		√		
A.1.5 m Long x 2.4 m Deep Loading Areas/Pads	√				x
B. 6.1 m Long x 1.8 m Deep Waiting Area	√		√		
Omit Steps Between The Sidewalk/Bus Pad And The Shelter		x		√	
C. 7.9 m Stop Area Along The Curbline Should Be Kept Free From Obstructions	√			√	
D. 1.2 m Deep Pedestrian Path	√				x
F. 0.6 m From The Curb Edge, 2.7 m Minimum Height Clear Area	√		√		
A Pitched Roof		x	√		
1.98 m (3 Persons) Seating Length	√			√	
Position Bus Stops And Benches To Facilitate View of Oncoming Transit Vehicles	√		√		

(continued)

Table 3 (continued)

Seat Dimensions: 50-60 cm Depth, 105 cm Length			x		√	
Back Support: Backing at a Height of 5-45 cm			x			x
Provide Seating With Armrests	√					x
Glass Panels	√			√		
Mark Glass Panels With Distinctive Pattern	√				√	
106 cm Panels Leaning Rails Above The Ground		x			x	
The Top Of A Panel On A Pole Should Be No More Than 1.5 m From The Ground.		x			x	
Position Bus Signage in Accoardance With City And Bus Operator Requirements	√			√		
Anchor Receptacle To Prevent Unauthorized Movement			x			x
Trash Receptacles At Least 60 cm From The Curb			x			x
Mimumum 114 lt. Trash Receptacle	√			√		
Do Not Locate The Bin In Direct Sunlight			x			x
Sign Located Around 30 cm in Front of Bus Stops	√			√		
Bottom Of Signs Should Stand Min. 2m Up from the Ground	√			√		
Remove Curb Side Planter	√			√		
A Maximum Slope Of 1:50 (2%) For Water Drainage	√			√		
At Dimension Proportional with Sidewalk Width			x	√		
Positioning at an Adequate Pulling Distance			x	√		
Positioning Without Interrupting Pedestrian Flow			x	√		
Connection To Streets, Sidewalks Or Pedestrian Paths By An Accessible Route			x			x
Using the Furniture Without Conveying Physical or Psychological Pressure			x			x
The Furniture Arrangement Being Comfortable	√					X
4.VALUE						
Monetary Value	√					x
5. MAINTENANCE						

Litter Removal	√			√		
Convenient Cleansing	√			√		
Grafitti Resistant	√			√		
Furniture Inspection			x			x

(continued)

Table 3 (continued)

6.SUSTAINABILITY						
6.1. Sustainable Design						
Simplicity						
Design with Minimum Materials	√			√		
Avoiding Decorative Purposed Pieces	√			√		
Decreasing Support Structure Weight	√			√		
Miniaturization						
Reducing Furniture Size to Level of Standards	√			√		
Multi-Functionality						
Designing Urban Furniture to include functions other than the main function	√			√		
Avoid Installing Unnecessary Secondary Functions	√			√		
Being Modular	√			√		
6.2. Durability						
6.2.1. Natural Weather Conditions Use Vegetation To Help Control Sun, Wind, Snow and Direct Air Circulation			x			x
Heat						
Use of Non-Metallic Benches and Handrails			x			x
Furniture Elements Such As Permanent Benches Should Not Be Positioned Near Extensively Paved Areas Or Wall Surfaces	√			√		
Sufficient Tree Canopy Provides Shade, Facilitating the Reduction of Built Up Heat.		x				x
Light-Colored Paving Materials Reflect Solar Rays More Than Darker Materials, Diminishing Built Up Heat.			x	√		
Noise						
Site Furniture May Be Positioned Towards Entertaining Sources like The Sound Of Songbirds, Children, Buskers, etc.	√			√		
Barrier Walls, Earth Mounds, And Other Ways To Control Noise Should Be Considered.	√			√		
Rain						
Benches Should Discharge Water Well; They Should Be Produced From Non-Absorptive Materials To For Fast Drying.		x			x	
Materials Selected For Humid Climate Use Should Decay Naturally And Be Mold-Resistant			x			x
Wind						
Positioning Sheltering Furniture from Strong Prevalent Wind			x			x
Providing A Wind Screen Be Enclosed From At Least Two Sides	√				x	

Table 3 (continued)

6.2.2. Resistance to Accidents				
Impact Resistant	√		√	
Vandalism Resistant	√		√	
Fastener Durability	√		√	
Non-Slip, Solid, Smooth, Level And Well-Drained Paved Area		x	√	
6.2.3. Material				

Compliance with Urban Identity		x	√	
Applicability of End Product Format to Mass Production	√		√	
Adapting to Materials Used for Other Components in the Product	√		√	
Longevity Minimizes Resource Use	√		√	
Stainless Steel Fasteners (Screw, Bolts, Etc.)		x		x
6.3. Regeneration / Recycling				
Convenient Disassembly	√		√	
Wood, Steel, Metal Material Usage	√		√	
Reducing the Number of Materials	√		√	
Natural Material Instead of Composites		x		X
6.4. Technology				

(continued)

Table 3 (continued)

Bus Stop With AR Marker		
Perceiving Individuals	√	x
Energy Generating	√	x
Complimentary Photovoltaic Phone/Electronic Device Charging Points	√	x
Wifi Link	√	x
Public Services And Citizen İnformation Delivery On Continually Updated Screens	√	x
Scholastic Purpose Regarding Renewable Energy	x	x
Amassing Data To İmprove Public Services (e.g. Pedestrian Traffic, Use Of Public Facilities, Location Comparisons)	√	x
Customary 'Advertisement Function'	√	√

standard stop; periodic inspections of the stops were observed to be partial at both stops.

It has been observed that the smart stop is partially compatible with its surroundings and that the standard stop is compatible with its surroundings.

While the smart stop is functional, the traditional stop is partially functional. While the loading area, the 1.2 m pedestrian path that should be in front of the station and the 3 m seat that should be in the station, the armrests in the seats, are standard sizes in the smart stop, the standards were partially met in the traditional stop. While the level difference around the station was detected in the smart one, there was no problem in the traditional one. Seating size is more suitable in smart stop; It was determined that the backrests, the reclining bars, The top of a panel on a pole should be no more than 150 cm inches from the ground in both stops were not placed in accordance with the standards at both stops. On the other hand, it was seen that the smart stop was more comfortable in placing the furniture in general.

Smart stop design costs are more expensive than traditional stop. It was determined that the furniture was partially inspected at both stops.

It has been determined there is no significant difference in both stops regarding sustainable design and durability, including simplification, miniaturization, and multi-functionality. In terms of durability, it was determined that non rust-proof metal material was used at both stops and there was not enough vegetation to shade the vicinity of the bus stop. Again, it was determined that dark colored floor material was used around the smart stop and lighter colored, permeable floor material was used in the traditional stop.

Regarding Regeneration, it was determined that standards such as reducing the number of materials at both stops, using furniture that can be disassembled, reducing the use of materials, and the rule of using natural materials instead of composites were not observed at both stops. Other than that it fulfilled the advertisement function and the use of the augmented reality (AR) marker, it was determined that there was insufficient technology at the standard stop.

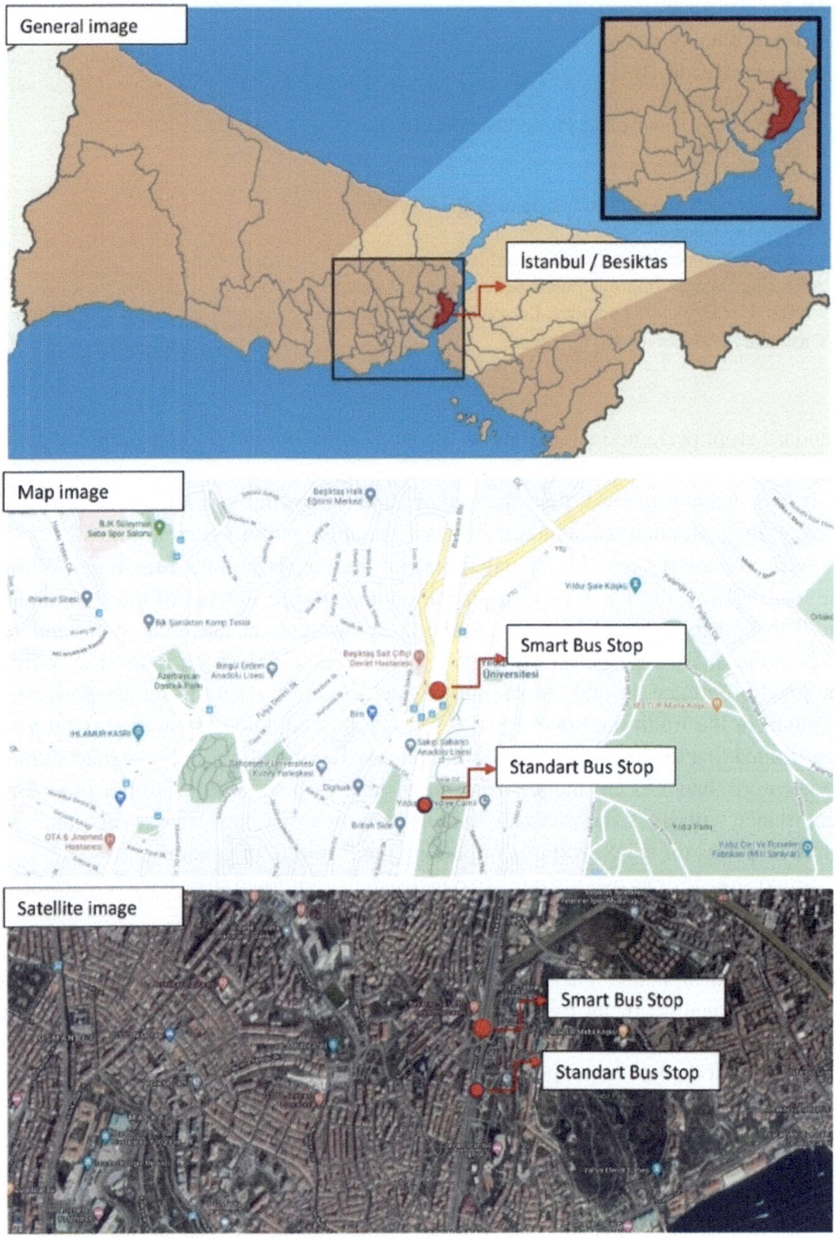

Fig. 2 Locations of study fields [48]

5 Discussion

Savio et al. [61], Imperial Country Transportation Commission [36], Hatai et al. [37???] stated in their studies that furniture found at the smart stop and its vicinity is more comfortable and better suited to standards than traditional stops. Moreover, Avila et al. [25], Kamal et al. [32] and Yu [37] all stated that the space around the station and in-station lighting would be more effective with the use of technology and indicated the waiting area. The results obtained in this study support the results of the literature.

In the working example, it is an important problem that the buses ae not easily visible due to the thick metal frame around the panel as the bus approaches the station. Although the smart stall is comfortable in itself, it is not placed in a sufficient pull distance and does not interrupt pedestrian flow when viewed at the scale of the area it is located in. This result may possibly be related to the standards not being taken into account during the stop's installation in the vicinity. In addition, the use of natural stone material around the smart stop contrasts with the modern look of the stop.

The results of studies carried out by Li [39], Ashwin et al. [31], and Barns [34] emphasized the advantages of using technology in the vicinity of the stop. As Yao [26] had also mentioned, considering the technological solutions of smart stops, smart stop costs are higher than standard stops.

The disadvantage in the use of the stop is that the side panels related to the creation of the technological infrastructure at the smart stop studied do not start higher than the ground and that the solar panels placed on the roof should be placed flat. This causes dirt and dust to accumulate in the stall, water on the roof does not discharge properly and is exposed to bird droppings.

It was determined that furniture maintenance was partially good at both stops. Particularly as Kaufman [41] stated in his smart station study, there is gradual fading in the colors of the colored floor. Despite the predominant use of metal material at both stops, it was observed that rust accumulated at the joints of the traditional stop.

The absence of a significant difference in simplification, miniaturization, multi-functionality and durability at both stops can now be associated with the prevalence of modern design of standard stops. On the other hand, the use of permeable materials, which is important in the use of sustainable materials, is met at the standard stop, while the use of concrete material around the smart stop is intriguing.

In line with the general approach in the designs around the stop on Regeneration, the use of uniform metal material at both stops is an important and meaningful consequence.

Smart stop is also an advantage to allow transportation cards to be loaded at the intermediate stops, compared to a normal stop. Wachira and Karthik [35] emphasized the importance of the kiosk installed within the stop, in terms of bus travel schedule and direction determination. Providing all of the electricity need in the smart stall by means of its own solar panels has made it a safe, eco-friendly stop. The voice command system established for visually impaired individuals also enabled the smart stop to encompass individuals with disabilities compared to the traditional stop. Moreover, as Kaya [30] also mentioned in design criteria for bus stops, there is a waiting area for the disabled person in the smart stop whereas its environment is arranged accordingly. In addition, thanks to GPS, a joint network has been created at the station and has been made controllable for maintenance monitoring or security. Free WIFI has also been standardized at these stops. The presence of AR marker use at standard stops is important in terms of providing information about the up-to-the-minute use of stops. It is seen that technology is used effectively at smart stops with the installed solar panels, WIFI, and the presence of furniture that can detect people and produce energy.

6 Conclusion

In this study, an attempt was made to standardize traditional and smart urban furniture designs, and then these standards were tested in order to conduct comparisons between two smart and traditional stops located near one other in Istanbul.

As a result of the study, when it comes to technology and security, the smart stop offers better quality than the traditional stop. It was determined there are no significant differences between form and character, function, value, maintenance, sustainability and regeneration of the Smart and Traditional stops.

References

1. Ingraham C (2019) People who spend more time outdoors lead more fulfilling lives, new research shows. https://www.washingtonpost.com/business/2019/06/19/people-who-spend-more-time-outdoors-lead-more-fulfilling-lives-new-research-shows/
2. Active Public Space (2020) http://activepublicspace.org/about/
3. Balc N (2018) IoT (Internet of Things) in public space. https://www.architects.nsw.gov.au/download/BHTS/IoT_[Internet_of_Things]_in_public_Space_Nicola_Balch_BHTS_2018.pdf
4. Galič M (2019) Surveillance, privacy and public space in the stratumseind living lab: the smart city debate, beyond data. Ars Aequi, special issue July/August. https://papers.ssrn.com/sol3/papers.cfm?abstract_id=3435518
5. Moughtin C, Oc T, Tiesdell S (1999) Urban design: ornament and decoration. Routledge. https://books.google.com.tr/books?id=EhVUAAAAMAAJ&q=Urban+design:+ornament+and+decoration.+Routledge.&dq=Urban+design:+ornament+and+decoration.+Routledge.&hl=en&sa=X&ved=2ahUKEwjP2sfklanqAhVv1qYKHcEYDtoQ6AEwAXoECAEQAg
6. Gurafikkusha Henshb̄u (1992) Elements & total concept of urban street furniture design, vol 6. Graphic Sha Pub Co
7. Carmona M, Heath T, Tiesdell S, Oc T (2010) Public places, urban spaces: the dimen-sions of urban design. Routledge, https://books.google.com.tr/books?id=EYtNvjLVisUC&printsec=frontcover&dq=Public+places,+urban+spaces:+the+dimensions+of+urban+design.+Routledge.&hl=en&sa=X&ved=2ahUKEwiT7Nyak6nqAhW1RBUIHRKODlMQ6wEwAHoECAUQAQ#v=onepage&q=Public%20places%2C%20urban%20spaces%3A%20the%20dimensions%20of%20urban%20design.%20Routledge.&f=fals

8. Eichner RB (1981) Urban streetscape analysis: interim report, literature review. BioTechnology, Incorporated.
9. Mueller T (2020) Smart urban street furniture solutions in smart cities. https://hub.beesmart.city/solutions/smart-urban-street-furniture-solutions
11. Nassar MA, Luxford L, Cole P, Oatley G, Koutsakis P (2019) The current and future role of smart street furniture in smart cities. IEEE Commun Mag 57(6):68–73
12. Crıstına Uţăi I (2020) Bringing inspiration to#worldchangers through storytelling. https://brandminds.ro/upgrading-our-cities-with-smart-street-furniture/
13. Froister A (2019). Bench 003. http://www.alexanderforestier.com/smart-street-furniture-1
14. Mars Architect 2020 Water bench|by mars architects—bench collecting & storing rainwater. https://www.jebiga.com/water-bench-mars-architects-bench-collecting-storing-rainwater/
15. Streetscape Guidance (2020) Technical guidance: street furniture. http://plangate.no/areal-transport/london/Streetscape/8%20-%20streetscape-guidance-2009-street-furniture-108.pdf
16. Willerton P (2016) Swiftmile: solar powered electric bike rental stations. https://electricbikereport.com/swiftmile-electric-bike-solar-charging-stations-videos/
17. CityManager (2020) Smart street lighting for smart(er) cities. https://www.tvilight.com/wp-content/uploads/2019/03/Brochure-Intelligent-Smart-City-Lighting-Control-Sensor-CMS-Light-Management-EN.pdf
18. Saimoğlu S (2016) Beş adımda Akıllı Aydınlatma. https://akillisebekeler.com/2016/12/12/bes-adimda-akilli-aydinlatma/
19. Minkave Construction & Development Co (2020) Energy and natural resources. https://www.euroasiaip.com/news/energy-and-natural-resources/23-our-news/47-minkave-construction-development-co-ltd-and-eip-sign-strategic-cooperation-agreement-for-urban-lighting-projects
20. Newyork City Furniture (2020) Furniture. https://www1.nyc.gov/html/dot/downloads/pdf/nycdot-streetdesignmanual-interior-05-furniture.pdf
21. EUtouring (2020) GPS Location maps of public toilets in Paris, https://www.eutouring.com/map_public_toilets_paris.html
22. City of Toronto (2006) Toronto's coordinated street furniture program design and policy guidelines, https://www.toronto.ca/wp-content/uploads/2017/11/96e3-Vibrant_Streets-Document.pdf
23. Haddi G (2013) The 'techno-pods' garbage bin design gives updates on LED screens, https://www.trendhunter.com/trends/garbage-bin
24. ASİS (2020) https://www.asiselektronik.com.tr/euts
25. Hassanein H (2017) Smart technical street furniture design: case study of new cairo administrative capital. https://www.researchgate.net/publication/321807999_Smart_Technical_Street_Furniture_Design_Case_Study_of_New_Cairo_Administrative_Capital
26. Gómez-Carmona O, Casado-Mansilla D, & López-de-Ipiña D (2018) Multifunctional interactive furniture for smart cities. In: Multidisciplinary digital publishing institute proceedings, vol. 2, no. 19, p 1212)
27. Avila M, Toledo J, Córdova F, Icaza D, & de los Angeles Tello M (2018) Intelligent multifunctional solar urban furniture: a multidisciplinary methodological vision of technology. In: 2018 International conference on Smart Grid (icSmartGrid). IEEE, pp 184–194. https://ieeexplore.ieee.org/stamp/stamp.jsp?tp=&arnumber=8634474
28. Yao P (2019) Smart objects for smart cities–the use of Internet of things in public spaces (No. 1372). EasyChair
29. İSBAK (2020). https://www.isbak.istanbul/catalog/
30. Akıllı Şehirler (2020). 2020–2023 Ulusal Akıllı Şehirler Stratejisi ve Eylem Planı, Akıllı Şehirler Beyaz Bülteni. https://webdosya.csb.gov.tr/db/cbs/akillisehirler/
31. MATRİSLED 2020. http://www.matrisled.net/my-product/istanbul-buyuksehir-belediyesi-10-adet-cift-tarafli-led-ekran-projesi/
32. Kaya E (2001) A study on street furniture design criteria for bus stops. Master's thesis, Izmir Institute of Technology

33. Ashwin D, Mounika V, Kommineni M, Swetha K (2019) Secure design for smart bus shelter using renewable energy. J Crit Rev 7(1):2020
34. Kamal M, Atif M, Mujahid H, Shanableh T, Al-Ali AR, Al Nabulsi A (2019) IoT based smart bus stops. In: 2019 3rd international conference on smart grid and smart cities (ICSGSC). IEEE, pp. 6–9. https://pdfs.semanticscholar.org/c2dc/c599fb6765cedc7645b38365599b44c9c95b.pdf
35. Alikhanova A, Kakimzhan A, Mukhanov A (2017) Design of a bus shelter based on green energy technologies for extreme weather conditions in Astana. http://nur.nu.edu.kz/handle/123456789/3585
36. Barns S (2017) Using smart technologies for climate change adaptation in Western Sydney: A CAPS Research Report. https://www.uts.edu.au/sites/default/files/CAPS_Research_Final.pdf
37. Wachira K, Karthik J (2016) Smart bus shelters: enhancing public information systems in bus shelters by integrating smart IoT solutions. http://hdl.handle.net/2043/21496
37. Imperial County Transportation Commission (2014) Bus stop design and safety guideline handbook. http://www.imperialctc.org/media/managed/news/Final-2014-ICTC-Bus-Stop-Design-and-Safety-Guideline-Handbook.pdf
40. Hatami F, Mahjoobin MH, Hatami F (2016) Design and construction of bus stop shelter using composite materials and new building technology. Turk Online J Des Art Commun 6:2672–2689. http://tojdac.org/tojdac/VOLUME6-NOVSPCL_files/tojdac_v060NVSE172.pdf
41. Yu S (2018) The bus waiting area design. In: 2018 International conference on advances in social sciences and sustainable development (ASSSD 2018). Atlantis Press
42. Smith V (2019) Safer stops: increasing public perceptions of safety through bus stop design in the city of greater sudbury
43. Zhang KJ (2012) Bus stop urban design. Doctoral dissertation, University of British Columbia
44. Li C (2020) Landscape design of the urban bus station based on the urban characteristics and eco logical environment of smart city, China. Feb Fresenius Environ Bull:3300
45. Kaufman, B. (2015). Re-Inventing the Bus Stop: Design Guidelines and Analysis for Transit-Friendly Parklets in Alameda County, https://escholarship.org/uc/item/8hd6x856
46. Sungur C, Babaoglu I, Sungur A (2015) Smart bus station-passenger information system. In: 2015 2nd International conference on information science and control engineering. IEEE, pp 921–925. https://ieeexplore.ieee.org/stamp/stamp.jsp?tp=&arnumber=7120749
47. Participants S, Daniel Sr G, Green B, June E, McBeth M, Redmond J, Macdonald C (2005) Assessment of alternative futures for transit in Tal-lahassee. https://its.fsu.edu/sites/g/files/imported/storage/original/application/4aaf8e747216dc82797ba71666e64aea.pdf
48. Fitzpatrick K, Hall K, Perkinson D, Nowlin L, Koppa R (1996) TCRP report 19: guidelines for the location and design of bus stops. Transportation Research Board of the National Academies, Washington, DC
49. Foothill Extension Bus Interface Plan (2011) Proposed bus/rail interface facility improvements, chapter 3. http://foothillgoldline.org/images/uploads/Chapter%203_DRAFT.pdf
50. TeamTarget (2020) Smart Bus Shelter, https://www.ttidubai.com/smart-bus-shelter/
51. Team Technoid (2007) Intelligent transportation system in city bus shelter. https://www.hackster.io/technoid/intelligent-transportation-system-in-city-bus-shelter-cf076e#overview
52. Team Technoid (2007) Intelligent transportation system in city bus shelter. https://www.hackster.io/technoid/intelligent-transportation-system-in-city-bus-shelter-cf076e#overview
53. Sadeghi L, Amerian Bani J, Zoogi A (2011) The role of urban planning in security improvement. J Kurdistan Regul Knowl, 2nd year (8)
54. Gupta N, Bhatti V (2015) Importance of street furniture in urban landscape. Int J Latest Trends Eng Technol (IJLTET) 5(3). https://www.ijltet.org/wp-content/uploads/2015/05/24.pdf
55. Minneapolis Coordinated Street Furniture (2007) Program Guidelines Report, https://nacto.org/wp-content/uploads/2015/04/Street-furniture-program-guidelines_minneapolis.pdf
56. Urban Design City Strategy-City of Ballarat (2013) Urban design manual, part A: Street and park furniture guidelines version 4.1. https://www.ballarat.vic.gov.au/sites/default/files/2019–04/street_and_park_furniture_guidelines.pdf

57. Smartcity (2018) Smart city furniture is the next urban trend. https://www.smartcity.press/smart-city-street-furniture-trends/
58. ARUP (2013). Smart streets principles, concepts and delivery. https://www.seng.org.au/sites/default/files/chapteradmin/vicfiles/Smart%20Streets%20Arup_110913.pdf
59. Najafidashtape A (2018) Ulaşım Aktarma Merkezlerinde Akıllı Kent Mobilyaları. Mimarlık ve Yaşam, 3(1), 63–74. https://dergipark.org.tr/tr/download/article-file/455980
60. Malt HL (1970). Furnishing the city. McGraw-Hill. https://books.google.com.tr/books?id=KYhbAAAAMAAJ&q=Furnishing+the+city.+McGraw-Hill.&dq=Furnishing+the+city.+McGraw-Hill.&hl=en&sa=X&ved=2ahUKEwiWyZX1lKnqAhVvwqYKHX-oAB8Q6wEwAHoECAQQAQ
61. PPS (2018) Bus Shelter. https://www.pps.org/article/busshelters
62. Transit BC (2020) Design guidelines for accessible bus stops. https://nacto.org/docs/usdg/design_guidelines_accessible_bus_stops_bc_transit.pdf
63. Delaware Valley Regional Planning Commission (2012) Septa bus stop design guidelines. http://www4.septa.org/strategic-plan/reports/SEPTA-Bus-Stop-Design-Guidelines-2012.pdf
64. Karslı UT, Öztürk ÖB (2018) Sürdürülebilir Çevrede Kent Mobilyaları. https://yapidergisi.com/surdurulebilir-cevrede-kent-mobilyasi/
65. Savio L, Cocina G, Gariano R, Giampetruzzi F, Pagani R, Pedrazzo MM, Pennacchio R (2017) Smart street furniture: innovation in the concept design process. In: International conference on applied human factors and ergonomics. Springer, Cham, pp 163–175

The manufacturer's authorised representative in the EU is Springer Nature Customer Service Centre GmbH, Europaplatz 3, 69115 Heidelberg, Germany. If you have any concerns regarding our products, please contact ProductSafety@springernature.com

Printed and bound by CPI Group (UK) Ltd, Croydon, CR0 4YY
30/09/2026
02225694-0011